Monograph of Living Chitons

Vol. 1. Order Neoloricata: Lepidopleurina

Monograph of Living Chitons

(Mollusca: Polyplacophora)

Volume 1
Order Neoloricata: Lepidopleurina

By

Piet Kaas

&

Richard A. Van Belle

E. J. Brill / Dr. W. Backhuys
Leiden · London · Köln · København

Edited by:
Leif Lyneborg

Printed by:
Vinderup Bogtrykkeri A/S
DK 7830 Vinderup, Denmark

ISBN 90 04 07414 7
90 04 07415 5

Contents

GENERAL PART

SYSTEMATIC PART

GENERAL PART

The present Monograph deals with the molluscan class Polyplacophora (bearing many plates), forming the third class of the subphylum Aculifera (spiny mollusks), the other two classes being the "wormlike" Caudofoveata and Solenogastres.

Definition

The Polyplacophora (Placophora, Loricata, Crepipoda), or popularly called "chitons" (coat-of-mail shells), are bilateral, untorted mollusks, dorso-ventrally flattened, oval or elongate oval in outline, and with head lacking tentacles as well as eyes. The back is shielded by eight longitudinally arranged articulating shell plates, called valves, surrounded by the girdle or perinotum, the epithelium of which segregates a cuticula as well as a variety of calcareous scales, or spicules, or corneous appendices. The ventral side is dominated by a broad flat creeping foot encircled by a groove in which a series of double feathered gills (ctenidia) is present at both sides. The terminal anus is placed on a papilla. There is a well developed radula with many transverse rows of generally 17 teeth.

External Characters of Body and Shell

As a rule Polyplacophora are easily recognizable by the eight valves, generally imbricating, allowing the animals to roll up like woodlice in order to protect the soft ventral parts when endangered. They vary in size from a few mm to c. 10 cm, but a few West American species attain a greater length, the largest being *Cryptochiton stelleri*, which may grow up to 30 cm in length. They are mostly more or less elongate oval in shape, though species of the genus *Cryptoplax* are narrowly elongate, almost worm-like.

The width of the girdle is rather variable; generally it occupies less than half of the total width of the animal. In *Placiphorella* and *Spongiochiton* the girdle is much enlarged anteriorly, in the Acanthochitonidae it is wide and fleshy, occupying about 50% or more of the total width, often encroaching at the sutures and more or less covering the shell plates, especially in *Notoplax*. In *Cryptoplax* the valves are much reduced and generally disconnected, especially the posterior ones. In *Cryptoconchus* and *Amicula* the valves are connected, but nearly wholly overgrown by the girdle, except for a narrow median strip, and in *Cryptochiton stelleri* the shell plates are totally internal.

The valves are built up by three or four layers. There is a very thin, transparent, organic membrane, the periostracum, covering the second layer or tegmentum, which is the dorsally visible part of the shell, mostly sculptured

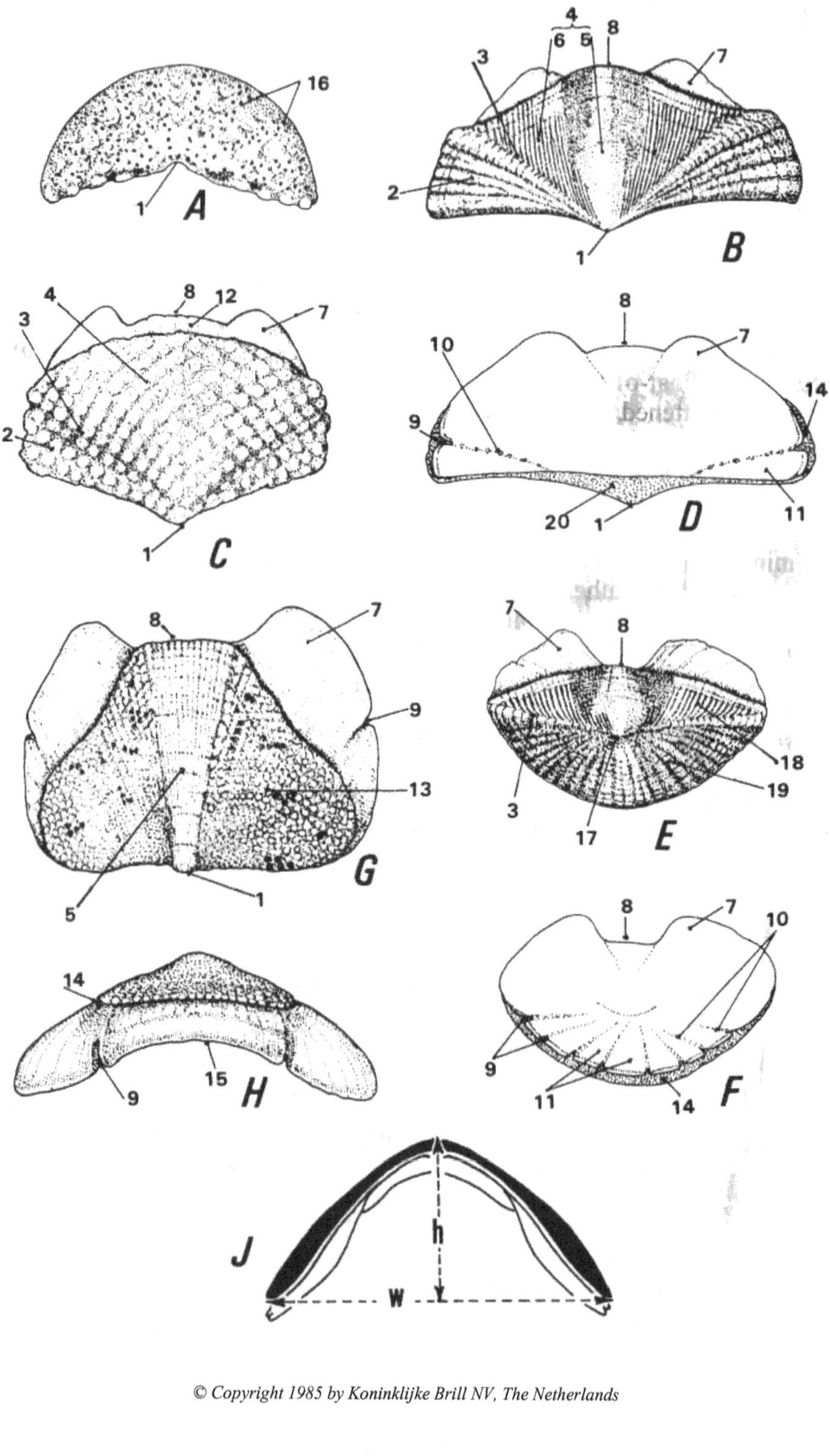
A
16
1
B
4
6 5
8
3
7
2
1
C
4
8 12
3
7
2
1
D
8
10
7
14
9
20 1
11
G
8
7
9
13
5
1
E
7
8
18
19
3
17
H
14
9
15
F
8
7
10
9
11
14
J
h
w

in various ways, often rich in pigment and consisting of organic as well as of anorganic material. The organic material is of a chitinous nature, forming a close network, the mazes filled up with anorganic substance, mostly calcium carbonate. Generally the median axis of the tegmentum is somewhat projected posteriorly, the apex of the valve, from which two low, often hardly discernible ribs radiate towards the anterior sides, dividing the tegmentum into three well defined triangular areas: a broad, obtuse-angled central area, and two narrow, more or less elevated sides or lateral areas. The sculpture of the central area often differs from that of the lateral areas. In the Acanthochitonidae the tegmentum is divided into a narrow dorsal strip, the jugal tract, and the two sides or latero-pleural areas.

The underlying layer, called articulamentum, contains less organic material and is somewhat nacreous. It projects at the sides, underneath the tegmentum, except in the Leptochitonidae and only in one or both of the end valves in the Hanleyidae, the overhanging edges of the tegmentum are called the "eaves", which may be either porous or solid. The projecting sides of the articulamentum are imbedded in the girdle tissue and are indicated as the insertion plates, mostly divided by slits into insertion teeth, which are either smooth or more or less pectinated by radiating grooves, especially at the outside. Slit rays may be developed as radiating rows of pores in the articulamentum from the slits towards the apices of the valves. Anteriorly the articulamentum of the second and subsequent valves shows two thin projecting processes beneath the preceding valve, the sutural laminae, or apophyses, which are inserted in the musculous tissue of the dorsal integument, enabling the animal to roll up. The apophyses are separated by the jugal sinus; in some genera (*Lepidozona, Connexochiton*) there is also a narrow projection of the articulamentum between the apophyses, a jugal lamina, which may be slitted or not.

The anterior or head valve is half cone-shaped, the base semicircular, the posterior margin straight, with a small inward curve at the apex; as a matter of course it is the only valve lacking apophyses. The intermediate valves are broadly rectangular, the posterior margin almost straight, or somewhat concave at both sides of the apex, which may be more or less strongly "beaked".

Fig. 2. Isolated valves of chitons.

A, valve I of *Tonicia*, dorsal view; B, intermediate valve of *Chiton*, dorsal view; C, intermediate valve of *Connexochiton*, dorsal view; D, intermediate valve of *Chiton*, ventral view; E, valve VIII of *Chiton*, dorsal view; F, valve VIII of *Chiton*, ventral view; G, intermediate valve of *Acanthochitona*, dorsal view; H, valve VIII of *Acanthochitona*, posterior view; J, valve IV of a chiton, anterior view, h = height of tegmentum, w = width of tegmentum. Dorsal elevation = h/w.

1, apex; 2, lateral area; 3, diagonal line; 4, central area; 5, jugal area; 6, pleural area; 7, apophyses; 8, jugal sinus; 9, slits; 10, slit rays; 11, insertion plates; 12, jugal lamina; 13, lateropleural area; 14, eaves; 15, posterior sinus; 16, rays of shell eyes; 17, mucro; 18, antemucronal area; 19, postmucronal area; 20, periostracum, folded under.

The anterior margin is hidden under the preceding valve. The posterior or tail valve is semicircular to half elliptical, with a top, called mucro, which is mostly about central, but it may be either anterior or posterior of the centre and even quite terminal. The tegmentum of that valve is divided by two more or less pronounced diagonal lines, or elevated ridges, radiating from the mucro, dividing the tegmentum into two distinct areas: the antemucronal area, which does not differ in sculpture from the central areas of the intermediate valves, and a postmucronal area sculptured like the head valve and the lateral areas of the intermediate valves. The postmucronal or posterior slope is mostly concave, but in some species it may be straight or even convex. If the mucro is placed terminal, it results in absence of a postmucronal area.

In a living state the girdle is, like the foot, firmly appressed to a hard, by preference smooth substratum, except for little anterior and posterior upward waves allowing a respiratory current to pass through the pallial groove. The girdle is always covered by a thick chitinous cuticula. Imbedded in it various kinds of armature are found. The simplest are small, oval or egg-shaped, closely set calcareous corpuscules, as they are found in the Lepidochitoninae. Among them occur small calcareous spicules, sheathed in stalked chitinous cups. All the girdle elements are segregated in epidermal pockets. In the Ischnochitoninae and Chitoninae the girdle is dorsally covered with imbricating scales, which may be smooth or dorsally sculptured with riblets. In the Hanleyidae the girdle is beset with much crowded erect spicules

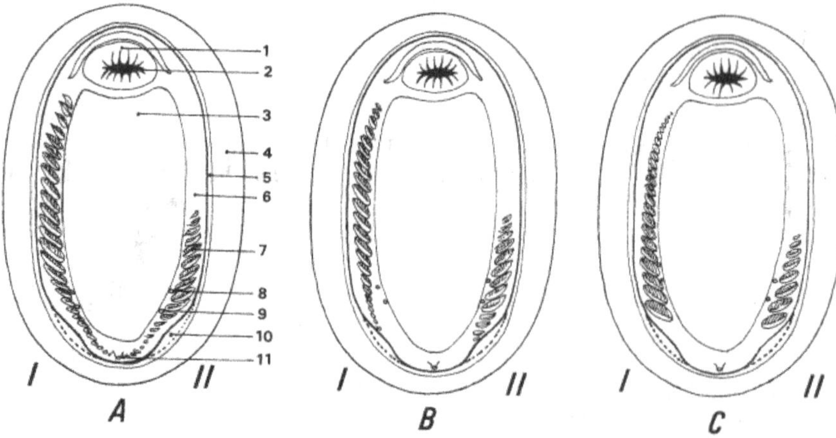

Fig. 3. Arrangement of the gills in chitons.

A, adanal without interspace; B, adanal with interspace; C, abanal. I, holobranchial; II, merobranchial.

1, head; 2, mouth; 3, foot; 4, girdle (ventral side); 5, pallial fold; 6, pallial groove; 7, gills; 8, gonopore; 9, nephridiopore; 10, pallial lappet; 11, anus.

of various sizes. This is also the case in the Acanthochitonidae, which bear also a series of pores near the sutures of the valves and four more around the head valve, eighteen in total, from which bunches of longer, straight or slightly bent spicules arise, the sutural tufts. Along the outer margin of the girdle often a fringe of longer spicules is present. In the Acanthopleurinae the girdle bears stout calcareous spines, and in the Mopaliidae and Chaetopleurinae cuticularized, often branching, bristly processes are found, which sometimes bear calcareous spicules at their tips.

The ventral side of the girdle is more simply clothed, either with rectangular flat scales in radiating series, or with pointed scales, or with spicules.

The pallial groove is bordered at the outside by a pallial fold, which forms pallial lappets at the posterior end, on both sides of the anus.

The gills are hanging down from the roof of the pallial groove in a longitudinal row. They vary in number from six to eighty or more pairs, but the number is not constant within a certain species and it is often different on the two sides. The number increases with growth of the animal, new gills being formed at the posterior end of a row, but in specimens of comparable size the number may differ considerably.

Plate distinguished six types of arrangement of gills. The arrangement is called holobranchial if the gills extend from the anterior to the posterior end of the foot, merobranchial if they only occupy the posterior half of the pallial groove. If the size of the gills increases posteriorly and the last one in each row is the largest, leaving a free space between the two rows and the anus, they are called abanal. In case the largest gills are followed by a few smaller ones towards the anus, they are called adanal. Are those rows continuous, they are adanal without interspace, are they ending up at square level with the anus, the arrangement is adanal with interspace.

The greater part of the ventral surface is occupied by the broad, strongly muscular foot; only in the Cryptoplacidae the foot is rather narrow. The epithelium is ciliated and contains many mucus cells; in some species of Lepidochitoninae a row of mucus glands is present near the anterior margin of the foot. The head is separated from the foot by a groove. In *Placiphorella* there is a fringe of precephalous tentacles, which probably acts as a veil to hold small animals like Amphipods, captured under the greatly enlarged anterior part of the girdle.

Musculature

The muscular system of chitons is rather complicated. The animal is able to roll up principally by contraction of the lateral longitudinal muscle, encircling the body just under the outer margins of the valves.

The valves themselves are directed by four pairs of muscles. The straight muscles are narrow bundles of fibers fixed to the anterior margin of the valves in the jugal sinus, and proceed parallel just underneath the jugum, where they insert on the body wall under the preceding valve. The oblique muscles also originate in the jugal sinus to the left and the right of the straight muscles, diverging forwardly underneath the diagonal lines to their insertion on the body wall under the preceding valve. The unpaired transverse muscles merely hold the valves together. They chiefly consist of dorso-ventral fibers, firmly attached to the underfolded posterior margin of the valve and to the apophyses of the next valve. The side margins of the valves are connected with the foot by two muscular bundles on each side, the anterior and posterior lateral muscles.

Fig 4. Musculature of chitons.

1, straight muscles; 2, oblique muscles; 3, anterior lateral muscles; 4, posterior lateral muscles; 5, transverse muscles; 6, lateral longitudinal muscle.

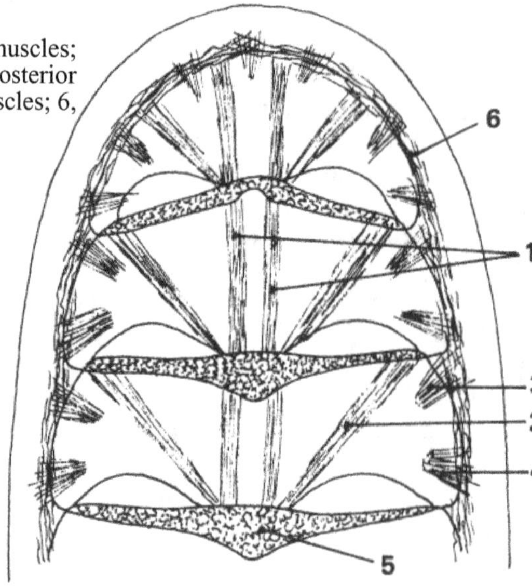

Digestion

The mouth opening lies in the centre of the ventral surface of the head and leads to the cavity of the mouth, which posteriorly forms a blind sack ending just beneath the paired subradular organ, which probably serves as an organ of taste. Where the mouth cavity passes into the gullet it receives the salivary glands, the function of which is not yet clear. They are rich in mucus. The gullet has a strong muscular wall, strengthened by chitine as a firm underlayer for the radula. From the sack-shaped radular sheath, extending ventrally of the digestive tract as far as the third or fourth valve, the radula distally emerges into the pharynx. By means of a very complicated muscular system the radula is moved back and forth across a pair of bolsters, allowing the animals to scrape off the food from the substratum on which they live.

The radula is formed at the base of the radular sheath as a long chitinous ribbon-like membrane, closely beset with numerous (40 or more) transverse rows of generally seventeen differently shaped teeth or plates. The first who gave a correct description and figures of chiton radulae was Lovén (1847: *Lepidochitona cinerea* and *Callochiton septemvalvis*). Dall (1879) described and illustrated the radulae of many American chiton species. In 1893 Thiele examined the radulae of many species and founded several new taxa on their features, but his system has proved to be artificial, so that the majority of his taxa are now out of use.

The radulae of chiton species are invariably built along the same pattern,

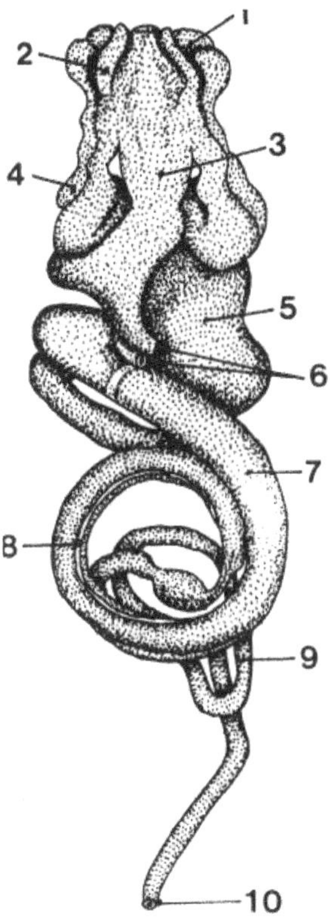

Fig. 5. Digestive tract of *Lepidochitona cinerea*, dorsal view (after Fretter, 1937).

1, buccal gland; 2, ciliated groove; 3, gutter; 4, salivary gland; 5, stomach; 6, entrances of midgut gland; 7, anterior intestine; 8, ciliated band; 9, posterior intestine; 10, anus.

though there are minor specific differences, which makes the radula an instrument of taxonomic value.

The designation of the different teeth in a row is rather confusing. The central or rhachidian tooth (Thiele's "Mittelplatte") is mostly small, simple, elongate, with a narrow, thickened cutting edge, called blade. It is accompanied by a row of eight teeth on both sides, the first of it (Thiele's "Zwischenplatte") is the minor lateral tooth in Dall's description, usually larger than the rhachidian tooth and often provided with a short blade. The next in the row, or major lateral tooth (Thiele's "Hakenplatte"), is by far the largest. It consists of a long chitinous shaft, crowned by a hard, black "cusp", containing silicium and iron, provided with one to four mostly sharply pointed denticles, which make them perfect scrapers. Then follow two boss-like and thickened uncinal plates and a spatulate uncinal which generally bears a long, inwardly curved and pointed blade. The last three, or external uncini in Dall's nomenclature, are scale-like, hardly or not having a blade. We prefer to call them marginal plates.

The gutter is separated from the stomach by a sphincter, or orbicular muscle. In primitive forms the stomach is hardly wider than the gutter, but generally it is a large sack deformed by the pressure of the very large midgut glands, often incorrectly called "liver". The right midgut gland is much smaller than the left one. It is located in depressions of the stomach, whereas the left gland occupies the spaces among the coils of the intestine. The two glands open separately into the stomach just before the latter passes into the intestine. They are mostly accumulation organs.

The intestine, about four or five times the length of the animal, is loosely coiled and divided by a short valve into an anterior and a posterior part. In some species the rectum is somewhat widened before it ends in the anus, which is situated in the median line, behind the foot, and usually placed on a papilla.

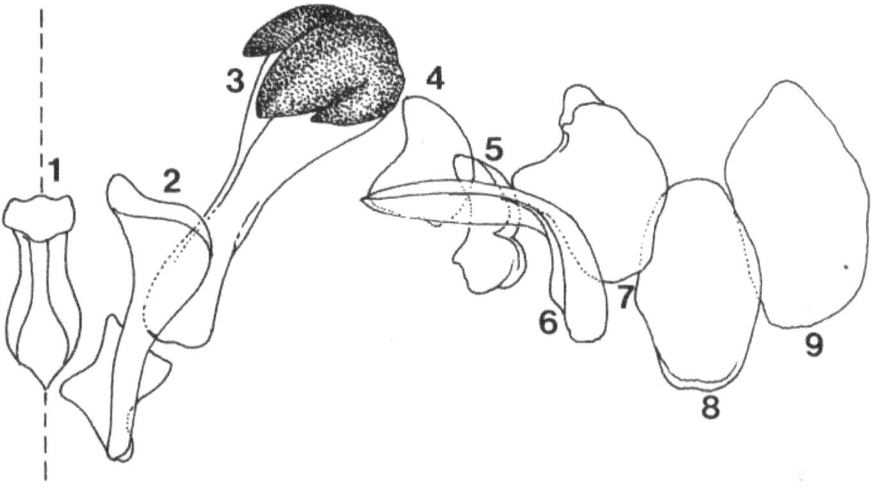

Fig. 6. Half of a radula row of *Lepidochitona cinerea*.

1, rhachidian or central tooth; 2, first or minor lateral tooth; 3, major lateral tooth; 4-5, uncinal teeth; 6, spatulate uncinal tooth; 7-9, marginal teeth.

Blood Circulation and Respiration

The amount of blood in chitons is relatively large, 40% of the wet weight of the animal (without the shell plates). The blood consists of colourless hemolymph, containing coloured granulae varying from pale green to light orange or brownish, (according to the colour of the animal, and blood cells or amoebocytes of two types: those with relatively little cytoplasm around a large nucleus, and others with more cytoplasm. The respiratory pigment is the for the Mollusca typical hemocyanin, dissolved in the hemolymph. It has only little affinity to oxygen, but it is sufficient for chitons as their need of oxygen is limited by their sluggish metabolism, the relatively large branchial surface and the easy ventilation, as they live in well aerated sea water, principally in the tidal zone.

The circulatory system is typically open and diffuse as the main blood channels have no definite wall. The heart is situated beneath the valves VII and VIII, enclosed in the dorso-ventrally flattened pericard. It consists of a median, tubular shaped ventricle, with a left and a right atrium or auricle. The ventricle ends blind posteriorly; anteriorly it continues into the dorsal aorta. The auricles are connected behind the ventricle. They open into the ventricle by generally one or two pairs (rarely three or four pairs) of ostia. The dorsal aorta proceeds anteriorly above the gonad, giving off a number of blood channels to the gonad and to the valve muscles. After passing through the diaphragma, a vertical partition of connective tissue separating the head from the rest of the body, the aorta opens into the head sinus. This gives off several longitudinal sinuses extending posteriorly to the foot, the viscera and the mantle. In the foot the venous blood enters into several lacunae to be collected

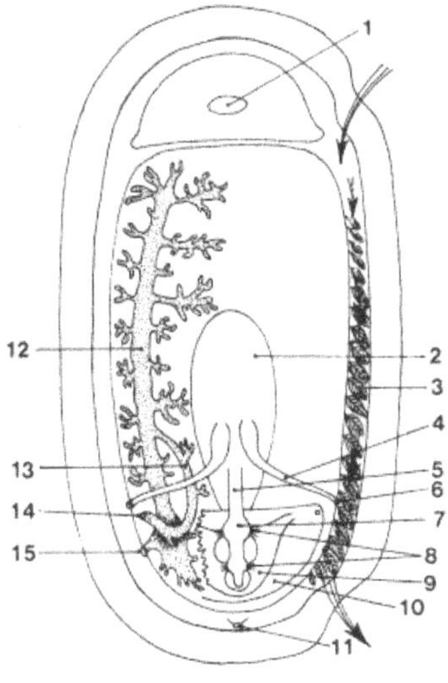

Fig. 7. Situation of the heart, the gonad, and the excretion system of a chiton (altered after Götting, 1974).

1, mouth; 2, gonad; 3, gills; 4, gonoduct; 5, dorsal aorta; 6, gonopore; 7, ventricle; 8, ostia; 9, auricle; 10, pericard; 11, anus; 12, nephridial canal; 13, renopericardial canal; 14, renopericardial pore; 15, nephridiopore.

in the median longitudinal sinus of the foot. Posteriorly this sinus sends to either side an afferent branchial sinus, which parallels the roof of the pallial groove, giving off branches into every gill. After aeration the arterial blood is carried through the efferent branchial sinuses, coming out through pores into the auricles and from there through the ostia into the ventricle.

The base of each gill is attached to the roof of the pallial groove. The axis of the gill carries two alternating series of oval, thin, ciliated lamellae, gradually decreasing in size towards the proximal end of the gill.

Ventilation of the gills is realized by a constant longitudinal current of water through the pallial groove, stimulated by movements of the cilia on the lammellae of the gills. The water enters anteriorly under the slightly uplifted girdle and flows out posteriorly in the same way.

Excretion

The excretory system of chitons consists of a pair of very large nephridia latero-ventrally of the visceral cavity. Probably by ultrafiltration the residual products in the hemolymph are delivered to the pericard, which has two openings at the anterolateral corners, the renopericardial pores, opening into the strongly ciliated renopericardial canals. In primitive forms these canals proceed anteriorly for only a short way, coming out into the main nephridial canal. This is a wide tube with many strongly branching diverticula, extending from the anterior end of the foot to the posterior end, where it widens to a sac. A short ureter leads from this sac to the excretion opening, the nephridiopore, on a papilla in the inner wall of the pallial groove among the largest posterior gills.

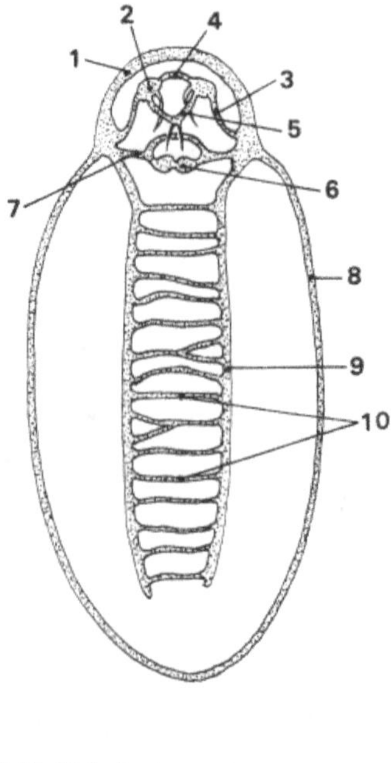

Fig. 8. Nervous system of *Acanthochitona fascicularis* (after Pelseneer, 1899).

1, cerebrobuccal ring; 2, buccal ganglia; 3, buccal connective; 4, dorsal buccal commissure; 5, ventral buccal commissure; 6, subradular ganglia; 7, subradular connective; 8, lateral nerve cord; 9, ventral or pedal nerve cord; 10, ventral commissures.

In most species the renopericardial canal continues forward along the inner side of the main nephridial canal before entering it at the anterior end, which gives the nephridium a hair pin shape, the outer and inner canal being of equal length. In some species the inner canal is also diverticulated.

Nervous System and Sense Organs

The nervous system is of a rather primitive nature, consisting of a cerebrobuccal nerve ring, giving off the paired longitudinal lateral and the also paired pedal or ventral nerve cords. Cerebral ganglia are wanting. The ventral nerve cords, connected by many commissures, proceed posteriorly in the median part of the foot, giving off numerous branches innervating the musculature of the foot and the ventral integument. The lateral nerve cords, just above the pallial groove, are sometimes connected at the posterior end, forming a wide loop. In some species they are also connected with the ventral cords by several commissures. The branches of the lateral cords innervate the lateral and dorsal body wall, the girdle, the aesthetes in the shell plates, the nephridia, gills and gonad.

The cerebrobuccal ring gives also rise to two buccal commissures, connected with the buccal ganglia, for the innervation of the anterior organs of the digestive tract.

Besides the subradular organ, which is possibly an organ of taste, chitons are provided with neurosensory cells in the ventral epidermis in the region of the mouth, in the cavity of the mouth and especially in the epidermis of the glandular tracts in the anterior part of the pallial groove.

Most peculiar in chitons are the sense organs which are imbedded in the tegmentum of the shell plates: the aesthetes. These are complex nerve cells, innervated by the lateral cords. Branches of these enter the articulamentum of the shell plates by the slits and the pores of the slit rays. They horizontally branch between the articulamentum and the tegmentum and then ascend vertically through pores in the tegmentum. There are two kinds of pores: small micropores and larger macropores, or megalopores, forming a specific pattern. Usually the macraesthetes, or megalaesthetes, terminating in the macropores, give off a certain number of long strands to the surface, the micraesthetes, ending in the micropores. The two kinds of aesthetes terminate at the surface of the tegmentum with a hard, clear, protective cap. The number of micraesthetes given off by a macraesthete is variable but fairly constant within a species. Their function is not quite clear, but it is generally believed that they are photoreceptors.

In the Toniciinae and the Acanthopleurinae, and also in *Schizochiton,* some macraesthetes are transformed into shell eyes, arranged in definite patterns in the end valves and in the lateral areas of the intermediate valves, usually in radiating rows, especially along the diagonal lines. The thin outer layer of the tegmentum forms a clear cornea, protecting the convex lens and the underlying fibrillar area, filling a cup which contains retina cells and a layer of black pigment.

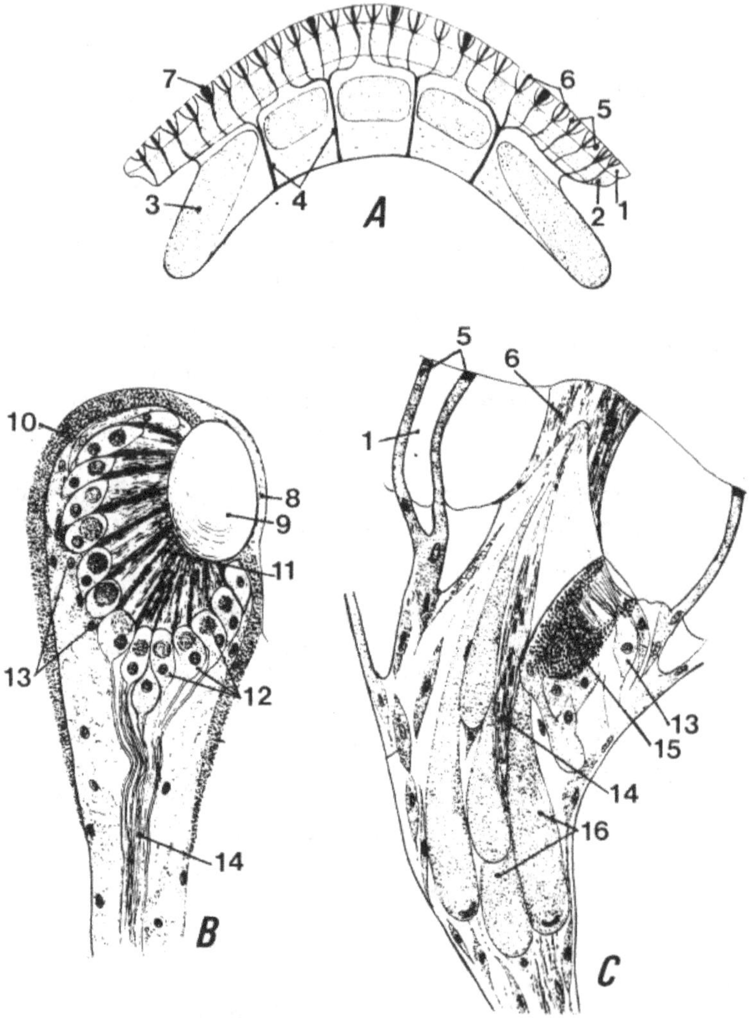

Fig. 9. Aesthetes and shell eyes of chitons (after Nowikoff, 1907, 1909).

A, schematic cross section through an intermediate valve of *Tonicia;* B, longitudinal section through a shell eye of *Acanthopleura;* C, longitudinal section through a shell eye of *Callochiton.*

1, outer layer of tegmentum; 2, inner layer of tegmentum; 3, articulamentum; 4, nerve fiber cords; 5, micraesthetes; 6, macraesthetes; 7, shell eye; 8, cornea; 9, lens; 10, pigment layer; 11, fibrillar area; 12, cells of cup; 13, retina cells; 14, branch of nerve cord; 15, pigment cup; 16, gland cells.

Callochiton species show a different type of eyes. Here some of the macraesthetes in the end valves and in a narrow band along the diagonal line in the lateral areas of the intermediate valves form ocelli. These are also provided with a small convex lens and a few retina cells in a pigmented cup. The macraesthetes also give off a number of micraesthetes. The number of eyes is variable in the various species and increases with the growth of the animal.

As the nervous system of chitons is hardly centralized it is highly probable that the "sight" of chitons will be limited to a vague perception of light and dark. Crozier & Arey (1918) established that young *Chiton tuberculatus* react photonegative, hiding in crevices of rocks and under stones, whereas full grown specimens are decidedly photopositive, living on exposed surfaces in the tidal zone. Half grown specimens are photopositive to weak light, but photonegative to a strong source of light. The same authors (Arey & Crozier, 1919) explain this inversion in the sense of chiton's phototropism by the gradual elimination of the shell photoreceptors by erosion of the shell plates, thus conditioning in the older individuals a lower specific stimulating power of the light.

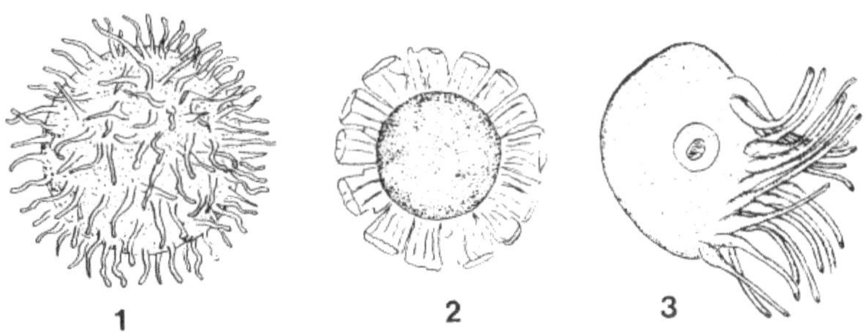

Fig. 10. Egg hulls of chitons [after Lyngnes (1-2) and Plate (3)].
1, *Ischnochiton albus;* 2, *Tonicella marmorea; 3, Chaetopleura peruviana.*

Reproduction and Development

Chitons are of separate sex, with only one exception: the Californian *Lepidochitona raymondi* is hermaphroditic. Externally the sexes are indistinguishable, except for the colour of the foot and mantle in a few species, caused by granulae in the hemolymph (*Chiton tuberculatus, Cryptoconchus porosus*). The gonads are paired in their inception, but they soon develop into one sack-shaped organ in the median area, anterior of the heart, extending between the third and sixth valves. Only in *Nuttallochiton* the gonads do not fuse. In all species the gonoducts are paired. The gonopores are located in

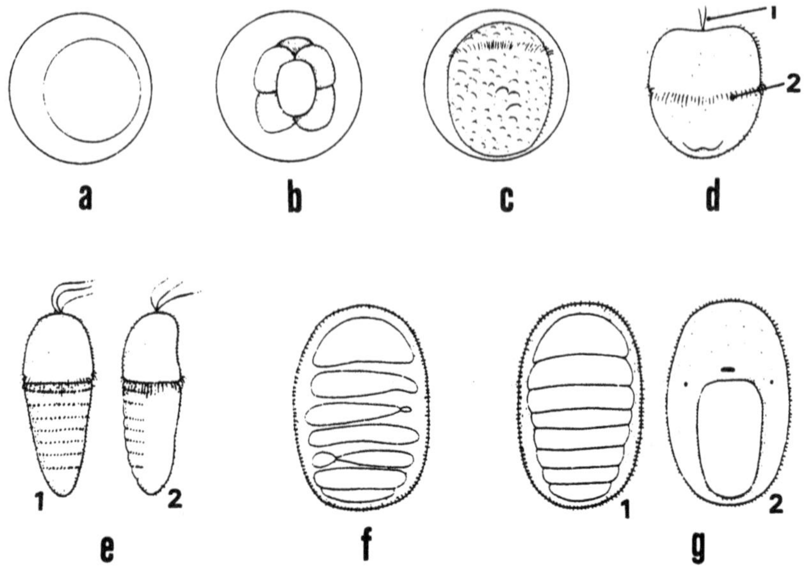

Fig. 11. Development of chitons (altered after Christiansen, 1954).

a, egg; b, eight-cell-stage; c, one or two hours before hatching; d, trochophora (1, apical tuft; 2, prototroch); e, nine days old larva (1, dorsal view; 2, lateral view), f, bottom stage; g, juvenile animal, two months old (1, dorsal view; 2, ventral view).

the pallial groove shortly before the nephridiopores, at the level of the suture of the sixth and seventh valves.

Copulatory organs are wanting in chitons; the sex cells are shed into the water, with the exception of a few brooding species. Spawning is generally initiated by emission of spermatozoids by one or more male individuals.

The eggs are always rich in yolk, enveloped in a thick membraneous hull, generally beset with all kinds of projections, often lobulate and typical for each species. The eggs are issued one or two at a time, sometimes free, but mostly held together by a slimy mass, either aggregated into clusters, or in long strings, which in *Cryptochiton stelleri* may attain a length of more than one meter. *Lepidochitona cinerea* emits one or two eggs every second during a quarter of an hour, making a total of about 1500 eggs in one or several mucous clusters.

Brooding of the eggs in the pallial groove is by no means a rare phenomenon, exhibited for instance by the Mediterranean *Lepidochitona corrugata*. Vivipary is only known in *Calloplax vivipara* from Chile.

Development takes place within the egg's hull. Fully developed larvae crack the hull and emerge as free living trochophores, equipped with a prototroch and an apical tuft of two to four flagella, or sometimes with a bunch of cilia. For *Leptochiton asellus* Christiansen (1954) established that the escape of the trochophores begins about twenty hours after fertilization of the eggs, but for *Lepidochitona cinerea* it takes three or four days, according to Hoffmann (1931).

Three or four days after the larvae are hatched metamorphosis begins with a dorso-ventral flattening of the trochophore and the appearance of six transverse grooves in the dorsal integument, marking the places where the valves I-VII are to be formed. The shell plates are rapidly developing, the tail plate appearing last. The larvae then settle on the bottom to start their chiton lives. After one or two years they are mature.

Ecology and Distribution

Chitons are mostly found in shallow coastal waters, preferably in the tidal and subtidal zones. They are well adapted to this life by their flattened shape and the broad creeping foot, which enables them to cling to rocks even in a strong surge. Only most of the Leptochitonidae and Hanleyidae, and a few species of other groups, like Ischnochitonidae and *Placiphorella,* are found in the coralline, or even in the abyssal zone. *Leptochiton alveolus* (Lovén), for instance, is found in a depth varying from 100 to almost 5000 m., while other species of this genus, like *L. asellus* (Gmelin) and *L. cancellatus* (Sowerby) may be found from the extreme low tide zone to a depth of 400 and 1000 m respectively.

Littoral species are living on a hard, smooth substrate, such as rocks, stones and old shells, feeding on algae, which they scrape off with the radula. Only *Chaetopleura angulata* appears to feed on *Balanus* spp. and Foraminifera. Deep water species may feed on sponges (some *Hanleya* and *Notoplax* species), on Hydroids, or on corals (Cryptoplacidae), but most *Leptochiton* species are true bottom dwellers, feeding on small organisms like Diatomeae, Radiolaria, Foraminifera, larvae of Crustacea and Polychaeta, and on organic particles in the bottom mud.

As far as we could ascertain all chitons are stenopolyhaline. Only *Lepidochitona cinerea* (L.), penetrating into the Baltic Sea as far as the Isle of Bornholm, tolerates a 15% salinity.

Deep water chitons are mostly uncoloured, the shell plates as well as the girdle generally creamy white, if not brownish or blackish by external incrustation. Most shallow water species show vivid colours in a variety of patterns, making them difficult to detect against their substratum. Moreover, some species are often overgrown with all kinds of marine growth. Algae, serpulid worms, nullipores, and barnacles often choose a chiton for their settling.

Quite a few species are lodging symbionts in their pallial groove, such as Protozoa (*Urceolaria korschelti*), sphaeromid Isopods (*Eusphaeroma crenulatum*) (fide Arey & Crozier, 1919), and mites (*Halixonis chitonis*) (fide Pelseneer, 1899). The huge *Cryptochiton stelleri* (Middendorff) may lodge the pea crab *Opisthomus transverses,* and also the polychaete *Acholoe vittata* among its gills (fide Ricketts & Calvin, 1939). The only known parasites of chitons are the Sporozoa *Haplosporidium chitonus* in the European *Lepidochitona cinerea* (L.), and *Chitonicum simplex* in the Chilean *Ischnochiton inca* (d'Orbigny) (fide Plate, 1901). In some species the shell plates are found to be infected with boring bivalves of the genus *Lithophaga,* as was established by Bullock & Bors (1971:1) in specimens of *Chiton tuberculatus* Linnaeus and *Chiton stokesi* Broderip, and by Watters (1981a:77) in *Acanthochitona hirudiniformis* (Sowerby).

In some cases it was observed that chitons are eaten by man. *Acanthopleura granulata* (Gmelin) was eaten by natives in the Antilles and in Bermuda. *Cryptochiton stelleri* (Middendorff), and *Katharina tunicata* (Wood) were eaten by the Indians around Puget Sound and in Alaska.

Chitons are little mobile animals. Creeping is slow, performed by anteroposterior contraction waves of the muscles in the foot, gliding slowly over the surface, eased by mucus excreted by the glands in the anterior part of the foot. Creeping is mostly done in the dark, when they leave their shelter to feed. At break of day they usually return to the same resting place. In the Dutch Wadden Sea, *Lepidochitona cinerea* (L.), living in the tidal zone on mud flats, seeks shelter under stones and loose valves of *Mya arenaria* at low tide, but becomes active on the exposed side of its hiding place as soon as it is washed by the incoming tide.

Chitons are found in all seas of the world. In cold and temperate regions a limited number of species is found. Leptochitonidae and Ischnochitonidae are cosmopolitan. Most species, however, are found in warm seas. In this respect Australasia and tropical West-America are richest, not only in species, but also in individuals. Most species are rather restricted in their distribution, many of them even endemics, probably due to their sluggish locomotion and their relatively short pelagic stage of development. Others are limited by their habitat, e.g. *Stenochiton cymodocealis* Ashby, which only feeds on the stems of the sea grass *Cymodocea* on the southern coasts of Australia.

Chaetopleura angulata (Spengler), a native from the coasts of Southern Brazil to Patagonia was involuntarily shipped to the coasts of Portugal and northern Spain by Iberian traders, certainly due to the animal's ability to climb rapidly anchor chains (fide de Rochebrune, 1889).

It is much more difficult to explain why some deep water species, like *Leptochiton alveolus* (Lovén), and *Ischnochiton* (*Stenosemus*) *exaratus* (Sars) are almost cosmopolitan. The former was described a dozen times from areas so wide apart that the authors never dreamed that the species they described might be identical with the European boreal species described by Lovén. *I. exaratus,* also originally described from Norwegian deep waters, is not only found on the East coast of the United States, south to Florida, but also in Magellan Straits and on the southern coast of Chile.

Collection and Preservation

From the information in the chapter on the ecology and distribution one can conclude that chitons must be looked for in the tidal or subtidal zone, and more specifically in those places where loose stones, old shells, or other movable hard substrates can be inspected. As most species are photonegative, specimens will primarily be found on the sides or the underside of the substrate. Real rocks are not often preferred by the chitons, except for a few species that can be found in the splash zone. In dredgings from deeper water chitons can be found on stones, old shells, sponges, corals, kelp holdfasts, seldom on living shells, although it can occur, f.i. *Callochiton septemvalvis euplaeae* on *Charonia nodifera*.

A minimum of equipment is needed for collecting chitons: a knife with a thin, sharp-pointed blade; boards of different sizes, i.e. rectangular pieces of plexiglass or any other rigid material that does not deform or discolour in the killing or preserving solutions; fine thread that easily can be broken off; a widemouthed, tightly sealable killing jar of sufficient size to hold quite a lot of boards of tied chitons, the jar must be filled about two-thirds with a 60% alcohol solution, and bear a label on which all pertinent collecting data can be noted; a plastic bucket, preferably with different sections, to store the above mentioned items and the possible recalcitrant specimens. Back home, several wide-mouthed, tightly sealable preserving jars, of different sizes, and filled with a preserving solution of 2 parts glycerine, 2 parts alcohol 98% and 1 part water must be available.

Once in the field, care should be taken to pick up or turn over stones very cautiously, so as not to crush any possible chiton clinging to the underside. When a specimen is spotted, gently depose the stone in such a way that the chiton stays visible and above water; then first select a board of somewhat larger size than the chiton, wet it, and keep it in one hand; with the other hand, carefully slide the point of the knife under the animal and remove it without damaging the girdle or the foot, place it immediately on the board and, with a slight pressure of the thumb, keep it flat for a few seconds, and the chiton will clamp down to the board. Now it can easily be tied, lengthwise only, to the board, after which it is placed in the killing jar. Personal experience has proven this to be the best method for getting perfect specimens.

If time or collecting conditions do not permit tying the chitons immediately to the boards, they can be placed in an open container partly filled with frequently renewed seawater, to be taken home alive. There they should be placed immediately on a dish, to relax, flatten and clamp down to the bottom of the dish. Following which the same procedure can be followed as explained above.

Chitons must remain in the killing solution for at least 24 hours, larger specimens preferably for several days. In the last case the solution should be replaced accordingly when necessary.

The next step depends on the purpose of the collection. For a wet collection, it will suffice to untie the chitons and place them, with short intervals, in several gradually more concentrated alcohol solutions, culminating with a 70% solution for definite conservation. For a dry collection, the untied chitons must be placed in the preserving solution for two or more weeks, depending on the size of the animals. A heterogenous series should be sorted out and placed in several preserving jars. Afterwards the chitons should be tied to the boards again and be dried in a dark, well ventilated place for a length of time at least equal to the time they have spent in the preserving solution.

To conclude these short notes on how to collect and prepare chitons for further study or permanent preservation, a piece of advice must be added: never use formaline, unless neutralized, for any purpose in connection with chitons; never cut out the internal parts for any reason; never clean the chitons, unless you know why and how, and even then only if it is for personal study. Be careful to provide all specimens with the correct collecting data; do not necessarily discard a specimen that, despite all precautions, has managed to curl up; it can be used for disarticulation and study of the internal characters of the plates, and for the preparation of perinotum and radula slides. When in the field, remember the general recommendations applicable to any shell collector; for a chiton student a curled up or damaged specimen is better than nothing, so, when chancing upon a chiton, even when personally not interested and not equipped for an appropriate treatment, take it home. Someone, somewhere, will be grateful to you for it.

Historical Review

Very little was known about chitons until 1758, when the 10th edition of Linnaeus' Systema Naturae appeared. In this work, Linnaeus accepted the generic name *Chiton* and gave a short diagnosis of four species: *hispidus, tuberculatus, aculeatus* and *punctatus*, all of which except *tuberculatus* are now considered nomina inquirenda. In 1764 he created a fifth species, *Chiton squamosum*, with a, for that time, exemplary description. Three years later, in the 12th edition of the Systema Naturae, he increased the number of species to 9, adding 4 European forms: *fascicularis, ruber, albus* and *cinereus;* these are still considered valid names.

In the following decades more scientists began to pay attention to these animals. Chemnitz, in volume 8 of the Neues systematisches Conchylien-Cabinet (1785), gave extensive descriptions and figures of 19 species. Unfortunately, this important work, being non-binominal, was placed on the Official Index of Rejected and Invalid Works in Zoological Nomenclature, so that his new names cannot be used. However, several of them were taken over by Gmelin in the 13th edition of Systema Naturae (1791), which included a short diagnosis of 28 species. In 1797, Spengler, who seems to have taken no notice of Gmelin's work, published a fine monograph on *Chiton* and described and partly illustrated 27 species.

Several important contributions towards a better understanding of the chitons were published by de Blainville. In 1816 he separated them from the other mollusks as a Class named Polyplaxiphores, and in 1825, in the article upon the chitons in the Dictionnaire des Sciences Naturelles, he made a serious attempt to split up the group into several divisions and subdivisions, and described, unfortunately without illustrations, some 65 species, about half of them new to science. Almost at the same time (1829), another effort towards a classification was published by Guilding; he recognized five genera.

Meanwhile, thanks to the discoveries of Frembly, Cuming and others, a large number of new species were brought to England. Most of these novelties were described by Broderip and Sowerby I, and in 1840, in the Conchological Illustrations, Sowerby II published remarkably good drawings of over a 100 species. This number was almost doubled when Reeve, in 1847, published a monograph in the Conchologia Iconica, giving good illustrations but inadequate descriptions of 189 species. The same year Gray, in a series of three articles in the Proceedings of the Zoological Society of London, proposed a system of classification for the Chitonidae, creating many new, and indicating the synonymy and types of all known genera. The system was based on external characters, more particularly the features of the insertion plates, and practically all his names are still in use today. Simultaneously, a very important work on the anatomy of the Siberian chitons was published by von Middendorff. He also proposed an extensive scheme of classification, but his groups, being very heterogeneous, have not been retained by later workers, except for *Cryptochiton*.

For a length of time no important work was issued on chitons, until P. P. Carpenter (1877) started a series of articles on the West American and Atlantic chitons. In preparation of a monograph of the entire group, he then reexamined all the important American and English collections, but he did not live to see the results of his studies printed. In 1882 an abstract of his classification, based on the outstanding features of the insertion plates, was published by Dall, who just previously (1879) had also published an extensive and valuable paper on the dentition of the group, giving descriptions and figures of the radulae of species representing most of the then known genera. A further important step in the evolution of the chiton study was taken in 1889, when Dall divided the order Polyplacophora into two superfamilies, with several families in each group.

An epoch in the history of the group was marked with the publication of Pilsbry's monograph of the Polyplacophora in Tryon's Manual of Conchology, volume 14: 1892-1893, and part of volume 15: 1893-1894. In that monumental work more than 500 species are described at length and eventually discussed, accompanied by an exhaustive synonymy and bibliography and in most cases illustrated by one or more drawings. A new classification was introduced with the following words: "The classification given in this work is based mainly upon characters furnished by the articulating surfaces of the valves" (vol. 14: xxiii). It is composed of 3 superfamilies, with 6 families, 5 subfamilies, 38 genera and 27 subgenera or sections. A phylogenetic diagram of the families and the subfamilies and an analytical key to the genera are proposed, while other chapters deal with the shell, the girdle, the head, foot and gills, the teratology, the collection and preservation, and the literature of chitons. So profound has been the study of Pilsbry, that it has become the Bible of most later chiton students, especially as no similar work has been published ever since.

This does not mean that no other authors did any important work on chitons. Thiele (1893) made a profound study of the dentition and, as a result, created 23 new genera. That from a taxonomical point of view the radula has proved to be inadequate is clearly demonstrated by the fact that none of Thiele's genera are used in modern classification, with exception of *Rhyssoplax* which, due to the non-radular characteristics of the type species, has subgeneric status. Plate (1898-1901) published an extensive study on the anatomy and phylogeny of the chitons. In 1909-1910, Thiele, after a revison of the system of the chitons, proposed a new classification and created the suborders Lepidopleurina and Chitonina. That classification was slightly changed and elaborated in his Handbuch der systematischen Weichtierkunde (1929).

The beginning of this century was also marked by many contributions towards the better knowledge of the local chiton fauna by Australasian workers such as Ashby, Bednall, Hedley, Hull, Iredale, Matthews, May, Suter and Torr. Their work culminated in the publication, by Iredale & Hull, of A Monograph of the Australian Loricates (1923-1927), and The Loricates of the Neozelanic Region (1929-1932).

All over the world new species were discovered and described in short papers, local monographs, or revisions of a definite taxon. Also the study of the fossil species became intensified and in 1955, on the occasion of an important study on the Swedish fossil chitons, Bergenhayn proposed a provisional synopsis of the systematic classification of the entire class of Polyplacophora. A thorough work was done on the same subject by A. G. Smith in Moore's Treatise on Invertebrate Paleontology (1960), followed by Van Belle in a series of Belgian publications (1975-1978).

As to the number of valid species, it has slowly increased to about 750, and even now, thanks to the study of dormant material in old collections, the more sophisticated dredging methods and the prospecting possibilities in formerly inaccessible places, new species are being discovered year after year. So, after recently having published a Catalogue of living Chitons, (1980), it has become the bold object of the authors to enlarge the catalogue with descriptions, illustrations and general and systematic notes, so as to present an updated and complete monograph of all living species for the interested students.

List of Abbreviations

AM	Australian Museum, Sydney.
ANS	Academy of Natural Sciences, Philadelphia.
BMNH	British Museum (Natural History), London.
CASG	California Academy of Sciences, Geological Department, San Francisco.
CASI	California Academy of Sciences, Department of Invertebrate Zoology, San Franscisco.
CM	Canterbury Museum, Christchurch.
IBUSP	Instituto de Biociências, Universidado de São Paulo.
IRSN	Institut Royal des Sciences Naturelles de Belgique, Bruxelles.
ITZ	Instituut voor Taxonomische Zoology, formerly Zoölogisch Museum, Amsterdam.
K	Private collection of P. Kaas.
KNVSM	Kongelige Norske Videnskabers Selskab, Museet, Trondheim.
LACM	Natural History Museum Los Angeles County.
MNHN	Museum National d'Histoire Naturelle, Paris.
NHMW	Naturhistorisches Museum, Wien.
NHRM	Naturhistoriska Riksmuseet, Stockholm.
NM	Natal Museum, Pietermaritzburg.
NMNZ	National Museum of New Zealand, Wellington.
NMW	National Museum of Wales, Cardiff.
QM	Queensland Museum, Fortitude Valley.
RMNH	Rijksmuseum van Natuurlijke Historie, Leiden.
RSM	Royal Scottish Museum, Edinburgh.
SAMA	South Australian Museum, Adelaide.
SAMC	South African Museum, Cape Town.
SIO	Scripps Institute of Oceanography, La Jolla, California.
TM	Tromsö Museum, Tromsö.
TMAG	Tasmanian Museum and Art Gallery, Hobart.
USNM	United States National Museum of Natural History, Smithsonian Institution, Washington D.C.
UUZM	Uppsala Universitets Zoologiska Museum, Uppsala.
VB	Private collection of R. A. Van Belle.
WAM	Western Australian Museum, Perth.
WEM	The Wood End Museum of Natural History, Scarborough, England.
WM	Wanganui Museum, Wanganui, New Zealand.
YPM	Yale Peabody Museum, Newhaven, Connecticut.
ZIAS	Zoological Institute, Academy of Sciences of the USSR, Leningrad.
ZIM	Zoologisches Museum und Zoologisches Institut der Universität Hamburg.
ZMB	Zoologisches Museum, Bern.
ZMHU	Zoologisches Museum an der Humboldt Universität, Berlin.
ZMK	Zoologisk Museum Universitet København.
ZMO	Zoologisk Museum, Oslo.
ZMR	Zoological Museum, Roma.

SYSTEMATIC PART

It must be remembered that only the recent species will be treated here. At the end of the work a synopsis of the complete classification, fossil taxa included, will be given.

Class POLYPLACOPHORA Gray, 1821

–Lond. med. Repos. 15:234 (validation proposed L. R. Cox, 1958. –Bull. zool. Nom. 15:543 – ICZN pend.).

=Polyplaxiphores de Blainville, 1816; Loricata Schumacher, 1817 (non Desmarest, 1804); Cyclobranchia Cuvier, 1817 (partim); Crepipoda Goldfuss, 1820; Lamellata Latreille, 1825; Placophora von Ihering, 1876; Polybranchiata Spengel, 1881; Lepidoglossa Thiele, 1893.

Mollusks protected by a series of dorsal calcareous plates, generally eight in number.

Key to orders

1a Primitive chitons lacking articulamentum . . . . Paleoloricata (†)
1b Chitons with articulamentum . . . . Neoloricata

Order NEOLORICATA Bergenhayn, 1955

–Kungl. Fysiogr. Sällsk. Handl., Lund 66(8):32.

Chitons with articulamentum projecting from beneath the tegmentum as sutural laminae only, or as sutural laminae and insertion plates on one or more valves.

Key to suborders

1a Valves without, or with unslitted insertion plates . . . . Lepidopleurina
1b Insertion plates on all valves, at least valve I slitted 2
2a Tegmentum always present, number of slits in valve I: 6 or more . . . . Ischnochitonina
2b Tegmentum absent, or, number of slits in valve I at most 5 . . . . Acanthochitonina

Suborder LEPIDOPLEURINA Thiele, 1910

–Zoologica Stuttg. 22:116.

=Eochitonia Dall, 1889 (partim); Eoplacophora Pilsbry, 1893; Protochitonina Ashby, 1929 (partim); Lepidopleurida, Chitonida (partim) Thiele, 1929; Helminthochitonida

Quenstedt, 1931: Gryphochitonida Quenstedt, 1932; Isoplacophora Cotton & Weeding, 1939 (partim); Afossochitonina Bergenhayn, 1955; Choriplacina Starobogatov & Sirenko, 1975; Cymatochitonina Sirenko & Starobogatov, 1977.

Valves without, or with unslitted insertion plates.

Key to families

1a No insertion plates . . . . Leptochitonidae
1b Insertion plates on at least some valves 2
2a Valve I without insertion plate . . . . Protochitonidae (†)
2b Valve I with insertion plate 3
3a Insertion plate on valve I, more or less developed on II – VII, or on I and VIII only . . . . Hanleyidae
3b Insertion plates on all valves . . . . Afossochitonidae

Family LEPTOCHITONIDAE Dall, 1889

–Bull. Mus. comp. Zool. Harvard Univ. 18:414.

=Leptoidea Dall, 1879; Holochiton (partim), Eochiton (partim) Fischer, 1885; Lepidopleuridae Pilsbry, 1892; Gryphochitonidae Pilsbry, 1900.

Animal ovate to elongate, tegmentum generally uncoloured and granulated all over, insertion plates wanting, sutural laminae small, widely separated, girdle clothed with various processes.

Key to subfamilies

1a Fossil species of small to large size; form of valves and sculpture of tegmentum markedly different from those of Recent Leptochitons. Lower Ordovician to Lower Jurassic . . . . Helminthochitoninae (†)
1b Of small size, tegmentum generally sculptured with minute pustules. Eocene to Recent . . . . Leptochitoninae

Subfamily LEPTOCHITONINAE

Animals small; valves generally brittle, tegmentum generally sculptured with minute pustules.

Key to genera

1a Valves solid, tegmentum heavily sculptured 2
1b Valves thin, tegmentum finely granulose 3
2a Sculpture of heavy concentric ridges . . . . *Lepidopleurus*
2b Sculpture of flat nodules . . . . *Oldroydia*
3a Granules of equal size . . . . *Leptochiton*
3b Granules getting larger towards outer margins . . . . *Hanleyella*

Genus ***Lepidopleurus*** (Leach MS) Risso, 1826

–Hist. nat. Eur. Mérid. 4:267.

Type: *Chiton cajetanus* Poli, 1791, SD Herrmannsen, 1846. Non: *Lepidopleurus* (Carpenter MS) Dall, 1879 (=*Lepidozona* Pilsbry, 1892).

Shell elongate, valves solid, lateral areas much elevated and heavily sculptured; girdle clothed with scales interspersed with hyaline spicules.

Distribution: Mediterranean Sea; Atlantic Ocean from the Bay of Biscay to the Canary Islands; Pacific Ocean: off Cape San Lucas, Lower California.

Lepidopleurus cajetanus (Poli, 1791)

(Fig. 12. Map 1)

Type: lost (?)
Type locality: Gaeta, Tyrrhenian Sea.

Chiton cajetanus Poli, 1791:10, pl. 4:figs 1-2.
Lepidopleurus cayetanus; Risso, 1826:267.
Chiton concentricus Sowerby, 1840:figs 72, 72a. Hab.? Nom. nud.
Lepidoplerus cajetanus; Leloup, 1934b:2, figs 1-6 (bibliography and synonymy); Leloup & Volz, 1938:8, figs 4-7; A.G. Smith, 1960b:75, pl. 17:figs 1-5; Malatesta, 1962:146, figs 1, 2; Sabelli & Spada, 1971a:35, figs 1-17; Sabelli, 1974a:71; Laghi, 1977:95, pl. 1:figs 13-20, text-figs a-d; Barash & Danin, 1977:5, fig. 1.

Description. Animal elongate, more than twice as long as wide, moderately elevated, the back rounded, not carinated, valves and girdle dirty white to light ochraceous.

Head valve thick, with 7-10 strong, concentric, terraced ribs, finely granulated all over. Intermediate valves with highly elevated lateral areas, also marked with thick concentric ribs, central areas with somewhat branching or anastomosing longitudinal chains of granules, the interstices rather wide and pitted; apices worn away, even in young specimens, apophyses triangular, strongly developed, separated by a rather wide, convex sinus. Tail valve with the prominent, raised and forwardly curved mucro a little posterior; the posterior slope convex. Antemucronal area forwardly produced, sculptured like the central areas, postmucronal area like the head valve; the apophyses wide, rounded.

Girdle very narrow, dorsally densely clothed with rounded to squarish, not imbricating, calcareous scales, ornamented with about 10 fine riblets, 80 μm long and wide. Among them slender, pointed, almost smooth spicules are found, up to 280 μm long, rising from chitinous cups. Marginal spicules straight to slightly curved, longitudinally finely grooved, varying in length from 60 to 150 μm. Ventral side of the girdle covered with close radiating rows of imbricating, elongate oval, flat scales, those near the outer margin distally sculptured with three or four short longitudinal riblets, 85 μm long, 30-35 μm wide.

Central radula tooth twice as long as wide, narrowed in the middle, with a thin blade; major lateral tooth with a strong tridentate head, the central denticle largest, the interior one smallest.

Length up to 30 mm, usually 20-25 mm.

Distribution. Mediterranean Sea: European coasts until Israel; on the North African coast only known from Oran (Algeria). Atlantic Ocean: from

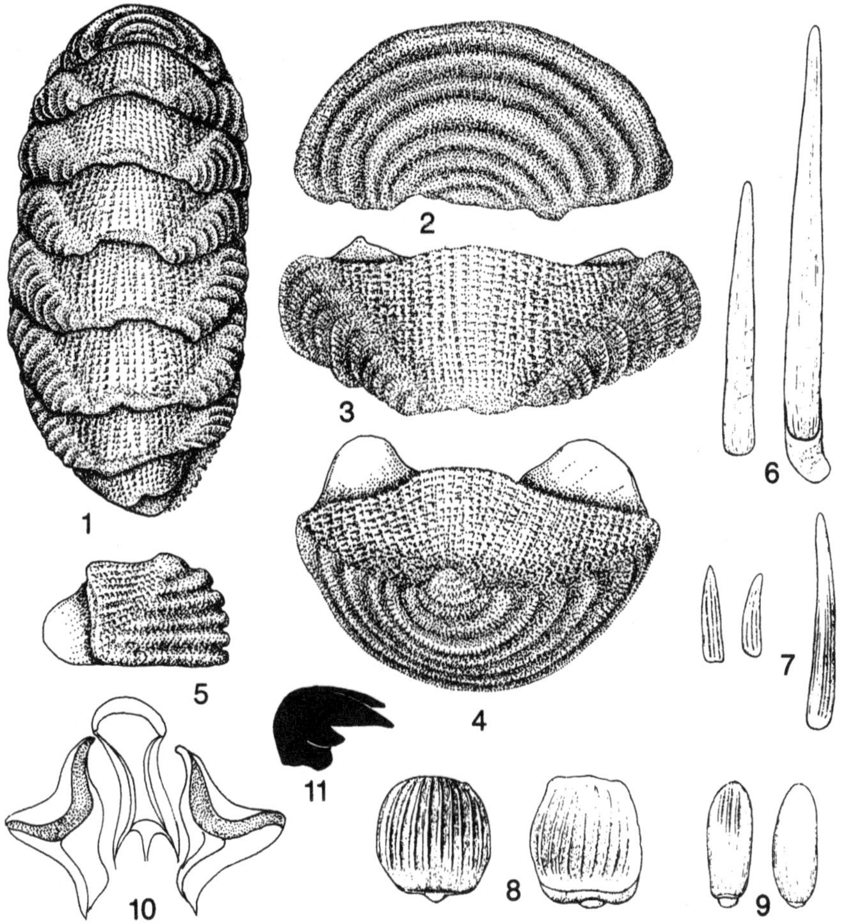

Fig. 12. *Lepidopleurus cajetanus* (Poli, 1791).

1, whole animal, × 4.5; 2, valve I, dorsal view, × 9; 3, valve II, dorsal view, × 9; 4, valve VIII, dorsal view, × 9; 5, valve VIII, lateral view, × 9; 6, dorsal girdle spicules, × 195; 7, marginal spicules, × 195; 8, dorsal girdle scales, × 195; 9, ventral scales, × 195; 10, central and first lateral teeth of radula, × 375; 11, cusp of major lateral radula tooth, × 190.

1, specimen from France, Alpes-Maritimes, Cap Ferrat, on porous stones, 40 m, VI.1951, Trégouboff leg., K 2754; 2-5, 10-11, specimen from Corsica, Calos harbour, on rocks, VII. 1956, H. van Haren leg., K 3866; 6-9, after Leloup, 1934.

the coasts of Spain and Portugal south to the Canary Islands. It was also recorded from Brittany: Quiberon and Gavre (Taslé, 1867), Roscoff (Vasseur, 1877), Lanninon (Daniël, 1883), but as it has never been rediscovered there in our century, it is highly probable that those specimens had been introduced with Portuguese oysters.

L. cajetanus is chiefly found in the littoral zone, living on stones, rocks and old shells. In the sublittoral zone it is generally found in a depth of at most 40 m.

Observations. Juveniles are hardly recognizable, as they lack the strong concentric ridges. At that stage they can easily be mistaken for young *Hanleya hanleyi,* from which they differ, however, by the pitted or scalloped interstices of the chains of granules.

Lepidopleurus scrippsianus Ferreira, 1980

(Fig. 13. Map 2)

Type: CASI, type series 00716 (holotype), 00717 (1 paratype); LACM 1700 (1 paratype); SIO (1 paratype).

Type locality: Mexico, off Baja California Sur, SW of Cabo San Lucas, 22°30.8'N 110°03.8'W through 22°37.2'N 110°15.5'W, 2507-2891 m.

Lepidopleurus scrippsianus Ferreira, 1980:55, figs 1-11.

Description. Animal of moderate size, elongate oval, less than twice as long as wide, valves rugose, subcarinated, colour white.

Head valve thick, semicircular, surface smooth except for coarse growth rugae. Intermediate valves broadly triangular, strongly beaked, apex somewhat eroded, central areas minutely granular, granules irregularly placed and variably spaced, lateral areas elevated and prominent, smooth except for coarse, irregular, concentric growth rugae. Tail valve semicircular, moderately inflated, mucro at the posterior third of the valve, postmucronal area with heavy concentric growth lines, strongly convex, sloping sharply.

Articulamentum chalky white, sutural laminae small, triangular, sinus very wide.

Girdle relatively thick and wide, creamy white, velvety in appearance, upper surface uniformly covered with juxtaposed spiculoid processes, 100 µm long, 20 µm wide, with pointed tips, often broken and jogged, undersurface completely devoid of any scalelike formations, reduced to a thin cuticle.

Central radula tooth about 70 µm long and 30 µm wide in the front where it bears a small blade, first lateral tooth with a socket-like formation on the outer side, major lateral tooth with a tridentate cusp, the central denticle largest, the exterior one smallest.

Gills adanal, posterior, about 20 per side, extending 40% of length of foot.

Holotype 12.1 × 8.2 mm, paratypes 20 × 10, 15 × 8 and 13 × 7 mm.

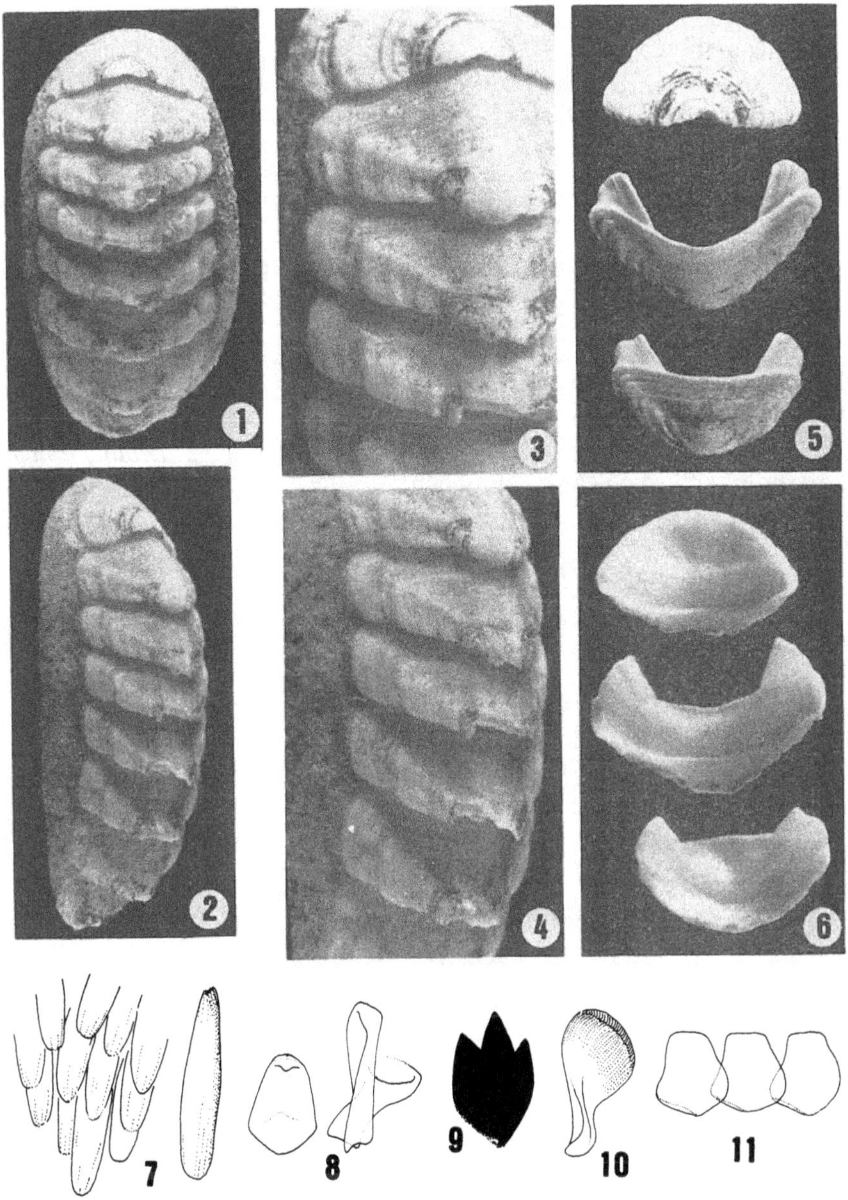

Fig. 13. *Lepidopleurus scrippsianus* Ferreira, 1980.

1, whole animal, dorsal view, × 3.7; 2, do, dorso-lateral view, × 4; 3, part of left side, dorsal view, × 7.5; 4, do, dorso-lateral view, × 6.5; 5, valves I, intermediate, VIII, dorsal view, × 4.8; 6, do, ventral view, × 4.8; 7, dorsal girdle spicules, × 200; 8, central and first lateral radula teeth, × 200; 9, cusp of major lateral radula tooth, × 200; 10, spatulate uncinal tooth, × 100; 11, marginal teeth of radula, × 100.

1-4, paratype; 5-11, holotype; all after Ferreira, 1980.

Distribution. *Lepidopleurus scrippsianus* is known only from the four specimens of the type lots.

Genus ***Leptochiton*** Gray, 1847

Proc. zool. Soc. Lond. 15: 127.

Type: *Chiton cinereus* Montagu, 1803; SD Gray, 1847. (=*Chiton asellus* Gmelin, 1791); non *Chiton cinereus* Linnaeus, 1767.

Shell ovate to elongate, valves thin, lateral areas hardly elevated, tegmentum finely granulose.

Key to subgenera

1a Tail valve normal, mucro median or submedian 2
1b Tail valve extremely long, mucro posterior . . . . *Parachiton*
2a Girdle with small scales, spicules, or both . . . . *Leptochiton* s.s.
2b Girdle with spicules and silky hairs . . . . *Pilsbryella*

Subgenus ***Leptochiton*** s.s.

=*Deshayesiella* (Carpenter MS) Dall, 1879; *Lophyropsis* Thiele, 1893; *Beanella* Thiele, 1893 (non Dall, 1882); *Terenochiton* Iredale, 1914; *Xiphiozona* Berry, 1919; *Belchiton* Ashby & Cotton, 1939.

Mucro submedian, girdle clothed with small scales, sometimes interspersed with spicules, or wholly spiculose.

Distribution. Cosmopolitan, with concentrations in North Atlantic, Australasian and Japanese waters.

Cosmopolitan species

Leptochiton (L.) alveolus (M. Sars MS, Lovén, 1846)

(Fig. 14. Map 6)

Type: NHRM, type collection 104.
Type locality: Bohuslän, Sweden.

Chiton alveolus M. Sars MS, Lovén, 1846:159.
Leptochiton alveolus; Dall, 1879:317; Ferreira, 1979b:152, figs 9-16, 37, 38 (bibliography and synonymy); Kaas, 1979a:15; 1981:223, figs 8, 10 A-B.
Leptochiton belknapi Dall, 1878:1.
Leptochiton benthus Haddon, 1886:10, pl. l:fig. 1, pl. 2:figs 1a-m.
Lepidopleurus similis E. A. Smith, 1894:167, pl. 4:figs 9-12.
Lepidopleurus mesogonus Dall, 1902:505.
Lepidopleurus halistreptus Dall, 1902:556.
Lepidopleurus halistreptus abbreviatus Dall, 1908:354.
Lepidopleurus luridus Dall, 1902:556.
Lepidopleurus farallonis Dall, 1902:557.
Lepidopleurus giganteus Nierstrasz, 1905:3, pl. l:figs 1,2, pl. 2:figs 39-42.
Lepidopleurus simplex Nierstrasz, 1905:4, pl. l:fig. 3, pl. 2:figs 44-47.

Lepidopleurus opacus Dall, 1908:354.
Lepidopleurus japonicus Thiele, 1909:11, pl. 1:figs 1a-c.
Parachiton opiparus Iredale & Hull, 1925:345, pl. 39:fig 20.
Lepidopleurus japonicus aequivalvus Bergenhayn, 1933:4, figs 1a-c.
Lepidopleurus aequispinus Bergenhayn, 1933:6, figs 1d-g, pl. l:fig. 1, pl. 2:fig. 20.

Description. Animal rather large for the family, up to 3, exceptionally to 4 cm long, elongate, more than twice as long as wide, rather elevated, back evenly rounded to subcarinate, side slopes convex, girdle narrow.

Head valve semicircular; intermediate valves of about equal width, highly arched, rather long, the anterior margin decidedly curved inward, posterior margin almost straight, without a distinct apex. Lateral areas not or hardly raised, always with a few concentric growth marks; posterior valve semicircular, the front margin curved inward, mucro central, not prominent, posterior slope a little concave directly behind the mucro.

Tegmentum sculptured with neatly separated roundish to oval, raised granules arranged irregularly quincuncially on the central and antemucronal areas, quincuncially on the head valve, the lateral and postmucronal areas, where they are less prominent.

Colour of tegmentum and girdle white or creamish, sometimes stained with black or brown by environmental deposits.

Apophyses small, sharply triangular, widely separated by a concave sinus in the intermediate valves, trapezoid in the tail valve.

Girdle dorsally covered with not imbricating, elongate, bluntly pointed scales, sculptured with about 6 raised riblets, narrower than the interstices, 60-65 μm long, 30-35 μm broad; the sutural scales larger, sharply pointed, with 4-5 riblets, 100 μm long, 35 μm broad. Scattered among the scales finely striated, slender needles are found, up to 120 μm long, 10 μm thick. There are also small groups of intersegmental spicules, somewhat stronger striated and relatively thicker, 150 μm long, 16 μm thick. Ventral scales more or less wedge-shaped, with 6 weak riblets, 85 μm long, varying in width from 16-30 μm.

Radula with a small, short, weakly cusped central tooth; cusp of major lateral tooth with one long, sharp denticle, 180 μm long.

Length 20 mm (average) to 40 mm (exceptional); width 7-8 mm (average) up to 14 mm.

Distribution. *L. alveolus* is a deep water species, hardly ever found in a depth of less than 100 m; in the Bay of Biscay dredged in 4825 m. It is the only cosmopolitan species occurring in the three oceans. It was originally described from the west coast of Scandinavia (Bohuslän, Sweden, and Bergen, Norway), but the species was reported under a dozen different names from elsewhere.

Atlantic Ocean: from the Barents Sea all along the Norwegian and the Swedish west coast; Iceland; Bay of Biscay; NW Portugal; Gulf of St. Lawrence: between Cape Rosier and the SW point of Anticosti Id; Gulf of Maine; St. George's Bank.

Pacific Ocean: along the west coast of North America, from Point Barrow S. to Panamá Bay; in NE Asia: Kurile Is; Okhotsk Sea; Japan Sea S. to Korea; Central Pacific: Philippines and Indonesian archipelago, South Australia.

Indian Ocean: Sri Lanka; Kerguelen Is (southernmost record).

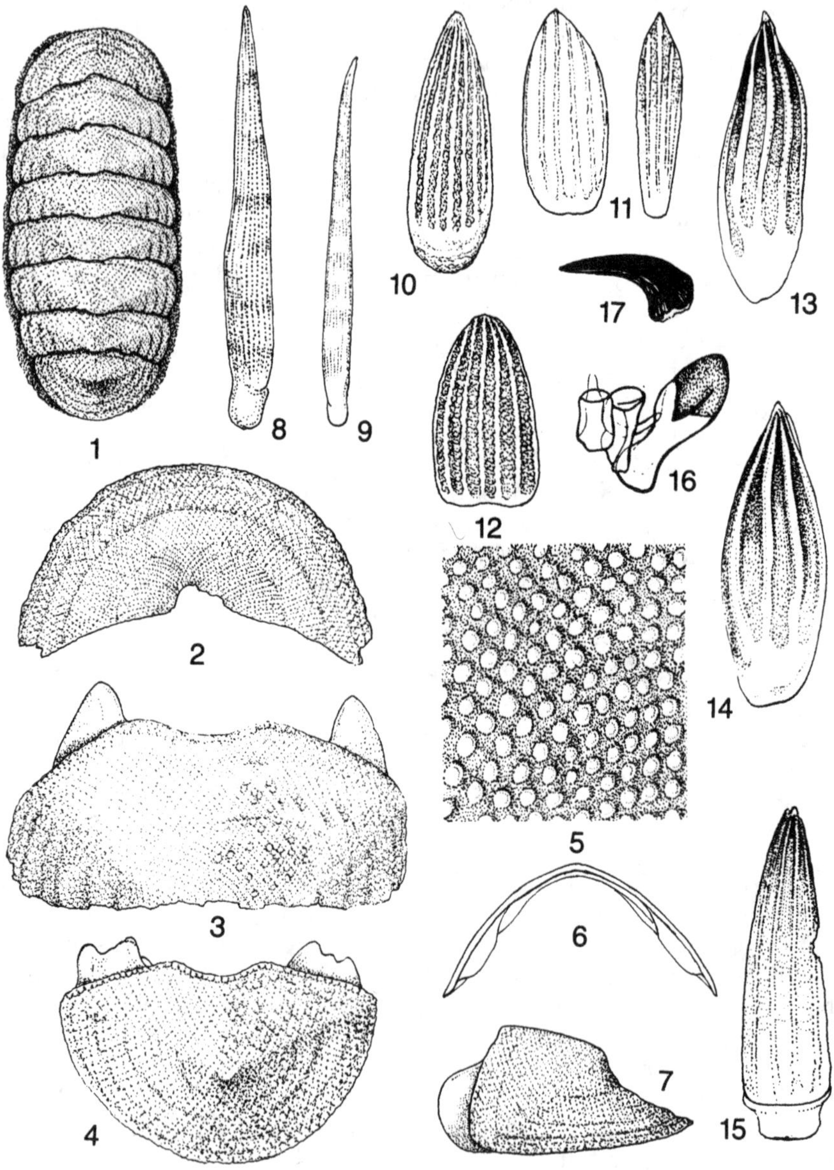

Observations. The type is a disarticulated specimen in alcohol deprived of its shell plates. A printed label reads: "Bohusl. Lovén," similar to the print in the "Index Molluscorum Scandinaviae". It is, however, easily recognizable by its girdle. Ferreira (1979b:152), who studied many types of *Leptochiton* species, particularly those described by W. H. Dall, was the first who gave a full account of the synonymy of the present species. To his list may be added *Lepidopleurus similis* E. A. Smith, 1894, from Sri Lanka. The type is BMNH 1894.9.11.13. It only differs from typical *L. alveolus* in the lateral areas of the intermediate valves being somewhat raised. *Parachiton opiparus* Iredale & Hull, 1925, from the holotype (AM, C 3026), collected 40 mi. S of Cape Wiles, South Australia, appeared to be another synonym of *alveolus*.

In *L. japonicus aequivalvus* Bergenhayn, 1933 (type: UUZM 136), and *L. aequispinus* Bergenhayn, 1933 (type: UUZM 142) the valves are more carinated, having a small but distinct apex. These minor differences may well fall within the range of intraspecific variation.

North East Atlantic Species

Leptochiton (L.) asellus (Gmelin, 1791)

(Fig. 15. Map 9)

Type: "Ex Museo Spengleriano", ZMK (lectotype + 1 paralectotype, design. Kaas, 1981).
Type locality: Norway, Telemark, Kragerö.

Chiton asellus Gmelin, 1791:3206; Spengler, 1797:99-100, con var. *c.*; non var. *a*, non var. *b*.
Chiton cimicinus Spengler, 1797:79, ex parte.
Chiton cinereus; Montagu, 1803:3 (non Linnaeus, 1767).
Chiton minimus; Wood, 1815:19 (non Gmelin, 1791; nec Spengler, 1797).
Chiton fuscatus Leach, 1852:231 (non Brown, 1827).
Chiton scoticus Leach, 1852:232.
Lepidopleurus asellus; Leloup, 1934a:2, 1 map (bibliography and synonymy); Knudsen, 1949:2; Jakovleva, 1952:51, figs 11:1, 2, 13, pl. 1:figs 1a-b; Christiansen, 1954:52, figs 1-31; Muus, 1959:37, fig. 21; Malatesta, 1962:150, not fig. 5; Knudsen, 1970:2; McKay & Smith, 1979:1.
Leptochiton asellus; Kaas, 1981:217, figs 1-3.

Fig. 14. *Leptochiton (L.) alveolus* (M. Sars MS, Lovén, 1846).

1, whole animal, dorsal view, × 4.8; 2, valve I, dorsal view, × 12; 3, valve IV, dorsal view, × 12; 4, valve VIII, dorsal view, × 12; 5, detail of sculpture close to front margin of intermediate valve, × 52; 6, valve IV, anterior view, × 8; 7, valve VIII, lateral view, × 12; 8, intersegmental spicule, × 400; 9, dorsal girdle spicule, × 400; 10, ventral girdle scale, × 400; 11, two other types of ventral scales, × 400; 12, dorsal scale, × 400; 13-14, dorsal scales near the suture, × 400; 15, marginal spicule, × 400; 16, central, first lateral and major lateral radula teeth; 17, cusp of major lateral radula tooth, × lOO.

1-15, 17, after Kaas, 1981; 16, after Thiele, 1893.

Description. Animal oval, up to 18 mm long, 10 mm broad, moderately arched, dorsal elevation 0.36, subcarinated, the side slopes almost straight, valves slightly beaked, lateral areas hardly raised.

Head valve almost semicircular, the tegmentum ornamented with 70-80

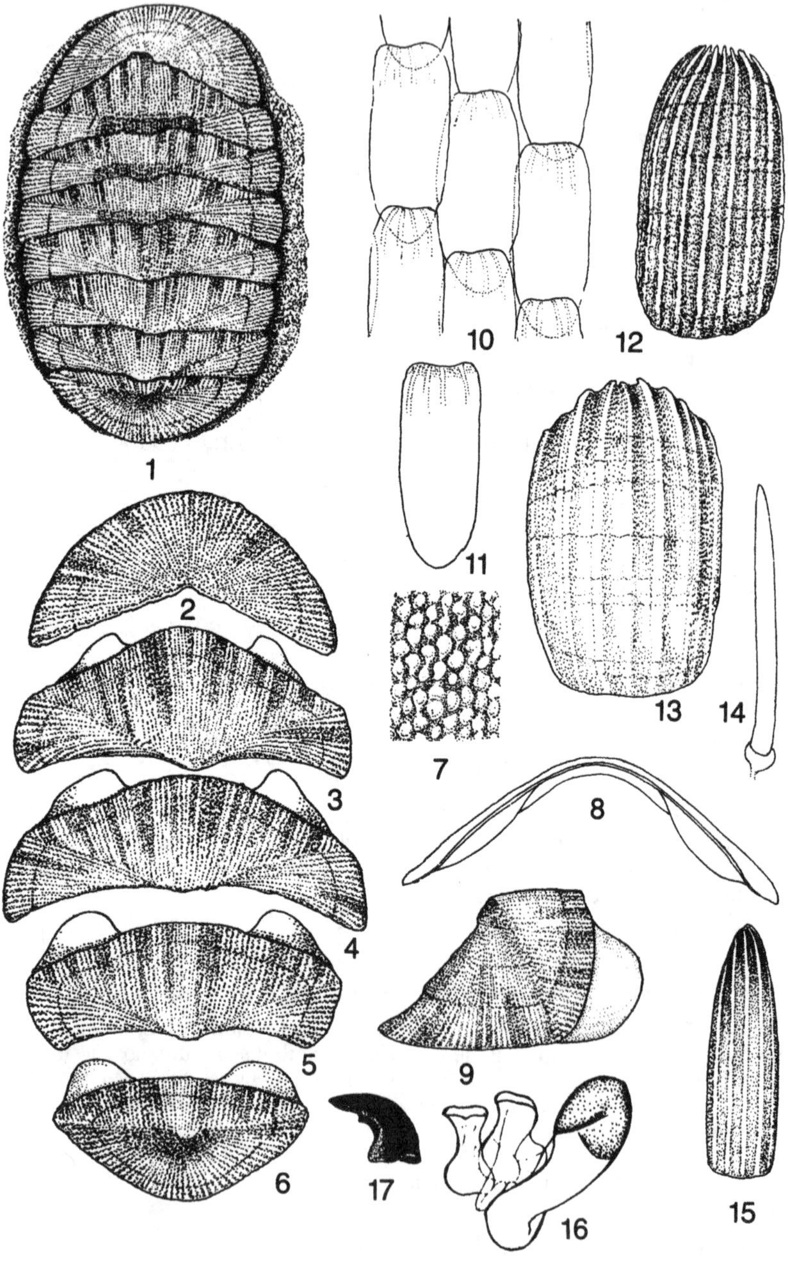

radiating series of small, little elevated, roundish to oval pustules. Intermediate valves with 60-70 longitudinal series of pustules, on the pleural parts somewhat converging towards the posterior margin, lateral areas distinctly separated by low diagonal lines, hardly or not elevated, sculptured with 15-20 radial series of pustules. The valves are more or less beaked, the apices always decidedly perceptible, the posterior margin slightly concave at both sides of the apex, the anterior margin evenly rounded, somewhat, produced in the middle of the second valve. Apophyses small, more or less triangular, rounded at the top, wide apart; the sinus wide, convex. Tail valve bluntly triangular, more than twice as wide as long, elevated, the mucro a little anterior, not swollen, the postmucronal slope rather steep, almost straight or a little concave; antemucronal area sculptured like the central areas, the postmucronal area like the head valve.

Colour of the tegmentum dirty white or light greyish, generally marked with short longitudinal stripes or narrow bands of greyish brown to olive green; the exposed parts of the valves sometimes entirely black or reddish brown by foreign deposits. Articulamentum weakly developed, white or greyish blue. The end valves and lateral areas always show a variable number of concentric growth marks.

Girdle narrow, dorsally clothed with much crowded, erect, slightly curved, not imbricating, elongate scales, sculptured with 8-10 fine riblets, narrower than the interstices, 130 µm long, 60 µm broad; near the sutures they become somewhat larger and relatively broader, 140 µm long, 90 µm wide. There is a short fringe of two types of spicules: slender, smooth, white, needle-like supra-marginal spicules, 120 µm long, 15 µm thick, in shortly stalked chitinous shafts, and much stouter, heavily sculptured infra-marginal spicules, 110 µm long, 30 µm thick, ornamented with 10-12 longitudinal ribs. Ventral scales more than twice as long as wide, distally bluntly pointed, the base truncated or a little emarginate, 70 µm long, 25 µm broad, arranged in imbricating radiating rows.

Radula with a weakly cusped central tooth, somewhat pinched in the middle; major lateral tooth with a strong cusp bearing one sharp denticle.

Distribution. Common in the NE Atlantic Ocean from Spitsbergen and the Barents Sea South along the Scandinavian coasts; Iceland; North Sea; all around the British Islands and Ireland, coasts of France, S. to Vigo, Spain.

Fig. 15. *Leptochiton (L.) asellus* (Gmelin, 1791).

1, whole animal, dorsal view, × 5; 2-6, valves I, II, IV, VII, VIII respectively, dorsal view, × 10; 7, detail of sculpture on central area of intermediate valve, × 47; 8, valve IV, anterior view, × 10; 9, valve VIII, lateral view, × 20; 10, arrangement of ventral girdle scales, × 400; 11, isolated ventral scale, × 400; 12, dorsal scale, × 315; 13, do, close to suture, × 315; 14, marginal spicule, × 315; 15, do, × 315; 16, part of a radula row; 17, cusp of major lateral radula tooth, × 133.

1-15, 17, after Kaas, 1981; 16, after Thiele, 1893.

Probably also from the E. coast of Greenland. It lives from just below the low water mark to a depth of 250 m.

Observations. *L. asellus* is the most common *Leptochiton* species in northern European seas. It is the *Chiton cinereus* of Montagu, 1803 (not of Linnaeus, 1767), and of many subsequent authors prior to Sars (1878), which makes old records unreliable. Its true identity was, however, recognized by such authors as a.o. Brown, von Middendorff and Forbes & Hanley.

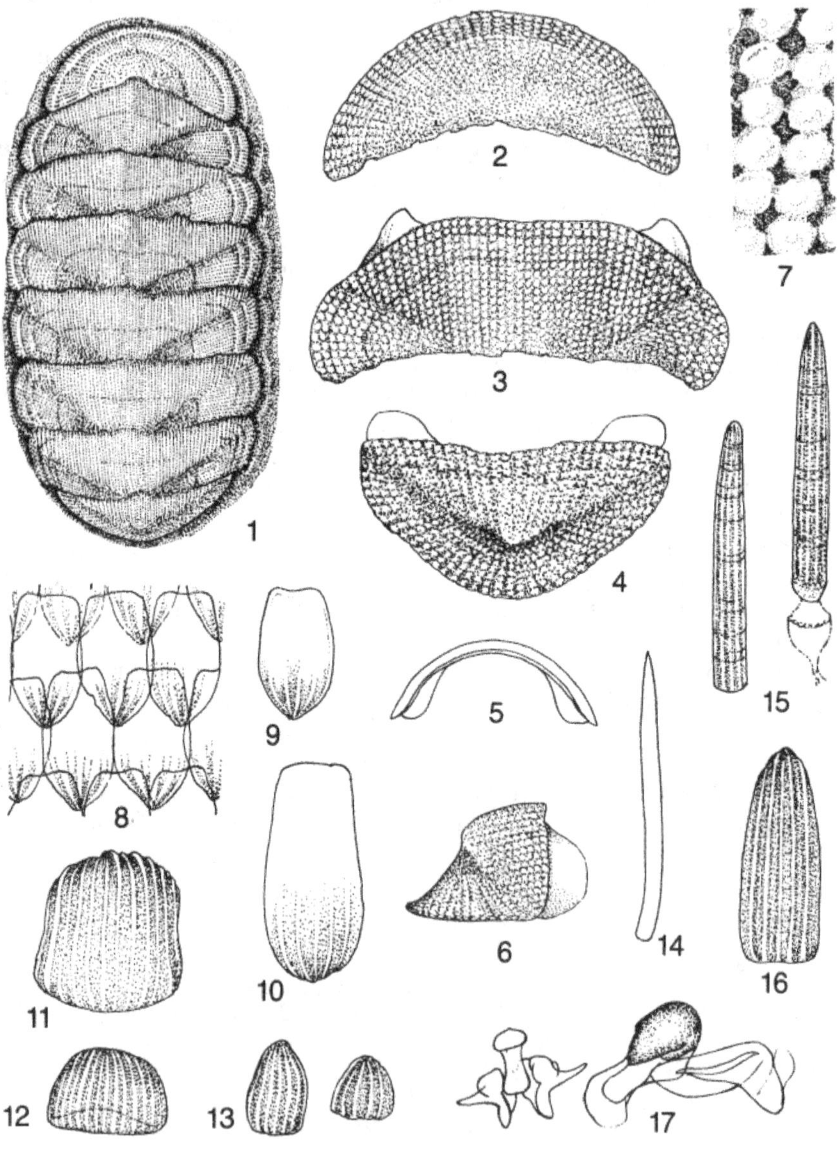

Leptochiton (L.) cancellatus (Sowerby II, 1840)

(Fig. 16. Map 4)

Type: ? (probably lost).
Type locality: ? (coast of Great Britain, probably Oban, Scotland).

Chiton cancellatus Jeffreys, 1839:39 (nomen nudum), Sowerby II, 1840:figs 104, 104a-b, 105; 1840a:4; Nobre, 1937:69.
Chiton tuberculatus Leach, 1852:230 (non Linnaeus, 1758).
Lepidopleurus cancellatus; Leloup, 1934a:12, 1 map (bibliography and synonymy); Leloup & Volz, 1938:10, figs 8-10, 47; Jakovleva, 1952:52, fig. 14, pl. 1:figs 2a-b; Muus, 1959:38, figs 19g, 22; Malatesta, 1962:147, figs 2-4; Laghi, 1977:98, pl. l:figs 1-3; Anadon, 1979:123; Kaas, 1979a:15.
Leptochiton cancellatus; Ferreira, 1979b:149; Kaas, 1981:217, figs 10 E-F.

Description. Animal rather small, up to 8 mm long, elongate oval, about twice as long as wide, moderately elevated, back evenly rounded, girdle narrow.

Head valve less than semicircular, posterior margin widely V-shaped. Intermediate valves of about equal width, roundly arched, broadly rectangular, anterior corners somewhat obliquely truncated, posterior margin about straight, apices not indicated. Lateral areas raised, mostly with a few more or less prominent growth ridges. Tail valve roughly triangular, mucro behind the centre when seen from above, swollen but not prominent, posterior slope a little concave directly behind the mucro.

Tegmentum evenly sculptured with very small roundish granules arranged in radiating rows on the head valve, the lateral areas of intermediate valves, and on postmucronal area of tail valve, in longitudinal series on central and antemucronal areas, interstices very narrow.

Colour white, mostly stained with dirty yellow, brown, or black by environmental deposits; girdle generally a shade darker.

Articulamentum white, apophyses transparent, small, triangular, widely separated, trapezoid in tail valve, jugal sinus slightly concave.

Girdle dorsally covered with not imbricating, distally rounded, curved scales, 44-56 μm long, sculptured with 12-16 raised riblets, narrower than

Fig. 16. *Leptochiton (L.) cancellatus* (Sowerby, 1840).

1, whole animal, dorsal view, × 9; 2, valve I, dorsal view, × 20; 3, valve IV, dorsal view, × 20; 4, valve VIII, dorsal view, × 20; 5, valve IV, anterior view, × 9; 6, valve VIII, lateral view, × 20; 7, sculpture of central area of intermediate valve, × 200; 8, arrangement of ventral girdle scales; 9, isolated ventral scale, × 400; 10, do, close to outer margin, × 400; 11, dorsal scale near suture, × 400; 12, do, from middle of girdle, × 400; 13, do, supra-marginal, × 400; 14, dorsal spicule, × 400; 15, intersegmental spicules, × 400; 16, marginal spicule-scale, × 400; 17, part of radula row.

1, Specimen from France, Brittany, Locquémeau, under stones, extreme low tide, VII.1972, R. A. Van Belle leg., K 4851; 2-16, specimen from France, Brittany, Baie de Morlaix, Château de Taureau, 10 m, 6.IX.1954, J. Stock leg., K 3614; 17, after Thiele, 1893.

the interstices; the sutural scales relatively larger, squarish, 56 × 54 µm, the supra-marginal ones small, 20-30 µm long, bluntly pointed, with 6-8 riblets. Scattered among the scales some slender needles, up to 100 × 7 µm, are found. There are also somewhat longer, striated, intersegmental spicules, stalked in chitinous cups, up to 88 × 12 µm. There is a short fringe of scale-like spicules, ornamented with 5-6 narrow riblets, 74 × 28 µm. Ventrally the girdle is clothed with imbricating bluntly pointed, flat scales, arranged in quincunx, the tops weakly ribbed, 40 × 28 µm, getting more elongate towards the outer margin, up to 76 × 36 µm.

Central tooth of radula with a well developed blade, first lateral tooth narrowly aliform, bladeless, major lateral with unicuspid blade, a tuberiform denticle attached to inner face of the blade.

Gills merobranchial, abanal.

Distribution. NE Atlantic coasts, from the Shetland Islands down to Vigo, Spain, and near Madeira; rather rare in the Mediterranean Sea. It lives from extreme low tide line (Brittany) down to a depth of 920 m (Bay of Biscay).

Observations. *L. cancellatus* has been reported from Scandinavian coasts, as well as from the N. Pacific Ocean. A thorough study by Ferreira (1979b:149) revealed that all Pacific records refer to *L. rugatus* (Carpenter *in* Pilsbry, 1892). Kaas (1981:226) pointed out that Scandinavian records (Sars, 1878:111, and many others) refer to a closely related species, *L. sarsi* Kaas, 1981, which constantly differs from *cancellatus* in growing somewhat larger, in the shape of the tail valve, in the granulation of the tegmentum, and in the perinotum armature. *L. rugatus* has a quite different radula.

Leptochiton (L.) algesirensis (Capellini, 1859)

(Fig. 17. Map 5)

Type: ? (Possibly in Ecole des Mines colln, Paris -Ph. Bouchet, *in litt.*)
Type locality: Spain, Algeciras.

Chiton algesirensis Capellini, 1859:327, pl. 12:figs 3a'''-c'''.
Lepidopleurus granoliratus Carpenter *in* Pilsbry, 1892:14, pl. 2:figs 47-53; Bergenhayn, 1931a:2, pl. 1:figs 1-4, pl. 2:figs 43-46.
Lepidopleurus algesirensis; Pilsbry, 1893:62, pl. 14:figs 20-21; Nobre, 1940:497,

Fig. 17. *Leptochiton (L.) algesirensis* (Capellini, 1859).

1, whole animal, dorsal view, × 4.5; 2, valve I, dorsal view, × 19; 3, valve VIII, dorsal view, × 19; 4, valve VII, dorsal view, × 9; 5, valve IV, anterior view, × 9; 6, valve VIII, lateral view, × 9; 7, dorsal girdle scale from the middle of the girdle, × 375; 8, do, near the suture, × 190; 9, dorsal girdle spicule, × 190; 10, marginal spicule, × 190; 11, ventral girdle scale, near outer margin, × 375; 12, do, near inner margin, × 375; 13, central, first lateral and major lateral radula teeth, × 375.

1-13, specimen from France, Alpes-Maritimes, Cap d'Antibes, 0.3 m, V. 1977, R. A. Van Belle leg., K 4818.

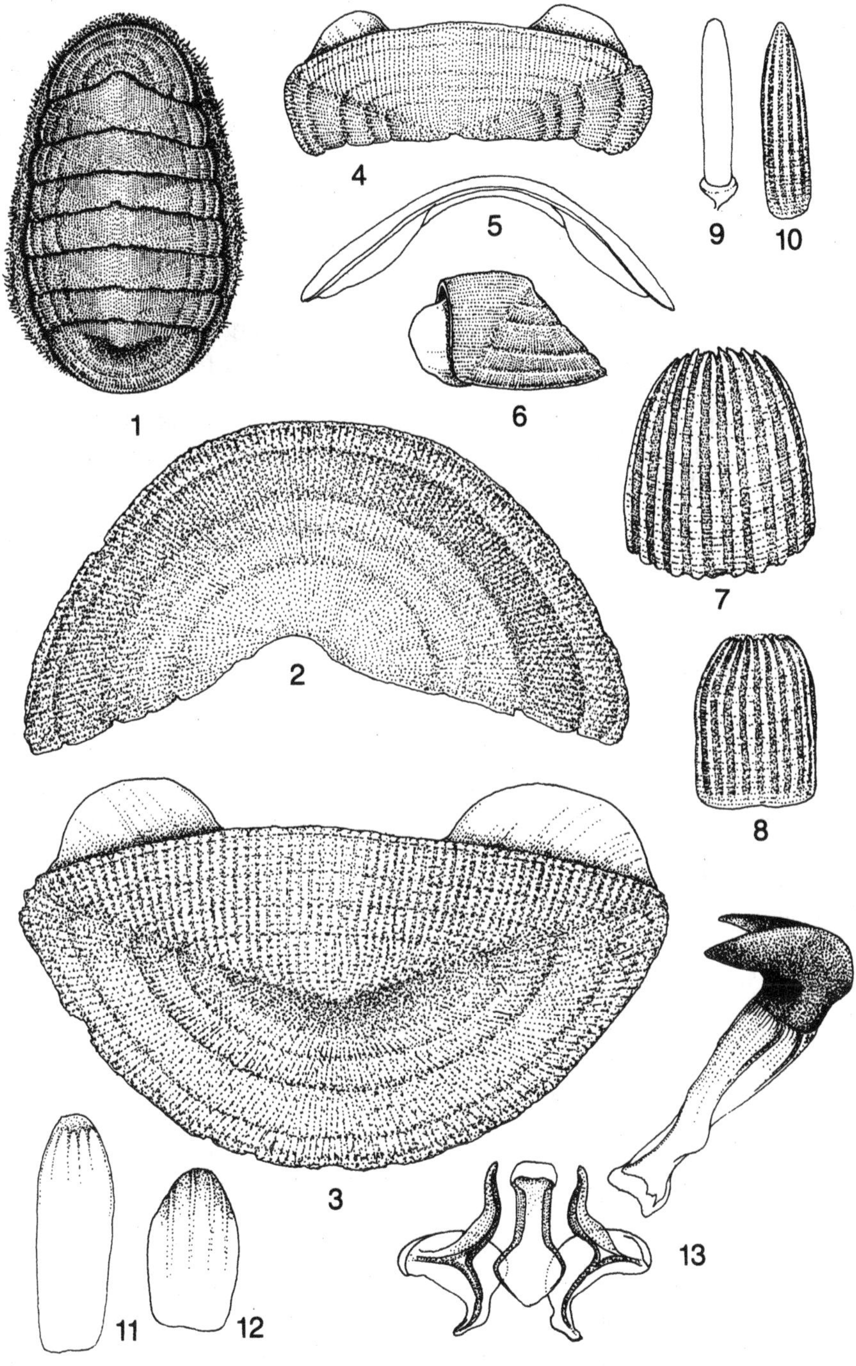
1
2
3
4
5
6
7
8
9
10
11
12
13

fig. 20 (bibliography); Malatesta, 1962:151, figs 7-8; Leloup, 1968a:7, fig 1, pl. 2: fig A.

Description. Animal of moderate size, up to 14 × 9 mm, of a peculiar oviform shape, moderately elevated, back evenly rounded, girdle rather narrow.

Head valve less than semicircular, posterior margin widely V-shaped, notched in the middle. Intermediate valves broadly rectangular, front and hind margins straight and perpendicular, except in valve VII where the anterior margin is slightly concave; apices not indicated. Valves gradually growing wider towards the sixth valve, lateral areas hardly raised, though clearly indicated. Tail valve semicircular, front margin regularly convex, mucro central, somewhat swollen but not prominent, postmucronal slope straight.

Tegmentum evenly sculptured with very small granules arranged in radiating series on head valve, lateral areas of intermediate valves and postmucronal area of tail valve, in longitudinal rows on central and antemucronal area, the rows becoming slightly converging towards the outer margins. Generally there are 4-5 obvious growth wrinkles encircling the shell completely.

Colour of tegmentum and girdle whitish to dirty yellow, often stained with dark brown, especially on the tail valve.

Articulamentum white, apophyses small, broadly triangular, widely separated, evenly rounded in tail valve.

Girdle dorsally clothed with not imbricating, squarish scales, sculptured with 10-11 raised riblets, about as wide as the interstices, ca 80 × 70 µm, somewhat larger towards the sutures, 124 × 92 µm. Along the margin there is a fringe of stronger, striated spicules, 134 × 34 µm. Ventrally the girdle is covered with elongated scales, truncated at the base, obsoletely striated near the rounded top, varying in length from 58 to 84 µm near the outer margin.

Radula with a weakly cusped central tooth, somewhat pinched in the middle; first lateral tooth slenderly pointed, not cusped, major lateral with a tricuspid blade, median denticle largest, outer one smallest.

Distribution. *L. algesirensis* is a sublittoral species, commonly distributed from the Berlengas Islands (Portugal) down into the western half of the Mediterranean Sea, and along the NW coast of Africa to Madeira and the Canary Islands.

Leptochiton (L.) arcticus (G. O. Sars, 1878)

(Fig. 18. Map 10)

Type: ZMO, type collection D 322a (lectotype, designation Kaas, 1981), D 322b (one paralectotype).
Type locality: Norway, Finnmark, Vardö.

Lepidopleurus arcticus G. O. Sars, 1878:112, pl. 7:figs 7a-h; Pilsbry, 1892:5, pl. 3:figs 59-63; Thiele, 1902:282; Nordgård, 1907:30; Thiele, 1909:9; Winckworth, 1926:13; Johnson, 1934:12; Knudsen, 1949:3; Jakovleva, 1952:52; Óskarsson, 1960:179.
Chiton arcticus; Jeffreys, 1882:668.
Chiton (Lepidopleurus) arcticus; Posselt, 1898:107.
Leptochiton arcticus; Norman, 1902:357; Kaas, 1981:221, fig 7.

Description. Animal of small to medium size, oval in outline, up to 12 × 7 mm, rather elevated (dorsal elevation up to 0.48), subcarinate to carinate, the sides convex. Valves not beaked, lateral areas hardly or not raised, girdle narrow.

Head valve less than semicircular, posterior margin widely V-shaped, notched in the middle. Intermediate valves broadly rectangular, anterior margin straight between the apophyses, laterally abruptly narrowing, posterior margin almost straight, not or very weakly umbonated, never beaked, lateral areas only marked by a difference in sculpture. Tail valve less than semicircular, posterior margin evenly rounded, anterior margin produced between the apophyses; length:width ratio 0.54; mucro central, not prominent, posterior slope straight to slightly convex.

Tegmentum evenly granulose. On head valve, lateral areas of intermediate valves, and postmucronal area of tail valve the granules are roundish, flat, only weakly raised, neatly quincuncially arranged. Central areas and antemucronal area of tail valve with longitudinal chains of oval granules, the chains hardly perceptible, narrow, and close set on the jugum, more and more pronounced and wider apart towards the side margins. There are always a few concentric lines of growth in the younger parts of the shell.

Colour of the tegmentum yellowish to ochraceous or light brownish, sometimes the older parts of the shell covered with a dark brown or blackish foreign deposit.

Articulamentum white, the apophyses well developed, bluntly triangular with a rounded top, wide apart; jugal sinus straight, produced.

Girdle dorsally covered with thick, opaque, not imbricating, conical scales, sometimes truncated at the top, always marked with 6-8 coarse ribs, as wide as the interstices, 45-65 μm long, 45 μm broad, interspersed with randomly disposed small, slender, slightly curved, striated, sharply pointed spicules, up to 200 μm long, 18 μm thick. At the sutures shorter and relatively stouter spicules are found, 140 μm long, 22 μm thick. There is a fringe of straight, bluntly pointed, weakly striated spicules, 245 μm long, 45 μm thick. Ventrally the girdle is covered with imbricating, flat, elongate ovoid scales, distally marked with 6 shallow grooves, the base somewhat emarginate, arranged in radiating series.

Radula with a slender central tooth, gradually tapering towards the base, with a narrow cusp; first lateral tooth broadly based, somewhat produced exteriorly, with a small, beak-like cusp. Major lateral tooth strongly developed, unicuspidate, with a very weak notch on the outer margin of the blade.

Distribution. Northern coast of Norway, from Folden in Nord Trøndelag to Varangerfjorden in Finnmark; Spitsbergen, Iceland, and (perhaps) Greenland; occasionally together with *L. asellus,* always in a depth of 10 m or more.

Observations. Thiele (1909:9), who examined type material of *arcticus,*

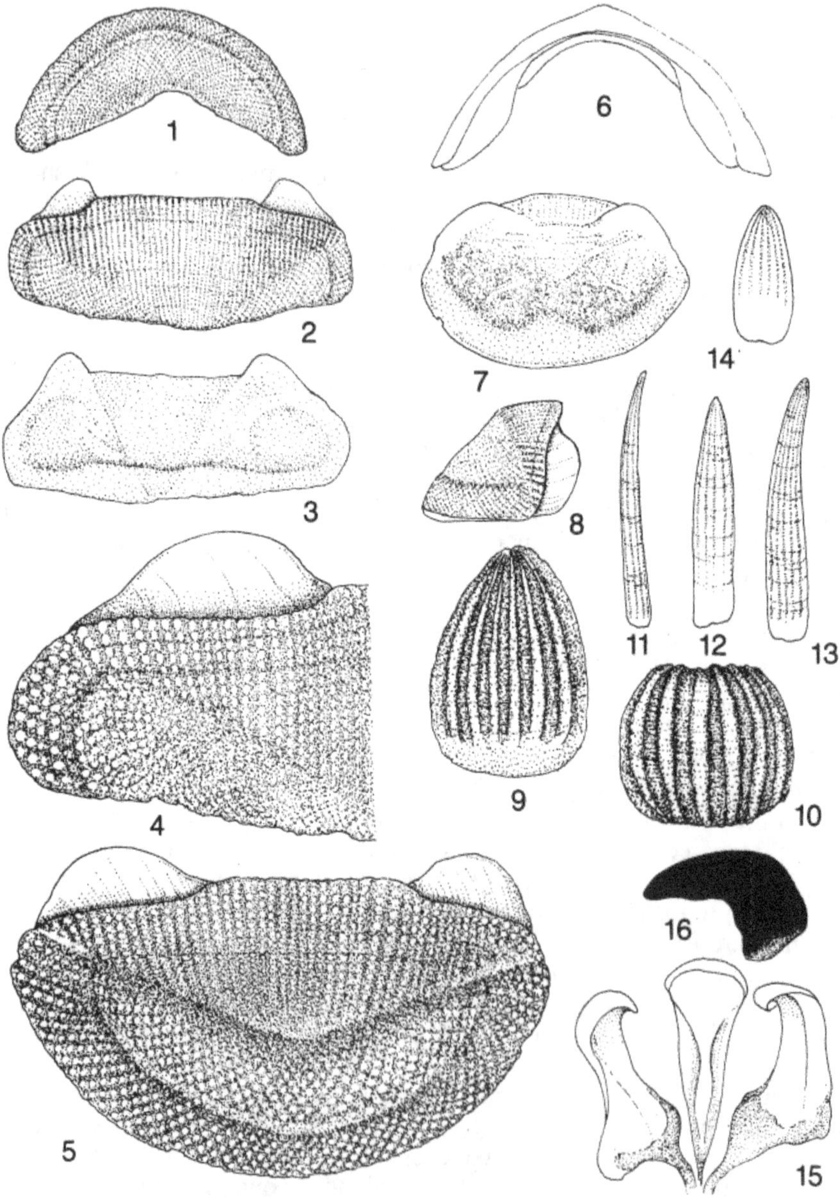

could not separate this from *asellus*, and his opinion was generally adopted. Winckworth (1926:13) regarded it as an extreme northern race of *asellus*. Kaas (1981:221) pointed out that both forms are regularly found in the same sample, which makes it impossible to regard *arcticus* as a subspecies of *asellus*. The two species are easily separable by the naked eye, *arcticus* being decidedly higher elevated, the sides more rounded, the back much weaker granulated, and also differing in colour. It also differs in the shape of the tail valve, and in the granulation, which is definitely radial in the lateral areas and end valves of *asellus*, quincuncial in *arcticus*.

Old records of both species, however, are to be reconsidered, as they can hardly be trusted.

Leptochiton (L.) scabridus (Jeffreys, 1880)

(Fig. 19. Map 11)

Type: USNM 177391/2
Type locality: England, Tor Bay; Jersey.

Chiton scabridus Jeffreys, 1880:33; Marshall, 1893:248; Warén, 1980:13.
Lepidopleurus scabridus; Sykes, 1894:35, pl. 3:figs 4, 7; Thiele, 1909:9, pl. 1:figs 5-10; Matthews, 1967:1; Van Belle, 1972:203, fig. 2.

Description. Animal rather small, elongate oval, exceptionally up to 8 mm long, width half the length, rather flat, back evenly rounded, but for a slight indication of a carina, girdle narrow.

Head valve semicircular, posterior margin widely V-shaped. Intermediate valves of about equal width, broadly rectangular, front and hind margins straight and perpendicular, except in valve II where the front margin is somewhat angulated anteriorly, apices not marked, lateral areas hardly raised but clearly indicated. Tail valve semicircular, front margin evenly convex, mucro about central, somewhat swollen, postmucronal slope concave.

Tegmentum evenly granulose, the roundish granules getting larger towards the anterior and outer margins, arranged in 36-48 radiating rows on the head valve, 7-8 rows on the lateral areas of the intermediate valves, and 30-36 on the postmucronal area of the tail valve, in 24-30 longitudinal moniliform rows on the central areas and the antemucronal area, tending to become slightly converging near the outer sides.

Fig. 18. *Leptochiton (L.) arcticus* (G. O. Sars, 1878).

1, valve I, dorsal view, × 8.4; 2, valve V, dorsal view, × 8.4; 3, valve V, ventral view, × 8.4; 4, part of valve V, dorsal view, × 18; 5, valve VIII, dorsal view, ? 18; 6, valve V, anterior view, × 8.4; 7, valve VIII, ventral view, × 8.4; 8, valve VIII, lateral view, × 8.4; 9-10, dorsal girdle scales, × 640; 11, dorsal spicule, × 235; 12, marginal spicule, × 165; 13, sutural spicule, × 335; 14, ventral girdle scale, × 280; 15, central and first lateral teeth of radula, × 350; 16, cusp of major lateral radula tooth, × 165.

1-6, after Kaas, 1981.

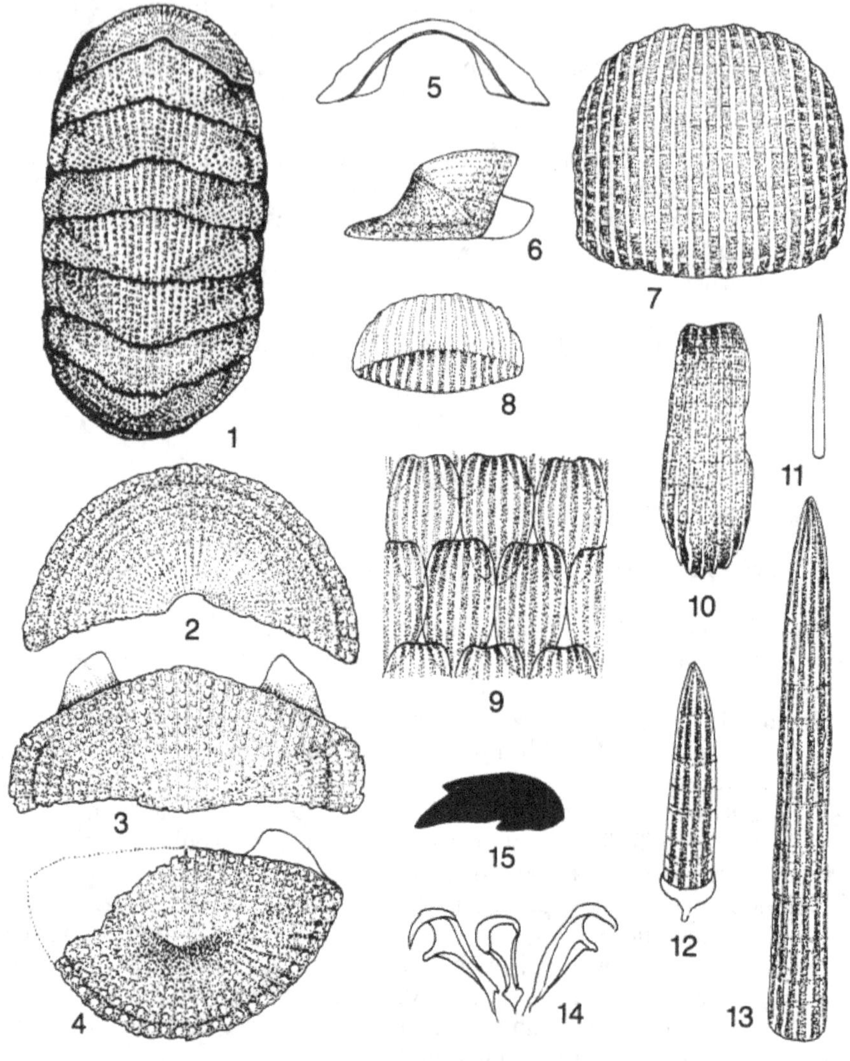

Fig. 19. *Leptochiton (L.) scabridus* (Jeffreys, 1880).

1, whole animal, dorsal view, × 20; 2, valve I, dorsal view, × 32, 3, valve II, dorsal view, × 32; 4, valve VIII, dorsal view, × 32; 5, valve II, anterior view, × 16; 6, valve VIII, lateral view, × 16; 7, dorsal girdle scale, dorsal view, × 400; 8, dorsal scale, ventral view, × 400; 9, ventral girdle scales, × 400; 10, ventral scale close to outer margin, × 400; 11, dorsal girdle spicule, × 400; 12, marginal spicule, × 400; 13, intersegmental spicule, ? 400; 14, central and first lateral radula teeth, × 800; 15, cusp of major lateral radula tooth, × 800.

1-15, specimen from Channel Islands, Herm, intertidal, on stones, Tomlin leg., K 4750.

Colour of tegmentum and girdle dirty white, often stained with yellow or light brown.

Articulamentum white, apophyses transparent, small, sharply triangular, widely separated.

Girdle dorsally covered with squarish scales, 92 × 84 μm, sculptured with about 18 riblets, narrower than the interstices, interspersed with small needlelike spicules, 48 × 4 μm, and near the sutures some heavily ribbed spicules, 80 × 16 μm, rising from chitinous cups. Ventrally the girdle is covered with alternating rows of elongated, obtusely pointed scales, sculptured with some 7 riblets, truncated at the base, 34 × 24 μm, larger towards the outer margin, up to 85 μm long.

Central radular tooth distally widening to a well developed blade, first lateral tooth pointed, without blade, major lateral bearing a tridentate cusp, median denticle largest.

Distribution. Endemic, sublittoral species, restricted to SW England, the Channel Islands, and the coast of Britanny. Bergenhayn (1931a:9, pl. 2 fig 7) reported one specimen and an odd valve from the Canary Islands.

Leptochiton (L.) rarinota (Jeffreys, 1882)

(Fig. 20. Map 7)

Type: BMNH 85.11.5.1439.

Type locality: not designated: NE Atlantic Ocean, either 48°31'N 10°03'W ("Porcupine" Exp. 1870, sta. 3a), or 40°06'N 09°44'W (do, sta. 14), 500 and 469 m respectively.

Chiton rarinota Jeffreys, 1882:668, pl. 50:fig 1; Pilsbry, 1892: pl. 50:fig 1; 1893:93; Nobre, 1931:258; 1936:193; Warén, 1980:13.

Description. "Shell oblong-oval, arched, rather thin and glossy: plates broader in the middle than at the sides; lateral areas indistinct: sculpture consisting of white tubercles, which are few in number and irregularly scattered; these are found in the middle of the shell, but become more raised and oval at the sides; under a microscope can be detected numerous and close-set lines or striae, which are arranged lengthwise: colour whitish: beaks none, except on the tail plate where they are nearly circular: inside glossy. L.0.1, B. 0.05" (J. G. Jeffreys); (L. 3.1, B. 1.8 mm).

Observations. The author added to the above description: "Although this is a very small species, and might be regarded as the young of some other species, I must observe that I have carefully compared both specimens with the young of all other European species of *Chiton* known to me, and some specimens of which last mentioned species are much smaller than those which I have now described. The peculiar character of having so very few and scattered tubercles is not presented by any other of those species. The girdle is membraneous and thin."

Jeffreys reported his "new species" from two stations of the "Porcupine" Exp. 1870 (see map 7). Warén (1980:13) could only locate one of them in the BMNH collections, the other was not to be found in the Jeffreys collection (in USNM) either. The only specimen is designated here as the lectotype. Despite our efforts we are unable to establish its true status, though we compared it with the types of other small species from the same habitat, viz *L. tenuis* Kaas and *L. gascognensis* n. sp., the only ones with which *rarinota* shows some resemblance. The latter, however, is much less elevated, somewhat flattened towards the outer margins. The valves are very thin and transparent, the articulamentum hardly developed. Apparently it is an immature specimen. In trying to study the girdle the tail valve got loose, so we had to abandon our effort for fear that the specimen might fall apart. The girdle (under a magnification × 50) appears to be covered with elongate, striated scales. There is a marginal fringe of small spicules and only a few long intersegmental spicules could be found, one of them has an estimated length of 400 µm.

The French "Thalassa" Expedition 1972 procured a number of very small *Leptochiton* from deep parts of the Bay of Biscay (Kaas, 1979), among which several immature ones that could not be identified. In our opinion *"Leptochiton rarinota"* is also unidentifiable, so it seems wiser to regard it as a nomen inquirendum.

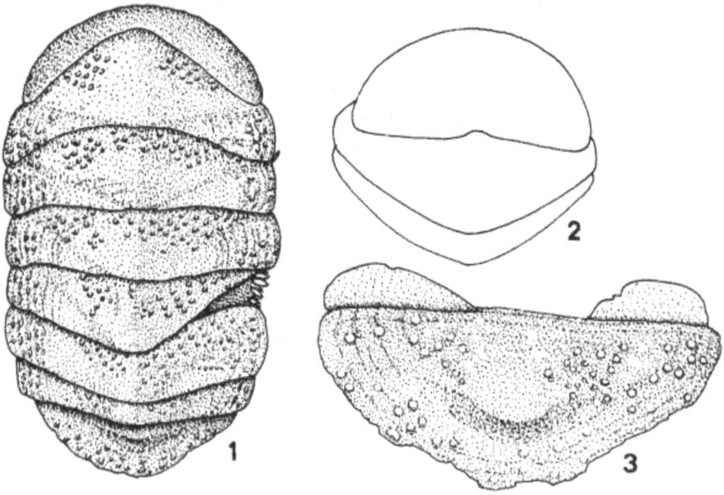

Fig. 20. *Leptochiton (L.) rarinota* (Jeffreys, 1882).

1, whole animal, dorsal view, × 20; 2, camera lucida sketch of valves I-III, showing dorsal elevation of III, × 20; 3, valve VIII, dorsal view, × 40.

1-3, lectotype.

Leptochiton (L.) odhneri (Bergenhayn, 1931)

(Fig. 21. Map 35)

Type: NHRM, type collection 1322.
Type locality: Canary Islands, Fuerteventura, Puerto Cabras.

Lepidopleurus odhneri Bergenhayn, 1931a:6, pl. 1:figs 5-8, pl. 2:figs 48-49.
Leptochiton odhneri; Altimira, 1978:182.

Description. Animal very small, maximum length 4.5 mm, oval, moderately elevated, back evenly arched to subcarinate, side slopes convex, girdle very narrow.

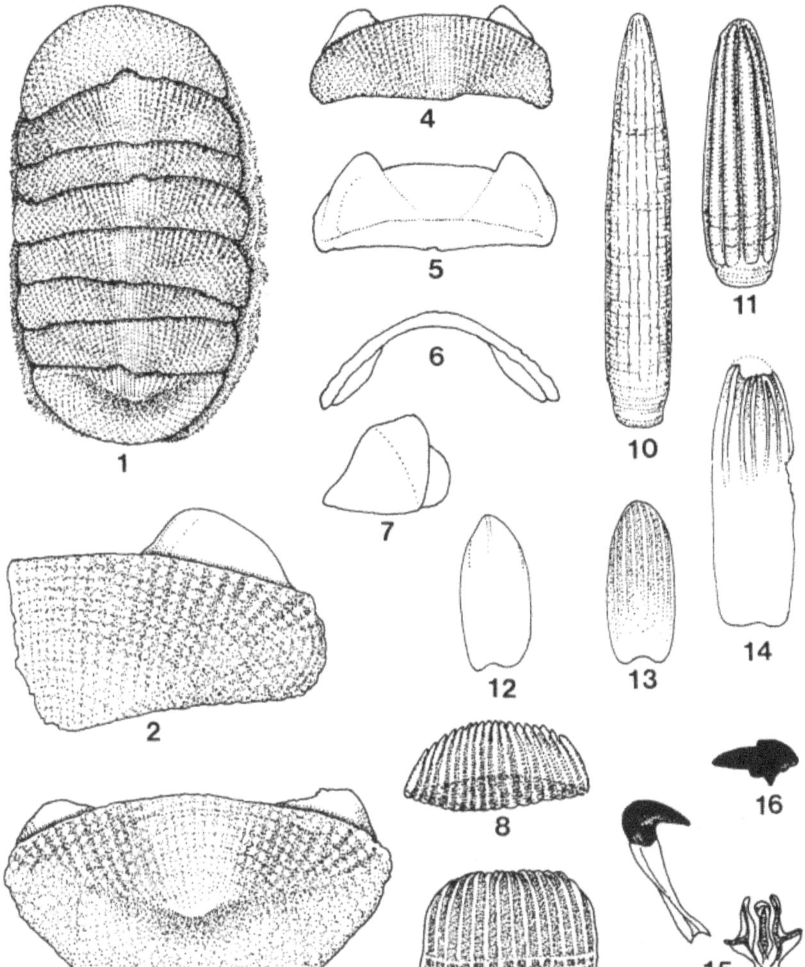

Fig. 21. *Leptochiton (L.) odhneri* (Bergenhayn, 1931).

1, whole animal, dorsal view, × 20; 2, part of valve IV, dorsal view, × 40; 3, valve VIII, dorsal view, × 40; 4, valve VII, dorsal view, × 20; 5, do, ventral view, × 20; 6, do, anterior view, × 20; 7, valve VIII, lateral view, × 20; 8, dorsal girdle scale from mid-girdle, × 400; 9, do, near suture, × 400; 10, intersegmental spicule, × 400; 11, marginal spicule, × 400; 12, ventral scale, × 400; 13-14, do, near outer margin, × 400; 15 central, first lateral and major lateral radula teeth, × 500; 16, cusp of major lateral tooth, × 500.

1-16, specimen from Canary Islands, S coast of Fuerteventura, Punta de Gran Taragal, calcareous algae, 49 m, 27. VIII. 1977, Cancap II-"Tydeman" Exp., sta. 44, K 4880.

Head valve semicircular, posterior margin widely V-shaped. Intermediate valves broadly rectangular, anterior margin convex, side margins rounded, posterior margin straight to very slightly beaked, apices not prominent, lateral areas little or not elevated. Tail valve semicircular, front margin evenly convex, mucro central, somewhat swollen though not prominent, post-mucronal slope slightly concave.

Tegmentum evenly sculptured with roundish granules, arranged in quincunx on the head valve, the lateral areas of the intermediate valves, and the postmucronal area of the tail valve, in longitudinal rows on the jugal area, the rows gradually becoming posteriorly converging on the pleural areas.

Colour of tegmentum and girdle dirty white to light yellowish.

Articulamentum white, apophyses very thin, small, broadly triangular, rather widely separated.

Girdle dorsally covered with short, curved, rectangular, ribbed scales, about 60 μm wide, 30 μm long, somewhat longer, up to 45 μm towards the sutures, the 15-18 riblets narrower than the interstices. There are heavy, obsoletely striated, intersegmental spicules, 135-150 μm long, and shorter, ribbed, marginal spicules, 90 × 27 μm. Ventrally the girdle is clothed with irregular radiating rows of obtusely pointed, elongated scales, 50 × 24 μm, the base somewhat curved inward, obsoletely ribbed, getting longer towards the outer margin, up to 92 μm, while the distal ribbing becomes more clearly visible.

The radula has a small central tooth, pinched in the middle, with a very narrow blade, first lateral tooth wing-like, tapering to a bladeless point, major lateral bearing a bidentate cusp, the outer denticle much smaller than the main one, which has an accessory protuberance on the inner side.

Distribution. Found at depths of 0-150 m along the NW coast of Africa, the Canary Islands, and Madeira.

Leptochiton (L.) intermedius (von Salvini-Plawen, 1968)

(Fig. 22. Map 36)

Type: NHMW, 77170.
Type locality: Adriatic Sea: off Rovinj, 4-7 m.

Lepidopleurus intermedius von Salvini-Plawen, 1968:251, pls 6-8, figs 44-57; Sabelli, 1974b:76.

Description. Animal small, up to 4.5 × 2 mm, elongate oval, only little elevated (dorsal elevation 0.29), the back roundly arched, not carinated, girdle very narrow.

End valves semicircular, the posterior one with a small, subcentral mucro, posterior slope slightly concave. Intermediate valves of equal width, broadly rectangular, not beaked, lateral areas not raised. Tegmentum of head valve and post-mucronal area of tail valve sculptured with well raised, roundish granules arranged in quincunx; on the lateral areas they are arranged at random. Central areas and ante-mucronal area of tail valve with alternating, widely separated, large granules in longitudinal series, converging towards the posterior margin.

Colour of tegmentum yellowish brown.

Articulamentum with small apophyses, wide apart, triangular; jugal sinus wide, somewhat convex.

Girdle narrow, dorsally covered with small, not imbricating, transparent, colourless scales, longer than wide, with 3-5 longitudinal sulci, though these are not always present. There is a marginal fringe of small, smooth, bluntly pointed spicules. Ventral scales of different sizes, opaque, longer than wide, distally pointed and very finely longitudinally striate.

Major lateral radula tooth with a tridentate cusp, the median denticle largest, the inner one smallest.

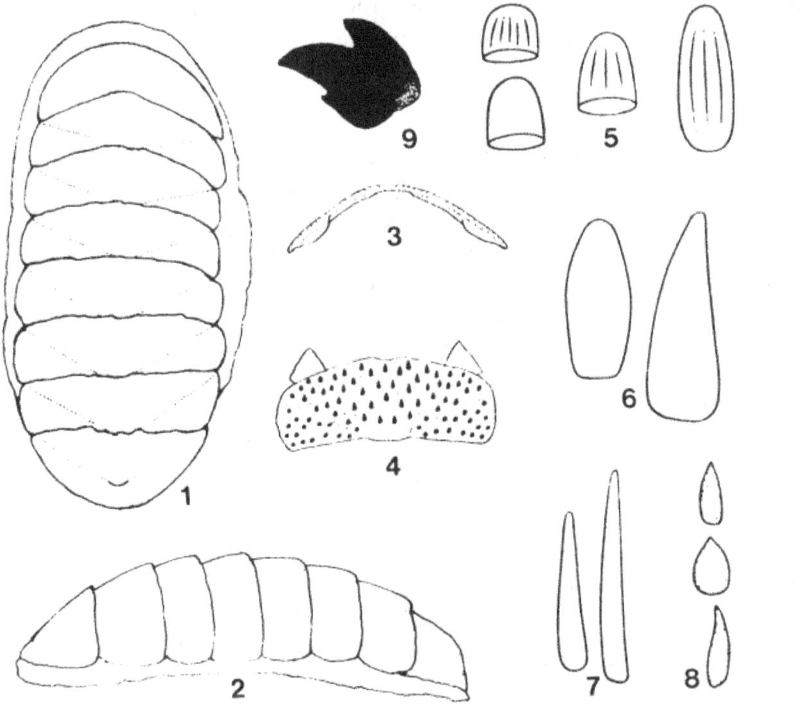

Fig. 22. *Leptochiton (L.) intermedius* (von Salvini-Plawen, 1968).

1, contours of a specimen, dorsal view, × c. 20; 2, do, lateral view, × c. 20; 3, valve IV, anterior view, × c. 20; 4, do, dorsal view, × c. 20; 5-6, dorsal girdle scales, × 290; 7, marginal spicules, × 290; 8, ventral scales, × 290; 9, cusp of major lateral radula tooth.

All after von Salvini-Plawen, 1968.

Gills merobranchial, adanal with interspace. There are up to 11 pairs of gills, extending backward from the front margin of the 6th valve.

Distribution. Only known from the type locality: Adriatic Sea, off Rovinj (Yugoslavia, Istria), on sandy bottom, in a depth of 4-7 m.

Observations. Despite our efforts we have not been able to examine the type specimen. The only material we could study was an immature, decalcified specimen. As the original illustrations are rather bad (the habitus photographs were apparently taken under a light-microscope), we could not do better than to reproduce the author's sketchy drawings, and the contours of his photographs of the specimen and its radula.

According to von Salvini-Plawen his new species should be intermediate between *L. asellus* (Gmelin, 1791) and *L. medinae* (Plate, 1899), principally on account of its anatomy. Especially the possession of a deep semicircular furrow dividing the head disc into a narrow outer edge and the actual disc containing the mouth, which Plate thought to be a unique characteristic of *L. medinae*, seems to be the main reason why von Salvini-Plawen compared his species with *medinae*. In fact it is a feature present in many other species of *Leptochiton*, such as *cancellatus, odhneri, asellus, arcticus, sarsi, alge-sirensis.*

In our opinion *L. intermedius* may be a good species, much alike *L. odhneri* (Bergenhayn, 1931), but for the girdle armature, it is not nearly related to either *asellus* or *medinae,* which also are mutually very different.

Leloup & Volz (1938), who made a profound study of the chitons of the Adriatic Sea, especially from the neighbourhood of Rovinj, reported, *L. cancellatus* from many locations. It is possible that this was a misidentification for *L. intermedius* because of the superficial resemblance. *L. cancellatus* differs markedly in the granulation of the valves, in the perinotum covering, and especially in the radula.

Leptochiton (L.) leloupi Kaas, 1979

(Fig. 23. Map 7)

Type: MNHN (holotype + 2 paratypes); K 4779 (1 paratype).
Type locality: Bay of Biscay, "Thalassa" sta. Z 407, 47°44'N 08°07'W, 1085 m.

Leptochiton leloupi Kaas, 1979a:16, pl. 1.

Fig. 23. *Leptochiton (L.) leloupi* Kaas, 1979.

1, whole animal, × 40; 2, part of intermediate valve, dorsal view, × 80; 3-4, dorsal girdle scales, × 500; 5-6, dorsal scales, near the suture, × 400; 7-8, ventral scales, × 400; 9, dorsal spine, × 400; 10, marginal spicule-scale, × 400; 11-12, do, × 400; 13-14, supra-marginal scales, side view, × 400; 15, cusp of major lateral radula tooth, × 400.

1, holotype; 2-15, paratypes, 1-14, after Kaas, 1979.

Description. Animal very small, up to 4 mm long, elongate oval, moderately elevated, non-carinated, the valves not beaked, chalky white, dull.

Head valve semicircular, sculpture of irregular warts or scalelike tubercles, quincuncially arranged in curved series running from the hind margin to the circumference in two directions, crossing each other rectangularly.

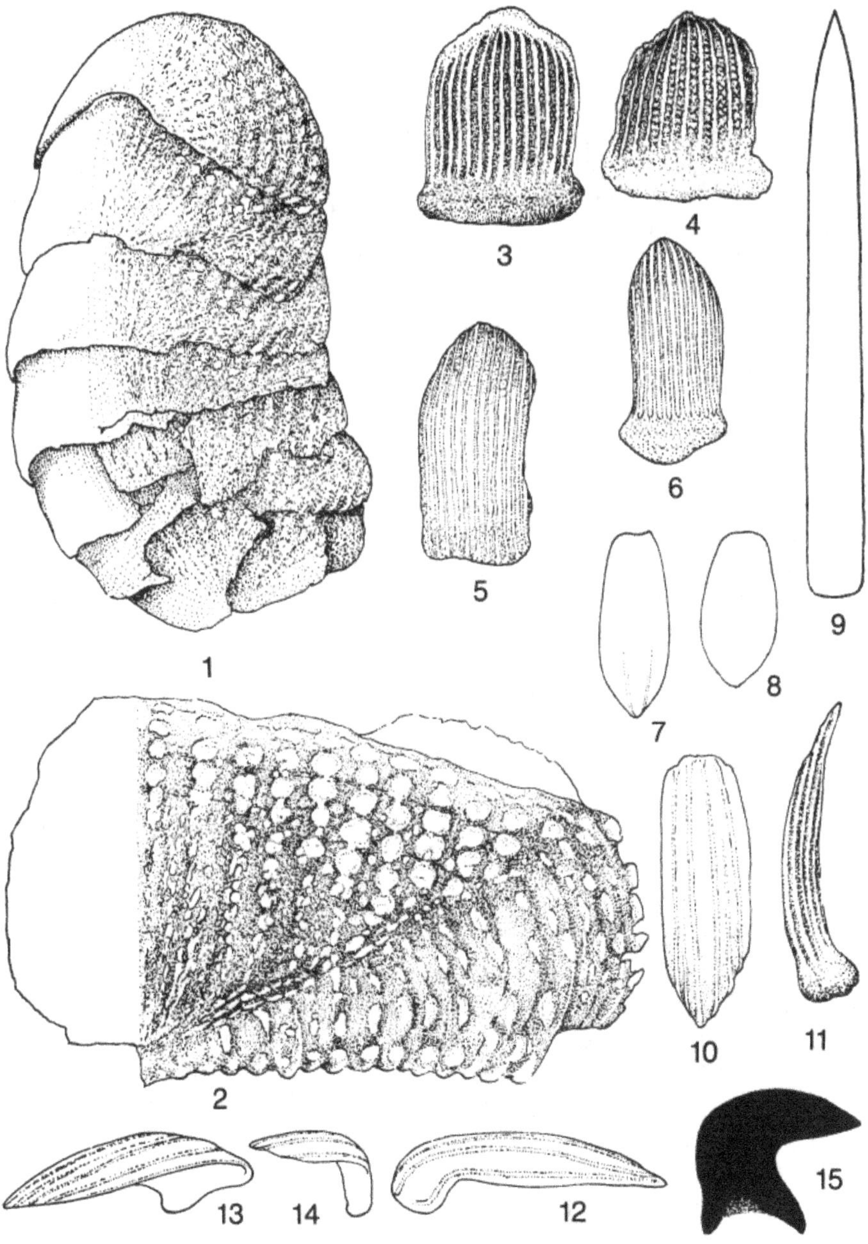

Intermediate valves about rectangular, regularly arched, central areas with rather widely separated longitudinal chains of irregularly shaped tubercles mostly composed of several granules cemented together, interstices finely and irregularly granulose, lateral areas decidedly raised, with about four radiating rows of scale-like tubercles, the hindermost dentating the sutures, at the same time arranged in concentric rows, separated by shallow grooves and lines of growth. Tail valve semicircular, mucro not prominent, about central, posterior slope slightly concave, postmucronal area sculptured like the head valve.

Articulamentum white, apophyses narrow, widely separated.

Girdle narrow, dorsally clothed with thin scales and very few spines; scales about 50 µm long, 40 µm wide, curved, squarish at the base, rounded at the top, with 12-14 narrow, sharp riblets, sutural scales about twice as long as wide, 75 × 35 µm, spines white, cylindrical, rather abruptly tapering to a sharp point, 192 × 17 µm. There is a marginal fringe of bluntly pointed spicules-scales, dorsally marked with 5-6 sharp longitudinal riblets, about 90 µm long. Ventrally the girdle is covered with radiating rows of imbricating flat, very thin and transparent scales, distally obtusely pointed, 50 × 25 µm.

Major lateral radula tooth with a strong, unidentate cusp, the base shaped like a fireman's helmet.

Length 3.9 mm, width 2.4 mm (largest paratype).

Distribution. *Leptochiton leloupi* seems to be restricted to the upper slope of the continental shelf in the Bay of Biscay. Besides the holotype, taken at "Thalassa" sta. Z 407, at a depth of 1085 m, 2 paratypes were collected at "Thalassa" sta. Z 417, 48°12'N 09°09'W, 865 m, and 1 paratype at "Thalassa" sta. Z 431, 48°38'N 09°47'W, 800 m.

Leptochiton (L.) tenuis Kaas, 1979

(Fig. 24. Map 7)

Type: MNHN (holotype + 6 paratypes); K 4780 (1 paratype).
Type locality: Bay of Biscay, "Thalassa" sta. Z 430, 48°37'N 09°52'W, 1080 m.

Leptochiton tenuis Kaas, 1979a:18, pl. 2.

Description. Animal very small, up to 3 mm long, elongate oval, very

Fig. 24. *Leptochiton (L.) tenuis* Kaas, 1979.

1, whole animal, × 20; 2, valve I, dorsal side, × 40; 3, valve IV, dorsal side, × 40; 4, valve VIII, dorsal side, × 40; 5, valve IV, anterior view, × 28; 6, dorsal girdle scales, × 350; 7, sutural scale, × 370; 8, ventral girdle scale, × 400; 9, ventral scale, near outer margin, × 370; 10, supra-marginal scale, × 370; 11, marginal spicules, × 350; 12, do, × 370; 13, dorsal spicule, × 350; 14, sutural spicule, × 370; 15, cusp of major lateral radula tooth, × 400.

1, holotype; 2-15, paratype; 1-14, after Kaas, 1979.

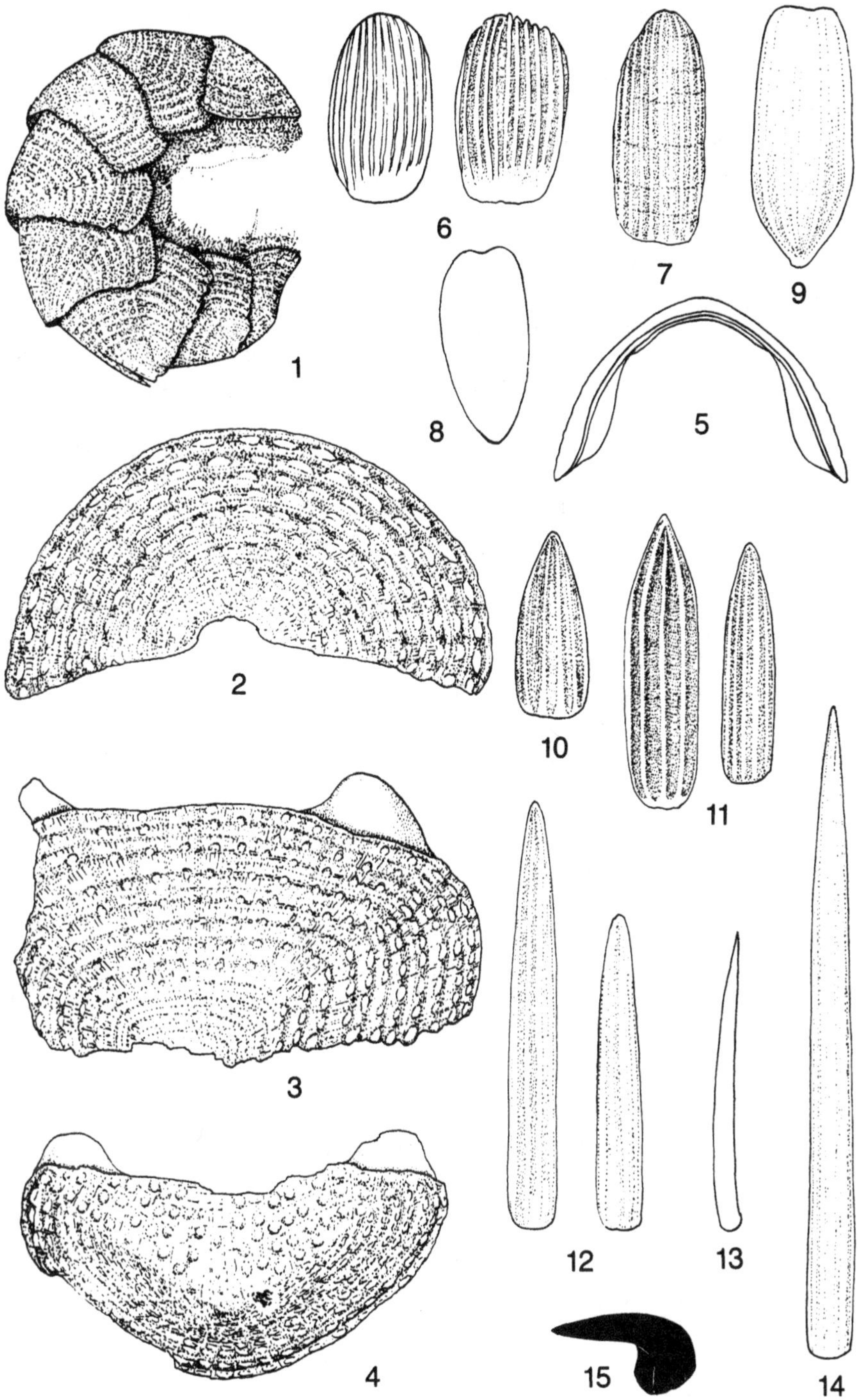
1
2
3
4
5
6
7
8
9
10
11
12
13
14
15

highly, angularly arched, not carinated, valves thin and brittle, only slightly mucronate, yellowish white.

Head valve semicircular, intermediate valves more or less rectangular, front margin a little emarginate, apex perceptible, but far from prominent, lateral areas not raised, posterior valve more or less triangular, twice as wide as long, much smaller than the other valves, mucro behind the centre, posterior slope steep, a little concave.

Tegmentum of head valve, lateral areas of intermediate valves, post-mucronal area of tail valve sculptured with concentric rows of elongate pustules, slightly dentating the sutures, continuing on the central areas, where they are also arranged in radiating rows, diverging from the apex towards the front margin. On the central areas the pustules are small, roundish, wide apart. All valves show fine, short, brown lines or scratchings in the interstices between the series of pustules.

Articulamentum poorly developed, apophyses small, triangular, widely separated.

Girdle rather narrow, dorsally clothed with very thin, striated scales, small, smooth spicules, and stout, ribbed spines in pairs. Scales elongate oval, 60-70 μm long, 35-40 μm wide, with 10-12 sharp riblets, sutural scales relatively longer, spicules sharply pointed, 100 μm long, 7-8 μm wide, spines much longer, irregularly disposed, 210-240 × 18 μm. Marginal fringe of stout, elongate spicules-scales, with an abruptly tapering point, marked with five strong longitudinal ribs, 100 × 20 μm, and more weakly ribbed spiculae. Ventrally the girdle is covered with radiating rows of very thin, imbricating scales, 58 × 30 μm, oval, obtusely pointed distally, somewhat longer near the outer margin.

Major lateral radula tooth with a simple, short, sharply pointed, unidentate cusp.

Length ca 3 mm, breadth 2 mm (largest paratype, somewhat curled).

Distribution. So far,*Leptochiton tenuis* has only been found on the upper slope of the continental shelf in the Bay of Biscay. Besides the holotype, taken at "Thalassa" sta. Z 430, at a depth of 1080 m, 1 paratype was collected at "Thalassa" sta. Z 431, 48°38'N 09°47'W, 800 m, and 6 paratypes at "Thalassa" sta Z 435, 48°40'N 09°53'W, 1050 m.

Leptochiton (L.) sarsi Kaas, 1981

(Fig. 25. Map 12)

Type: KNVSM (holotype + 2 paratypes); ZMO, type collection, D-33690 (1 paratype); K 4914 (1 paratype).
Type locality: Norway, Trondheimsfjord, Åsfjorden, 60-80 m.

? Der kleine iszländische chagrinirte Chiton, Schröter, 1786:509.
? *Chiton islandicus* Gmelin, 1791:3206 (ex Schröter, *l.c.*).
Chiton asellus varietas *a* Spengler, 1797:100.

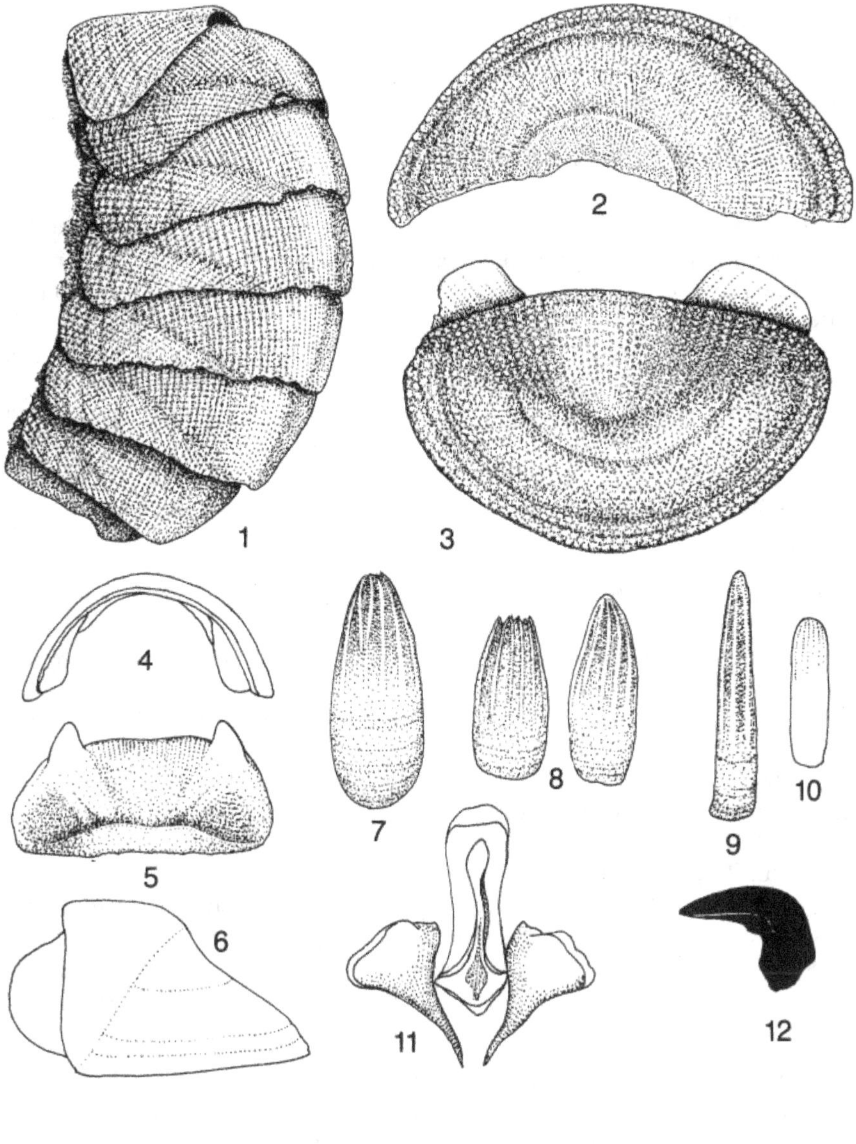

Fig. 25. *Leptochiton (L.) sarsi* Kaas, 1981.

1, whole animal, lateral view, × 10; 2, valve I, dorsal view, × 20; 3, valve VIII, dorsal view, × 20; 4, valve IV, anterior view, × 9.6; 5, valve IV, ventral view, × 9.6; 6, valve VIII, lateral view, × 20; 7, dorsal girdle scale, near suture, × 200; 8, dorsal girdle scales from mid-girdle, × 200; 9, marginal spicule, × 200; 10, ventral scale, × 200; 11, central and first lateral radula teeth, × 400; 12, cusp of major lateral radula tooth, × 240.

1-12, after Kaas, 1981.

Lepidopleurus cancellatus (non Sowerby II, 1840); G. O. Sars, 1878:111, pl. 7:figs 6a-h, et mult. auct.
Leptochiton sarsi Kaas, 1981:225, figs 4, 9, 10 C-D.

Description. Animal small, up to 10 × 5 mm, elongate oval, rather elevated (dorsal elevation 0.48), back evenly rounded to subcarinate, girdle very narrow.

Head valve a little less than semicircular, intermediate valves of equal width, twice as wide as long, not umbonated, posterior margin almost straight, anterior margins of valves II and VIII convex, of III-VII somewhat concave in the jugal part. Tail valve almost semicircular, mucro not prominent, central or a little anterior, posterior slope only slightly concave directly behind the mucro.

Tegmentum evenly granulose all over. On the head valve, the hardly raised lateral areas of intermediate valves and post-mucronal area of tail valve the granules are only little raised, roundish, arranged in quincunx, though towards the outer margin they appear to form radial rows, interrupted by many, very fine, concentric lines of growth. Central areas of intermediate valves and ante-mucronal area of tail valve with 50 longitudinal series of clearly cut, well separated, roundish granules, the interstices shallowly grooved.

Colour of tegmentum dirty white to light yellowish, often wholly or partly covered with brownish or black foreign deposits.

Articulamentum white, weakly developed, apophyses small, triangular, wide apart in the intermediate valves, trapezoid in the tail valve.

Girdle narrow, closely beset with elongate conical scales, dorsally ornamented with 5-6 weak, narrow riblets on the distal half, 125 μm long, 50 μm wide. Marginal spicules cylindrical, bluntly pointed, the visible part with 3 longitudinal riblets, 164 μm long, 25 μm thick; ventral scales elongate, round topped, the base truncated, with 6 longitudinal scratches towards the top, 96 μm long, 25 μm broad.

Central tooth of radula only little narrowed in the middle, with a weak cusp, the basal plate spoon-shaped, posteriorly broadly lozenge-shaped. First lateral tooth triangular, not or hardly cusped; major lateral strongly cusped, with a curved, sharply pointed main denticle and a very small exterior one.

Distribution. *L. sarsi* appears to be continuously distributed along the Scandinavian coast, from Bohuslän in Sweden up to Finnmark in Norway, in a depth of 40 to over 700 m, occasionally together with *L. alveolus*.

Observations. G. O. Sars, *l.c.,* and subsequent workers on Scandinavian chitons called the present species *C. cancellatus,* and they are not to be blamed for this, as the two species are closely allied. There are, however, a number of constant differences by which the two are to be separated: *sarsi* grows a little larger than *cancellatus,* the rows of granules on the central areas are more spaced, the granules larger and not connected, the lateral areas are less elevated, the length: breadth ratio of the tail valve amounts to 0.45 in *cancellatus*, 0.60 in *sarsi;* the granulation of end valves and lateral areas is decidedly radial in *cancellatus,* quincuncial in *sarsi.* The armature of the perinotum, moreover, is quite different.

Chiton islandicus Gmelin may be the present species. Gmelin's diagnosis is worthless, but he refers to Schröter's no. 19 (1786:509) "der kleine iszländische chagrinirte Chiton", which was found attached to a valve of *Cyprina islandica* (L.). Schröter's description may be appliable to *L. sarsi.* It reads: "Er ist sehr klein, an beyden Seiten schmal, oben und unten abgerundet, und hat beynahe einen cylindrischen Bau. Er ist stark gewölbt, hat keinen scharfen, sondern völlig ebenen Rücken. Seine Farbe ist schwarz. Mit blossen Augen betrachtet scheinet er ganz glatt zu seyn; nimmt man aber das Vergrösserungsglas, so ist er durchaus mit den feinsten erhabenen Puncten besetzt, und gleicht einem feinen Chagrin. Er hat acht Schilder, die Mittelschilder haben keine dreyseitigen erhabenen Absätze; der feine, kaum merkliche Saum des Thiers ist grau."

Spengler's *Chiton asellus* var. *a* (type in ZMK) is a rather badly preserved specimen of *sarsi,* covered with a black, tar-like environmental deposit.

Leptochiton (L.) gascognensis sp. nov.

(Fig. 26. Map 7)

Leptochiton sp. 1 Kaas, 1979a:21, pl. 3:figs 1-14.

Material. Northeastern Atlantic Ocean, Bay of Biscay, "Thalassa" sta. Z 397, 47°34'N 07°13'W, 511 m, MNHN (holotype), 3.9 × 2 mm, now disarticulated.

Diagnosis. Animal very small, elongate oval, length about twice the width, highly arched (dorsal elevation ca 0.55), back almost evenly rounded. Valves not beaked, lateral areas hardly or not recognizable, tegmentum evenly granulated all over, colour whitish. Girdle clothed with imbricating scales, scattered needle-like spicules, and pairs of intersegmental spines.

Description. Head valve semicircular, almost as wide as the intermediate valves, posterior margin slightly concave. Intermediate valves trapezoid, front margin somewhat emarginate, hind margin straight, apices inconspicuous, side slopes evenly rounded, lateral areas not elevated. Tail valve more or less triangular, length 3/5 of the width, mucro behind the centre, post-mucronal slope concave.

Tegmentum sculptured with small, roundish, widely separated pustules, showing a more or less concentric arrangement. On the central areas, however, they are at the same time decidedly arranged in longitudinal rows, slightly diverging towards the front margin, separated by shallow grooves.

Articulamentum white, apophyses very small, sharply triangular, rounded in the tail valve, widely separated by a slightly convex sinus, somewhat emarginate in the middle.

The dorsal side of the perinotum is clothed with thin, colourless, imbricating scales, strongly bent, marked with 14-16 sharp riblets, round topped, 36 × 36 μm; only the sutural scales are longer, 88 × 36 μm, with less

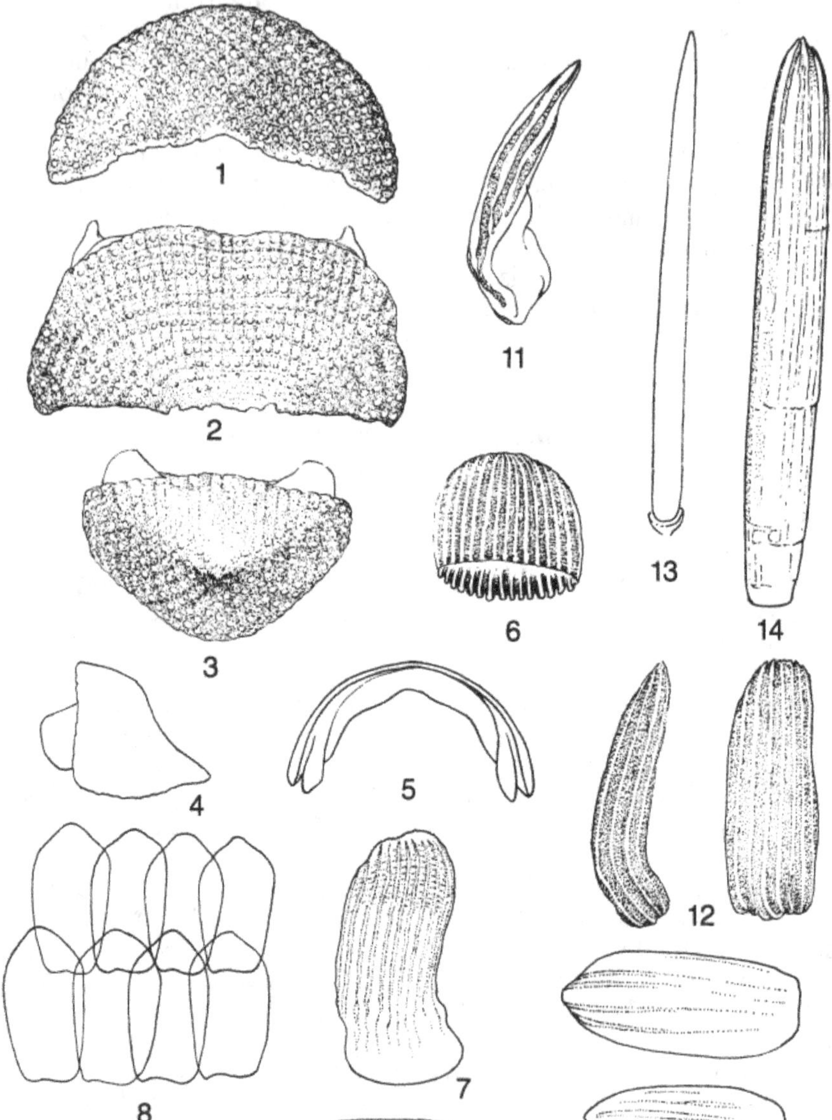

(10-12) riblets. Among the scales slender, white spicules are found, rising from chitinous cups, 192 × 18 μm. On the perinotum tongues between the valves pairs of longer, stout, intersegmental spines occur, rather weakly ribbed, 216 × 25 μm. There is a fringe of marginal spicules-scales, weakly 8-ribbed, somewhat angularly curved near the base, twice as wide as thick, 85-90 μm long, 32 μm wide. Ventral scales arranged in radiating, imbricating series, flat and thin, about oval in shape, obtusely pointed distally, truncated and a little emarginate at the base, 88 × 36 μm, toward the outer margin becoming relatively longer, sometimes with 5 or 6 hardly perceptible riblets, especially near the top.

Distribution. Only known from the holotype.

Observations. This species and the following one were originally described by the senior author (Kaas, 1979a) as *Leptochiton* sp. *1* and *2* respectively, without attaching an epitheton specificum to them, as it was hoped that more specimens might turn up. In the near future no new expeditions are to be expected in the region of the Bay of Biscay. As both species are quite different from all other known *Leptochiton* we have decided to recognize them as valid species and name them.

Leptochiton (L.) thalattius sp. nov.

(Fig. 27. Map 7)

Leptochiton sp. 2 Kaas, 1979a:23, pl. 4:figs 1-12.

Material. Northeastern Atlantic Ocean, off the coast of Portugal, "Thalassa" sta. Y 400, 40°45'N 09°19'W, 800 m, MNHN (holotype), ca 10 mm long, now disarticulated.

Diagnosis. Animal rather small, elongate oval, rather elevated (dorsal elevation c. 0.43), back rounded. Valves slightly beaked, lateral areas distinctly raised, tegmentum minutely granulated all over, colour pale brownish or roseate. Girdle clothed with not imbricating scales, scattered spicules, and intersegmental groups of longer spines.

Description. Head valve semicircular, sculptured with radiating rows of

Fig. 26. *Leptochiton (L.) gascognensis* sp. nov.

1, valve I, dorsal view, × 30; 2, valve V, dorsal view, × 30; 3, valve VIII, dorsal view, × 30; 4, do, lateral view, × 30; 5, valve V, anterior view, × 20; 6, dorsal girdle scale, × 550; 7, dorsal girdle scale, near the suture, × 390; 8, arrangement of ventral girdle scales, × 400; 9-10, ventral scales, near the margin, × 340; 11, supra-marginal scale, × 400; 12, marginal spicules-scales, × 400; 13, dorsal spicule, × 350; 14, intersegmental spine, × 350.

1-14, holotype, after Kaas, 1979.

roundish granules, interrupted by concentric lines of growth.

Intermediate valves somewhat rectangular, only slightly beaked, anterior margin a little emarginate, lateral areas decidedly raised, sculptured like the

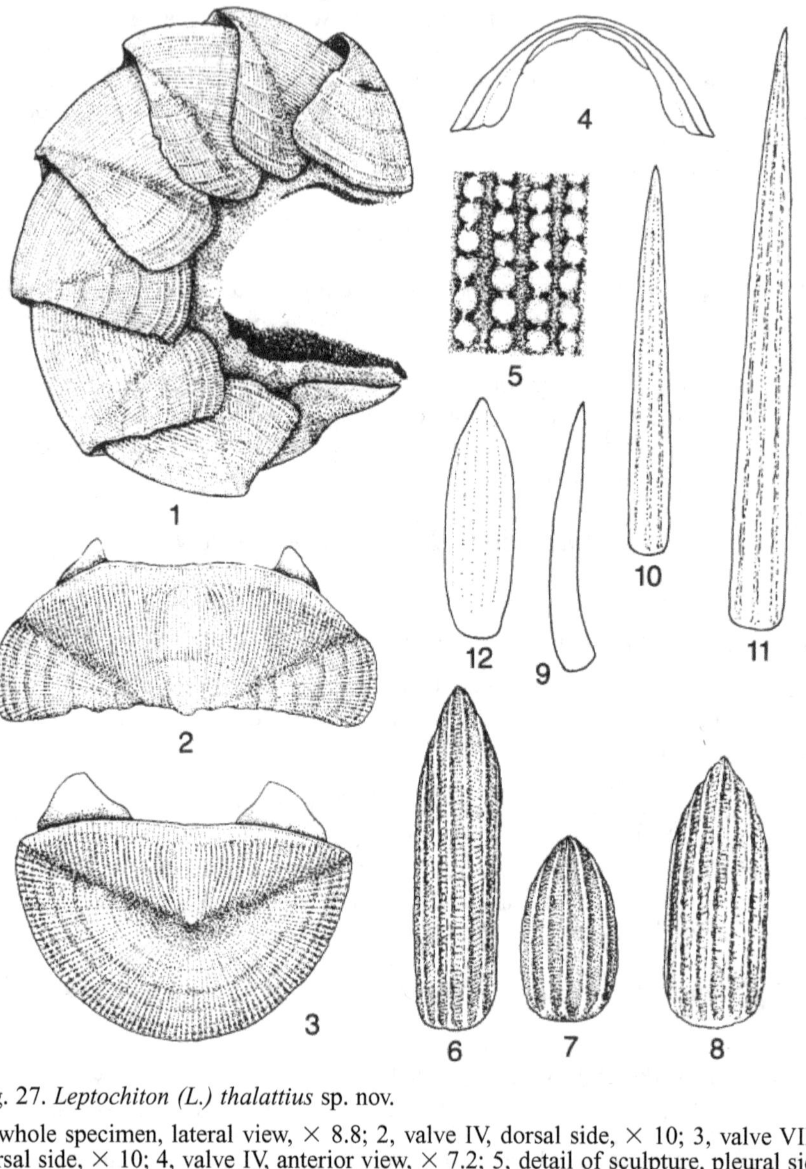

Fig. 27. *Leptochiton (L.) thalattius* sp. nov.

1, whole specimen, lateral view, × 8.8; 2, valve IV, dorsal side, × 10; 3, valve VIII, dorsal side, × 10; 4, valve IV, anterior view, × 7.2; 5, detail of sculpture, pleural side of central area, × 70; 6-8, three types of dorsal girdle scales, × 350; 9, dorsal spicule, × 350; 10, marginal spicule, × 350; 11, intersegmental spicule, × 350; 12, ventral scale, × 350.

1-12, holotype, after Kaas, 1979.

anterior valve; central area with longitudinal rows of roundish granules, stringed together, the grooves between the strings rather deep, about half as wide as the granules. On the jugal part of the valves there are about eight parallel rows, more weakly developed, the four or five adjoining rows on the pleurae are curved as a result of short interpolated rows, posteriorly as well as anteriorly; the remaining pleural rows are parallel again, somewhat diverging towards the anterior margin.

Tail valve rather semi-oval than semicircular, length about 5/8 of the breadth, mucro not swollen, a little before the centre, posterior slope concave just behind the mucro. Antemucronal area sculptured like the central areas, postmucronal area like the head valve.

Articulamentum white, apophyses rather small, sharply triangular, subtriangular in the tail valve, widely separated.

Girdle dorsally clothed with elongate, erect, calcareous scales of different sizes, not imbricating, with 5 (-7) rather strong ribs, obtusely pointed, 64-128 µm long, 30-36 µm wide. Among the scales, irregularly scattered, calcareous spicules occur, straight or somewhat curved, smooth, sharply pointed, 100 µm long, 16 µm thick. On the perinotum tongues between the valves groups of 2-4 weakly ribbed, gradually tapering, intersegmental spines are found, rising from chitinous cups, 220 µm long, 20 µm thick. There is a marginal fringe of equal, but smaller spines, 140 µm long, 16 µm thick. Ventrally the girdle is armed with elongate, flat imbricating scales, sometimes with traces of about 5 riblets, in rows perpendicular to the outer margin. The scales are truncated at the base, distally regularly tapering to a blunt point, 88 µm long, 25 µm wide.

Length: c. 10 mm when stretched. The holotype measures 7.2 mm, but it is strongly curled up.

Distribution. Only known from the holotype.

Observations. See our note under the foregoing species.

Leptochiton (L.) geronensis sp. nov.

(Fig. 28. Map 3)

Lepidopleurus alveolus; Van Belle, 1975a:57, figs 1-2, 4 (non Lovén, 1846).

Material. Spain, Gerona: off Llansá, on coralligenous rests, 200 m, summer 1974, C. Altimira leg., IRSN IG-26.356, type coll. 381 (holotype), 5.1 × 2.7 mm, dry; off Bagur, on coralligenous rests, 250 m, VI. 1974, C. Altimira leg., VB 2907a (paratype), 5.6 × 3 mm, dry, disarticulated.

Diagnosis. Animal small, up to 5.6 mm long, the width slightly more than half the length, elongate oval, moderately elevated (dorsal elevation c. 0.37), subcarinated, side slopes little convex. Valves thin, not beaked, lateral areas not raised, tegmentum evenly granulated all over, colour dirty white.

Girdle narrow, clothed with elongate scales and scattered needle-like spicules.

Description. Head valve semicircular, posterior margin almost straight. Intermediate valves rather short, anterior margin straight between the apophyses, somewhat backwardly directed towards the sides, side margins evenly rounded, posterior margin practically straight, apices inconspicuous, lateral areas hardly perceptible. Tail valve a little less than semicircular, width at least twice the length, mucro antemedian, not prominent, postmucronal slope straight.

Tegmentum sculptured with roundish, clearly separated granules. On the head valve, the lateral areas of the intermediate valves and the postmucronal

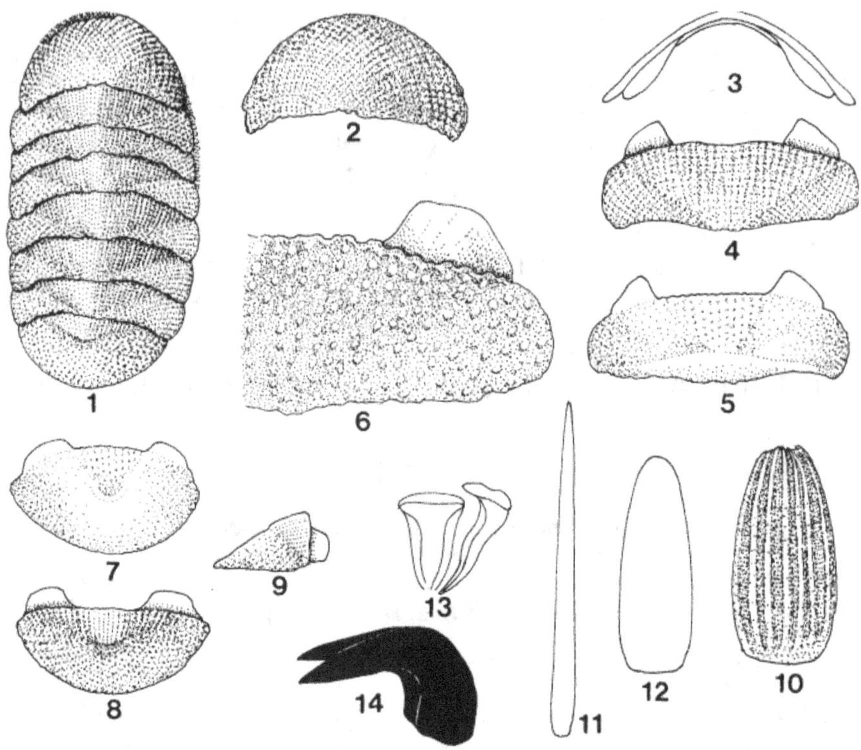

Fig. 28. *Leptochiton (L.) geronensis* sp. nov.

1, whole animal, dorsal view, × 9:6; 2, valve I, dorsal view, × 20; 3, valve IV, anterior view, × 20; 4, do, dorsal view, × 20; 5, do, ventral view, × 20; 6, do, right half, dorsal view, × 40; 7, valve VIII, ventral view, × 20; 8, do, dorsal view, × 20; 9, do, lateral view, × 20; 10, dorsal girdle scale, × 400; 11, dorsal needle, × 400; 12, ventral scale, × 400; 13, central and first lateral radula teeth, × 800; 14, cusp of major lateral radula tooth, × 800.

1, holotype; 2-14, paratype.

area of the tail valve, the granules are arranged in a quincuncial pattern, which, depending on the lighting, gives the impression of radial rows near the outer sides. On the central areas, the granules are arranged in longitudinal rows, ca 40 in number on valve V.

Articulamentum very weakly developed, the tegmental granules shining through, apophyses small, broadly triangular, trapezoid in the tail valve, widely separated by a more or less flat sinus.

Girdle dorsally clothed with elongate scales, 68 × 36 μm, with 6-8 longitudinal ribs, narrower than the interstices, and scattered, smooth, straight needles, 110 × 8 μm. The scales are somewhat longer near the sutures, shorter along the exterior margin. Ventral scales white, thin, flat, elongate, round-topped, truncated at the base, 70 × 24 μm.

Radula with the central and first lateral teeth broadening to a well-developed, flat blade, major lateral tooth with a curved, bidentate cusp, the sharply-pointed denticles of about equal size.

Distribution. So far only known from the holotype and a paratype, both dredged off the coast of Gerona, Spain. Apparently a deep-water species.

Observations. Initially considered as *Leptochiton alveolus* by Van Belle (1975:57), a renewed and thorough study of the type-material has revealed many important differences with that species, just as with all other species of the genus *Leptochiton.*

West Atlantic Species

Leptochiton (L.) pergranatus Dall, 1889

(Fig. 29. Map 17)

Lectotype: USNM 95171; d. Kaas, 1972.
Type locality: off Dominica, "Blake" sta. 192, 260-362 m.

Leptochiton pergranatus Dall, 1889b:414.
Lepidopleurus pergranatus; Pilsbry, 1892:6; Dall & Simpson, 1901:451, pl. 58:fig. 1a-c.
Lepidopleurus (Deshayesiella) pergranatus; Kaas, 1972:16, figs 1-6.
Non: Lepidopleurus pergranatus; Götting, 1973:247, pl. 8:fig. 1a-b = *Leptochiton binghami* (Boone, 1928).

Description. Shell small, elongate oval, moderately elevated, carinated, side slopes angularly convex, flattening towards the girdle. Colour whitish or greyish white, not glossy.

Head valve semicircular, front slope a little concave. Intermediate valves not or only very slightly beaked, lateral areas not raised, hardly perceptible. Tail valve with a slightly swollen mucro, a little anterior, back slope decidedly concave.

Tegmentum uniformly sculptured with minute, clearly cut, roundish granules forming longitudinal chains on the central areas and the antemucronal area of the tail valve, 30-40 on either side of the dorsal ridge, arranged in quincunx on the lateral areas, the head valve and the post-mucronal area of the tail valve.

Articulamentum white, apophyses small, more or less triangular, sinus wide and shallow.

Girdle rather narrow, dorsally clothed with minute, blunt, coarsely ribbed spicule-like scales, 50-65 µm long, and scattered long, slender, pointed glassy spicules, 170-210 µm long, very finely striate, emplanted in a chitinous shaft. Ventral side covered with close set, short, finely striated, pointed glassy spicules, about 80 µm long, directed towards the outer margin, marginal ones somewhat longer, forming a short fringe.

Gills 12 on each side, reaching forward to about the middle of the sixth valve.

Length 12 mm, width 6.5 mm.

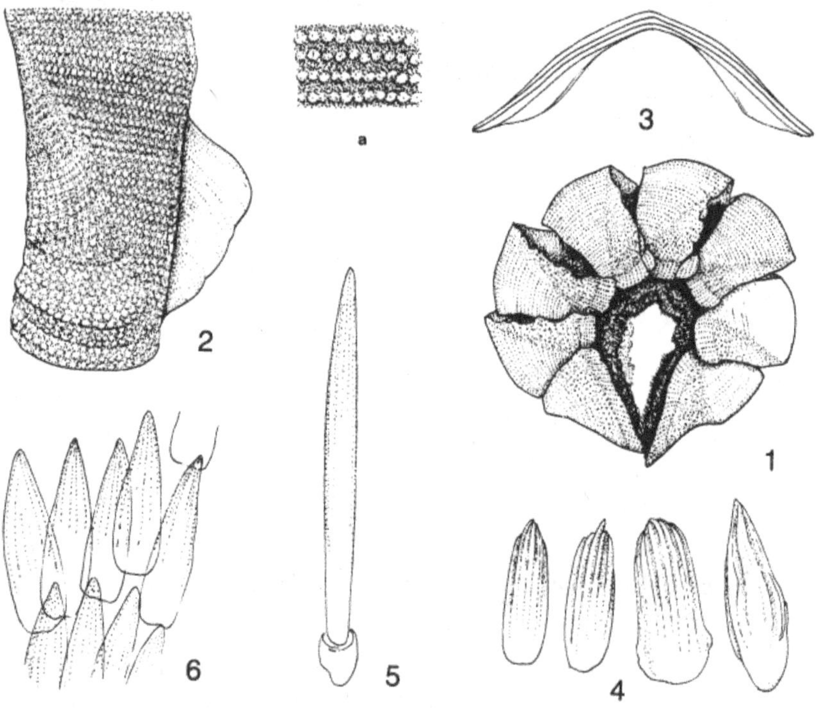

Fig. 29. *Leptochiton (L.) pergranatus* Dall, 1889.

1, whole animal, lateral view, × 6.4; 2, part of valve IV, × 16; 3, valve IV, anterior view; 4, dorsal girdle scales, × 400; 5, dorsal spicule, × 320; 6, ventral spicules, × 320.

1-6, lectotype, after Kaas, 1972.

Distribution. Near Dominica, 260-362 m; Florida: off Florida Keys and West Florida, Gulf of Mexico, 215-2225 m; Porto Rico: off San Juan.

Leptochiton (L.) binghami (Boone, 1928)

(Fig. 30. Map 17)

Type: YPM 15636.
Type locality: off British Honduras, N. of Glover Reef, 912 m.

Lepidopleurus binghami Boone, 1928:14, fig. 2.
Lepidopleurus (Deshayesiella) binghami; Kaas, 1972:18, figs 7-12.
Lepidopleurus pergranatus; Götting, 1973:247, pl. 8:fig. 1a-b (not of Dall, 1889).

Description. Animal rather small, elongate, moderately elevated, carinated, side slopes nearly straight. Colour yellowish white, girdle whitish.

Head valve very slightly concave. Intermediate valves decidedly beaked, having a small but distinct apex, lateral areas only slightly raised, stronger

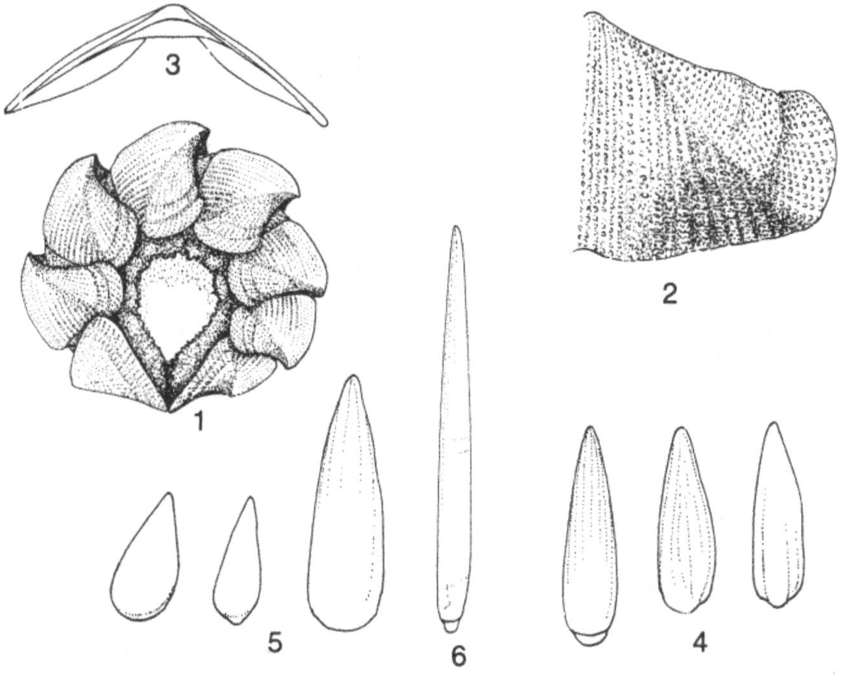

Fig. 30. *Leptochiton (L.) binghami* (Boone, 1928).

1, whole animal, lateral view, × 5.6; 2, part of valve IV, × 8.8; 3, valve V, anterior view; 4, dorsal spicules, × 560; 5, ventral girdle scales, × 410; 6, isolated girdle spicule, × 280.

1-6, holotype, after Kaas, 1972.

towards the side margins. Mucro of the tail valve somewhat swollen, blunt, slightly anterior, post-mucronal area strongly concave.

Tegmentum uniformly granulose, the granules fine and even, not much raised, not clearly cut, roundish, and arranged in quincunx on the head valve, the lateral areas of the intermediate valves and the post-mucronal area of the tail valve; on the jugal area the granules are very fine, hardly elevated, arranged in longitudinal series which are close together; on the pleurae they become much larger, more irregular, like double or treble longitudinal rows of which the granules are fused transversely, the rows separated by rather wide grooves, deepest and widest at the anterior margin, becoming shallower posteriorly, and ending shortly before the diagonal ridge, quite unlike any other *Leptochiton* species.

Girdle dorsally covered with close set, erect, short, conical, white spicules, very delicately striate, about 50 µm long, and scattered, much longer, slender spicules, 170-200 µm long. Ventral side clothed with elongate conical scales or spicules, up to 84 µm long. No marginal fringe of longer spicules present.

Length 15.3 mm, width 8 mm.

Distribution. Off the coast of British Honduras: N. of Glover Reef, 912 m; Colombia: N. coast, W. of Guajira Peninsula, 300-400 m.

Observations. Götting (1973) described and figured three specimens from the Caribbean coast of Columbia under the name of *Lepidopleurus pergranatus* (Dall, 1889). His figures, however, clearly show that they were misidentified. Actually the specimens belong to the present species.

Leptochiton (L.) darioi (Righi, 1973)

(Fig. 31. Map 14)

Type: IBUSP ZU100 (holotype); ZU101 (one paratype).
Type locality: Brazil, São Paulo: NE of Vitória Island, 50 m.

Oldroydia (?) *darioi* Righi, 1973b:259, figs 1-11.

Description. Animal very small, spool-shaped, up to twice as long as wide. Shell white, sometimes with irregularly spread rust-coloured deposits.

Head valve semicircular. Valves II to VIII wide and arched, not carinate, without a distinct apex, lateral areas not raised, angle of divergence in valve V 100°. Tail valve with a somewhat projecting, slightly anterior mucro, posterior slope decidedly concave. Valves I-VII with well marked lines of growth.

Valves decorated with small round granules, juxtaposed, arranged in distinct rows. Head valve with about 60 radial series of granules. Central areas of the intermediate valves and antemucronal area of valve VIII with longitudinal rows, shorter and converging towards the sides. Lateral areas clearly marked by the radial orientation of the rows of granules. On the post-mucronal area of valve VIII the granules are arranged in concentric series.

Apophyses very small, sharply triangular and wide apart. Valves I and VIII with a weakly developed non-slit insertion plate, perforated at the ventral side; intermediate valves without insertion plates.

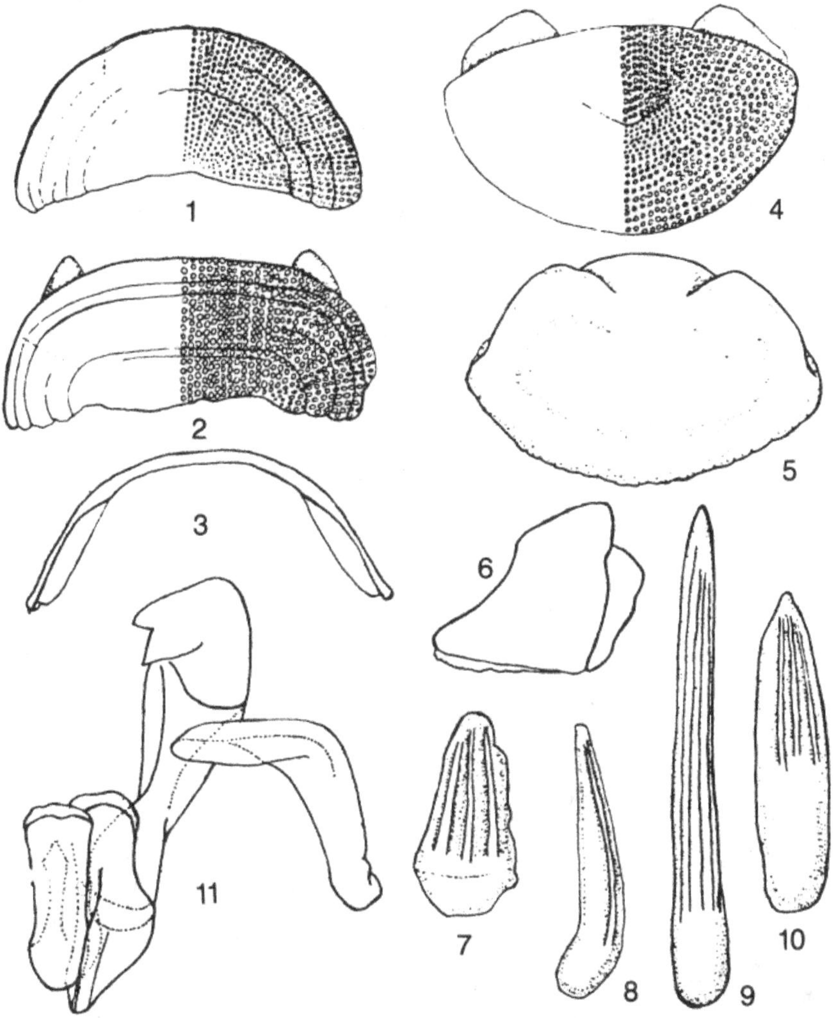

Fig. 31. *Leptochiton (L.) darioi* (Righi, 1973).

1, valve I, dorsal view, × 20; 2, valve V, dorsal view, × 20; 3, do, anterior view; 4, valve VIII, dorsal view, × 20; 5, do, ventral view, × 20; 6, do, lateral view, × 20; 7, dorsal girdle scale, dorsal view, × 600; 8, do, lateral view, × 700; 9, marginal spicule × 350; 10, ventral girdle scale, × 550; 11, radula, × 350.

1-11, after Righi, 1973.

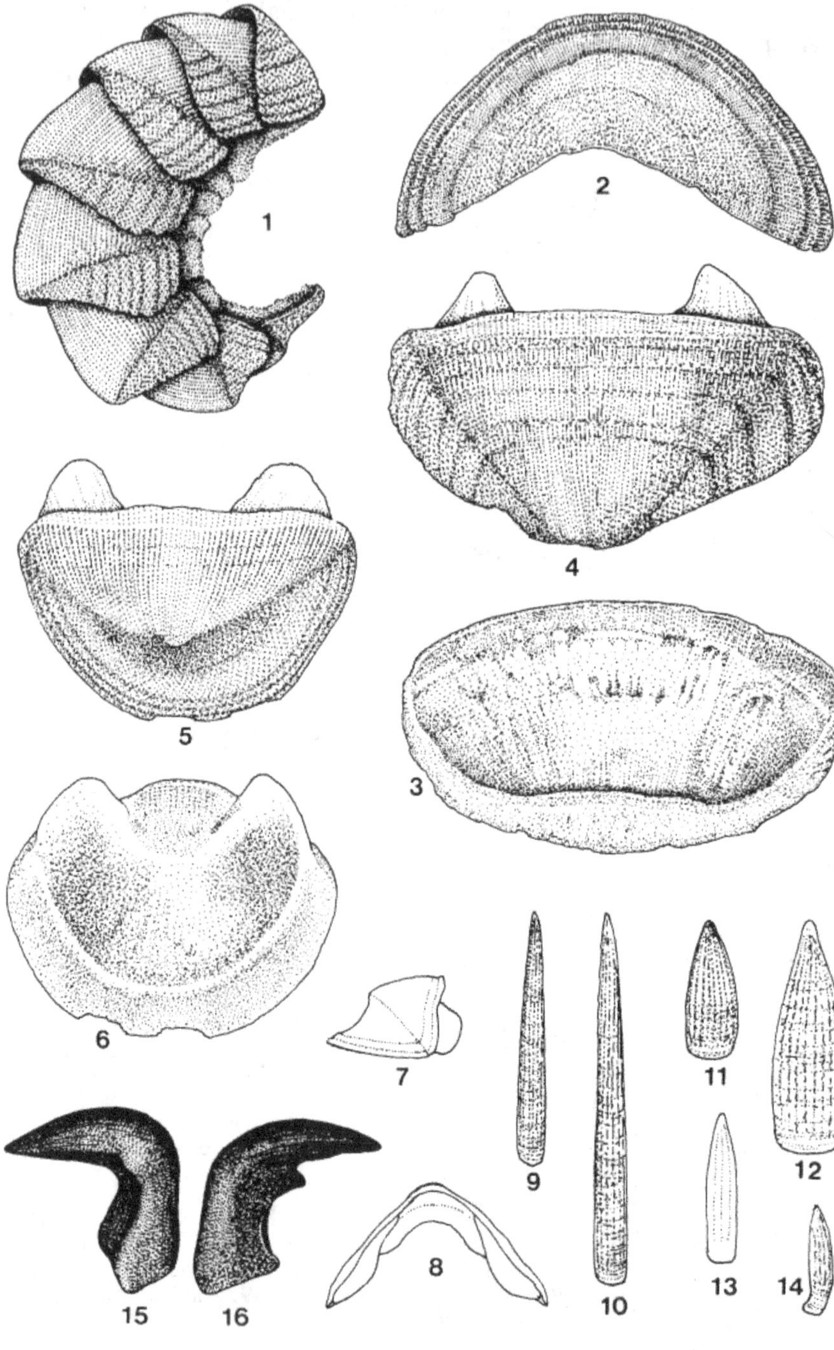
1
2
4
5
3
6
7
9
10
11
12
13
14
15
16
8

Girdle white, with chestnut spots at the sutures. Dorsal side covered with imbricate scales and scattered spicules. Scales 42 to 52 µm long, ornamented with 3-4 longitudinal ribs; spicules 125-183 µm long, longitudinally weakly grooved, more frequent at the sutures and near the outer margins. Ventral side of the girdle with imbricating scales, 57-74 µm long, in radiating rows, distal half decorated with 1-6 longitudinal ribs. Marginal fringe formed by ventral scales as well as dorsal scales and spicules.

Central and first lateral teeth of the radula with a broad cusp, major lateral tooth tricuspidate, first uncinal tooth smooth edged.

Gills merobranchial, adanal without interspace, 8 to 10 on each side, from the middle of valve VII to the anal papilla.

Length 6.1 mm, breadth 2.9 mm (specimen from Rio de Janeiro).

Distribution. Off the eastern coast of Brazil, São Paulo; NE of Vitória Island, 50 m ("Prof. W. Besnard", coll.); Rio de Janeiro, 23°22'S 44°26'W, 50 m ("Prof. W. Besnard", coll.).

Observations. The species was placed, with some doubt, in the genus *Oldroydia* by the author. It has, however, nothing to do with that genus. Unslit insertion plates in the head and tail valves only are unquestionably the main characteristics of the genus *Hemiarthrum*. It is clear that the present species does not fit in that genus either. Apart from the insertion plates in I and VIII the species shows all the features of a true *Leptochiton*, so we have decided to regard it as an outstanding member of this genus.

South African Species

Leptochiton (L.) sykesi (Sowerby III, 1903)

(Fig. 32. Map 13)

Type: SAMC A-5342.
Type locality: South Africa: False Bay, Cape Point Lighthouse, 384 m.

Chiton (Hanleya) sykesi Sowerby III, 1903:225, pl. 5:fig. 13.
Lepidopleurus sykesi; Thiele, 1906:328, pl. 29:figs 6-8; Ashby, 1931a:15, pl. 2:figs 16-19; Barnard, 1963:331.

Description. Animal of moderate size, elongate, more than twice as long

Fig. 32. *Leptochiton (L.) sykesi* (Sowerby III, 1903).

1, whole animal, × 4.5; 2, valve I, dorsal view, × 9; 3, do, ventral view, × 9; 4, valve II, dorsal view, × 9; 5, valve VIII, dorsal view, × 9; 6, do, ventral view, × 9; 7, do, lateral view, × 4.5; 8, valve II, anterior view, × 4.5; 9, dorsal girdle spicule, × 190; 10, marginal spicule, × 190; 11, dorsal scale from mid-girdle, × 190; 12, do, near suture, × 190; 13, ventral scale, × 190; 14, dorsal scale, lateral view, × 190; 15, cusp of major lateral radula tooth, posterior view, × 190; 16, do, anterior view, × 190.

1-16, specimen from S Africa, NW of Table Bay, 347 m, "Pieter Faure", SAMC.

as wide, highly arched (dorsal elevation 0.55), carinated, side slopes almost straight, valves heavy, apices small, pointed, hardly projecting, in older shells often worn away, lateral areas perceptibly raised.

Head valve almost semicircular, sculptured with numerous, closely set, radiating series of very minute, almost microscopical, round, flat granules, obsolete in older specimens. There are 3-5 concentric growth ridges close to the outer margin. Underfolded part of tegmentum forming a wide and thick ridge on inside of posterior margin. Intermediate valves relatively narrow, valve II longer than the others, the posterior margin widely V-shaped, almost straight in III-VII, slightly concave at both sides of the apex. Central areas with 70-80 very fine, longitudinal, beaded threads, somewhat converging posteriorly. Lateral areas raised, sculptured with equal minute granules arranged in quincunx. There are 5-8 concentric growth marks, continuing on the central areas but there less pronounced. Tail valve semicircular, mucro pointed, posterior, at two thirds of the valve's length when seen from above; antemucronal area sculptured like central areas, postmucronal area like lateral areas; posterior slope deeply concave.

Articulamentum strongly developed, white, strongly calloused towards outer margins of end valves. Apophyses small, triangular, wide apart; jugal sinus straight.

Colour of tegmentum dirty white to light yellowish brown.

Girdle narrow, dorsally clothed with elongate, bluntly pointed scales, marked with generally 8 faint longitudinal grooves, those near to the sutures largest, 160 μm long, 50 μm broad, on the middle of the girdle they measure 95 μm × 36 μm. Scattered among the scales needle-like calcareous spicules occur, longitudinally finely striate, 180 μm long, 22 μm thick. Ventral scales arranged in close radiating rows, imbricating, spatula-shaped, thin, flat, with three very faint grooves, 100 μm long, 22 μm broad, relatively longer towards outer margin.

Central tooth of radula rounded and widened at the base, distally with a well developed cusp; major lateral with a long and strong bidentate cusp, the inner denticle long, strongly bent, sharply pointed, the outer one small but distinct.

Gills merobranchial, adanal, with 10 gills on both sides.

Length 23 mm, width 9.5 mm.

Distribution. South Africa, Cape Province: Cape Point, west coast of Cape Peninsula, Table Bay, from 70-433 m.

Leptochiton (L.) chariessa (Barnard, 1963)

(Fig. 33. Map 13)

Type: SAMC A-9338.
Type locality: South Africa, off Cape St. Blaize, 225 m.

Lepidopleurus chariessa Barnard, 1963:332, fig. 29h.

Description. Animal of moderate size, elongate oval, almost twice as long as wide, elevated (dorsal elevation 0.45), carinated, side slopes a little convex; valves with a small, pointed apex, not beaked, lateral areas hardly raised.

Head valve less than semicircular, tegmentum ornamented with 60-70 fine radiating rows of round-topped, oval granules and 4-5 faint, concentric growth marks near the outer margin. Intermediate valves about twice as wide as long, with a rather sharp back; anterior margin sinuate, concave in the jugal area, convex at the pleurae which are regularly narrowing towards the sides. Posterior margin only little concave at both sides of the small, pointed apex. Central area with 40-45 longitudinal rows of oval, well-raised pustules, the interstices almost as wide. Lateral areas marked by the radially directed rows of pustules, numbering 10, towards the margin interrupted by 4 or 5 faint lines of growth. Tail valve elliptical in outline, the length two thirds of the width, mucro slightly posterior, almost central, when seen from above, bluntly pointed; hind slope regularly concave. Antemucronal area sculptured like central areas, with 30-35 longitudinal rows of granules, the rows closely set on the jugal area, wider separated and rather strongly converging

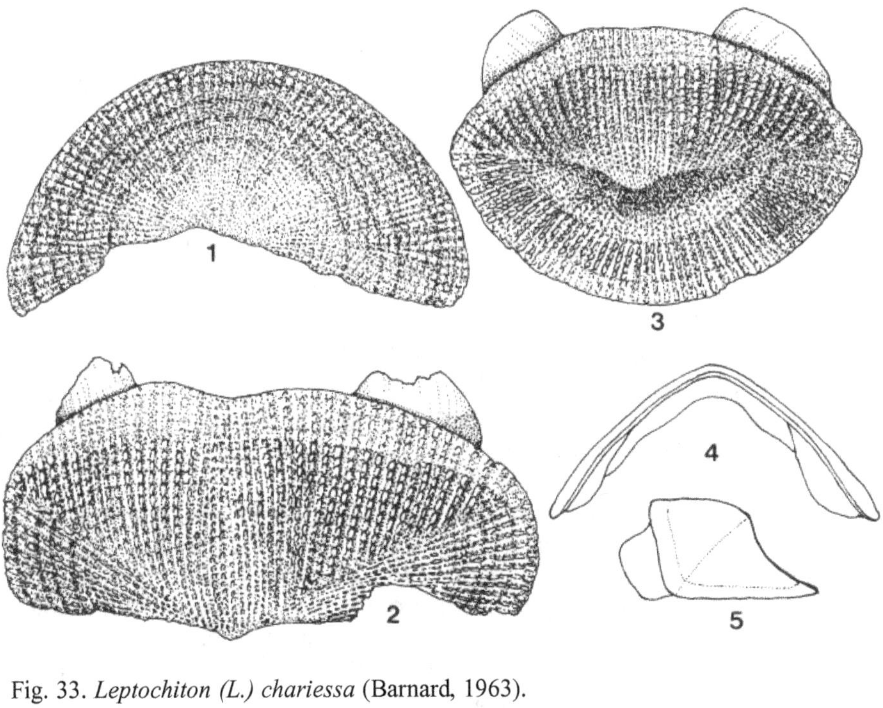

Fig. 33. *Leptochiton (L.) chariessa* (Barnard, 1963).

1, valve I, dorsal view, × 19; 2, valve IV, dorsal view, × 19; 3, valve VIII, dorsal view, × 19; 4, valve IV, anterior view, × 9; 5, valve VIII, lateral view, × 9.

1-5, holotype.

posteriorly towards the sides; postmucronal area with 50 radial rows of granules and a few faint, concentric growth marks.

Articulamentum rather weakly developed, white, the apophyses wide apart, subtriangular; jugal sinus concave in the intermediate valves, convex in tail valve.

Colour of tegmentum yellowish white.

Girdle narrow, with fine spicules ("but not in good condition", according to Barnard).

Length (by estimate) 13 mm, width 7 mm.

Distribution. Two specimens were procured, the animals mostly decomposed, "enough flesh remained to hold the valves together, but no radula was obtained" (Barnard).

Of the type only the 8 shell plates are preserved.

Antarctic Species

Leptochiton (L.) kerguelensis Haddon, 1886

(Fig. 34. Map 31)

Type: ?
Type locality: Kerguelen Island, Royal Sound, Balfour Bay, "Challenger" Sta. 149c: 49°32'S 70°E, 108 m.

Leptochiton kerguelensis Haddon, 1886:12, pl. 1:fig. 3, pl. 2:figs 3a-c.
Leptochiton pagenstecheri Pfeffer, 1886:107, pl. 3:figs 3a-b. [Type: ZIM 6971 (lectotype), South Georgia.]
Lepidopleurus kerguelensis; Carcelles, 1953:158; Leloup, 1956b:12, fig. 1 (bibliography and synonymy); Dell, 1964a:113; ?Righi, 1973a:239, figs 1,2.
Non *Lepidopleurus kerguelenensis;* Hedley, 1916 = *Leptochiton fairchildi* (Iredale & Hull, 1929).

Description. Animal small, elongate oval, almost twice as long as wide, up to 5 mm long, rather elevated (dorsal elevation 0.42), the valves roundly arched, neither carinated nor beaked, smooth to the naked eye, lateral areas slightly but distinctly raised, girdle narrow, clothed with elongate scales and long needles around the sutures.

Head valve almost semicircular, apparently smooth, with 4 or 5 concentric lines of growth, especially towards the outer margin. Under a magnification of at least 25 × the tegmentum appears to be beset with roundish, flat, hardly perceptible granules, arranged in quincunx. Anterior margin of intermediate valves straight or slightly concave in the middle, the pleural parts convex; posterior margin almost straight, the apices worn away, even in young specimens. Central areas sculptured with 36-40 very weak furrows between longitudinal rows of roundish granules, on the jugum only discernible when highly magnified, more clearly, but still very weakly shown on the pleural areas; lateral areas little raised, separated from the central area by a slight ridge, granulose all over, the granules hardly perceptible, arranged in quincunx. Posterior valve semicircular, the mucro about central, not swollen, posterior slope only little concave directly behind mucro, antemucronal area sculptured like central areas, postmucronal area like head valve.

Colour of tegmentum white, older parts of valves often stained with light brown or blackish by foreign deposits.

Articulamentum weakly developed, white, apophyses small, wide apart, sharply triangular in intermediate valves, trapezoid in tail valve.

Girdle narrow, white, dorsally closely paved with not imbricating, elongate,

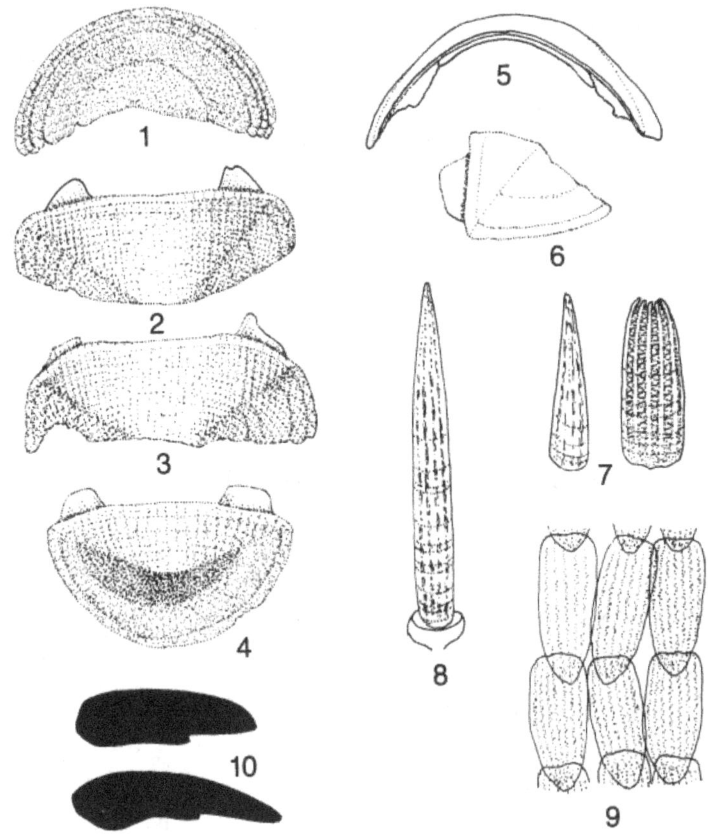

Fig. 34. *Leptochiton (L.) kerguelensis* Haddon, 1886.

1-4, valves I, II, V, VIII, dorsal view, × 20; 5, valve V, anterior view, × 20; 6, valve VIII, lateral view, × 20; 7, dorsal girdle spicule and scale, × 400; 8, intersegmental spine, × 400; 9, ventral scales, × 400; 10, cusps of major lateral radula teeth, × 800.

1-10, specimen from South-Georgia, Kochtopfbucht, 30.V.1902, Swedish Antarctic Exp. 1901-1902, NHRM.

rectangular or bluntly pointed scales, 56 μm long, 20 μm wide, with 4-6 narrow riblets, often dentating the top of the scale, interspersed with short, conical, striated spicules of equal length, 10 μm wide. Along the sutures long, slender, sharply pointed, strongly longitudinally grooved needles are found at regular intervals, rising from ochraceous chitinous cups, 104 μm long, 12 μm thick. Several of these needles are placed together on the encroachments of the girdle between the valves, forming small intersegmental bunches. Ventral side of girdle, closely paved with flat, imbricating scales in radiating rows, 54 μm long, 20 μm wide, weakly longitudinally grooved, the top evenly rounded, the base truncate.

Radula with numerous close rows of very thin, weakly developed teeth, of which only the bifid cusps of the major laterals are quite striking; they are narrow, elongate, rather bluntly pointed, with a very small outer denticle at about two thirds of the length of the cusp.

Gills merobranchial, adanal; 8-10 ctenidia on both sides.

Length 5 mm, breadth 3 mm, generally smaller.

Distribution. Between S. latitudes 40° and 70° all around the mainland of Antarctica: Tierra del Fuego, Falkland, S. Georgia, S. Shetland, and S. Orkney Islands, off Enderby Land, Ross Sea, Crozet and Kerguelen Islands. The northernmost habitat is Gough Island (Dell), at 41°S 10°W. The species has been recorded from depths of 12-125 m.

Righi (1973a:239) reported a specimen from Ilha Grande, Rio de Janeiro, but his figures of short, multistriate girdle scales indicate that it was wrongly identified.

Leptochiton (L.) medinae (Plate, 1899)

(Fig. 35. Map 37)

Type: ZMHU.
Type locality: Magellan Strait: Harbour of Punta Arenas, 15-19 m.

Lophyropsis imitatrix Thiele, 1893:388, pl. 31:fig. 36, nom. imp.pro *imitator* (non E. A. Smith, 1881); Thiele, 1909:4.
Lepidopleurus medinae Plate, 1899:82, pl. 5:figs 204-206; Thiele, 1909:4; Car-

Fig. 35. *Leptochiton (L.) medinae* (Plate, 1899).

1, valve I, dorsal view, × 20; 2, valve IV, dorsal view, × 20; 3, do, ventral view, × 20; 4, valve VIII, dorsal view, × 20; 5, valve IV, anterior view, × 20; 6, valve VIII, lateral view, × 20; 7, dorsal girdle scale, near suture, × 400; 8, do, near outer margin, × 400; 9, dorsal spicule, × 400; 10, marginal spicule, × 400; 11, ventral girdle scale, near inner margin, × 400; 12, do, mid-girdle, × 400; 13, do, near outer margin, × 400; 14, half radula row, × 120; 15, cusp of major lateral radula tooth, exterior view, × 190; 16, anterior half of the animal, ventral view, × 4.

1-13, 15, specimen from Chile, Golfo de Ancud, ESE of Isla Tac, 42°26'.40 S, 72°59'W, 250-300 m, 23.I.1949, Lund Univ. Chile Exp. 1948-1949, sta. 41, NHRM 792; 14, 16, after Plate, 1899.

celles, 1950:47; Leloup, 1956b:13, fig. 2 (bibliography); Dell, 1971:216.
Lepidopleurus (Leptochiton) agesilaus Dall, 1919:501 (lectoype + 1 paralectotype in USNM, 96227, Str. of Magellan, 110 m); Carcellas & Williamson, 1951:243; von Salvini-Plawen, 1968:258; Ferreira, 1979b:162, fig. 28.

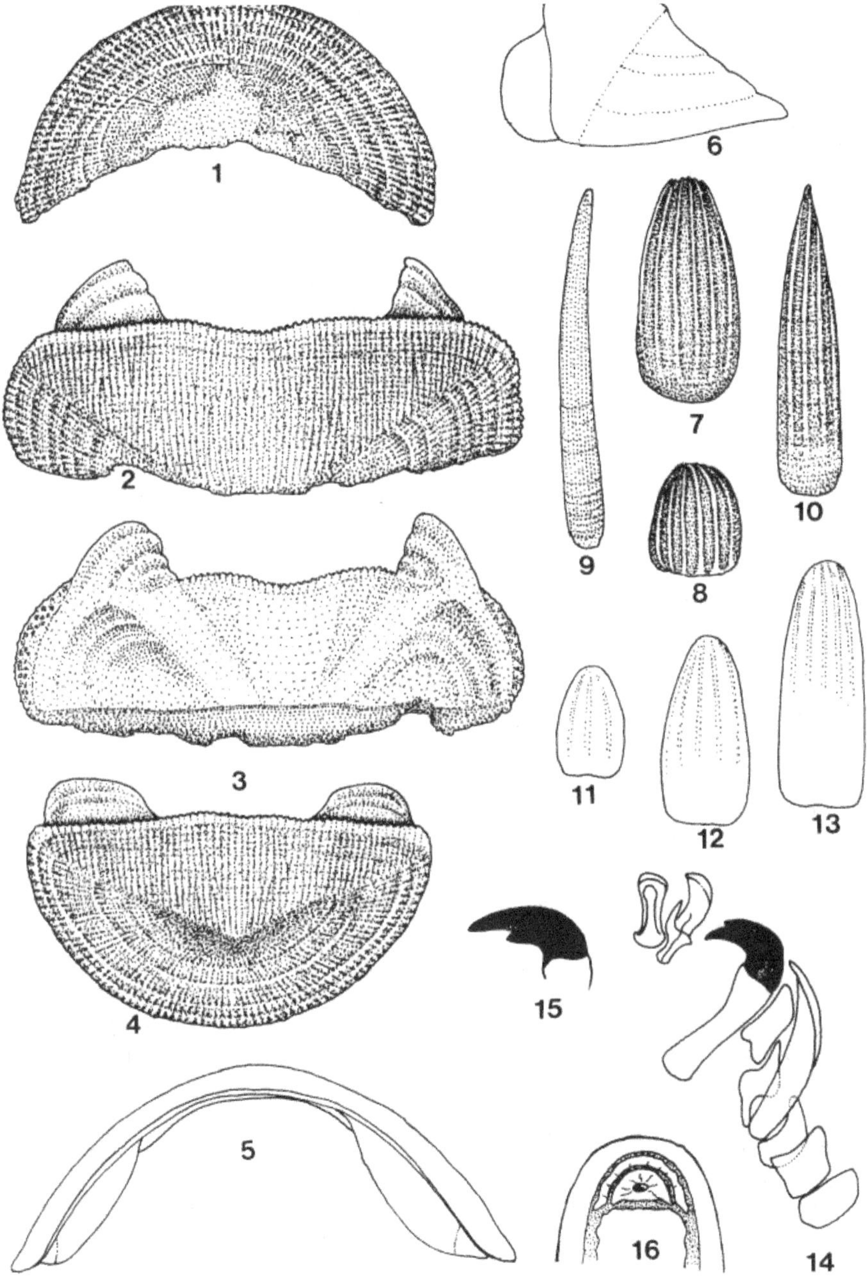

Lepidopleurus agesilaus; Boudet, 1945:135; Carcelles, 1950:47; Leloup, 1956b: 12; Castellanos, 1956:466; Dell, 1964a:113.

Description. Animal elongate oval, twice as long as wide, rather elevated (dorsal elevation 0.38), valves evenly rounded, neither carinated nor beaked, lateral areas hardly raised.

Tegmentum of head valve sculptured with numerous (80) radial rows of small, roundish, closely set granules, crossed by many growth marks, becoming stronger towards the outer margin which is semicircular in outline. Intermediate valves short and wide, anterior margin curved inward between the apophyses, posterior margin almost straight, the umbo not swollen, nor protruded. Central areas sculptured with 70 closely set, longitudinal rows of very small roundish to oval granules, weakly pronounced in the jugal area, stronger towards the side margin. Lateral areas only little raised, sculptured with 15 radial rows of similar granules, interrupted by 7-9 concentric growth marks. Anterior and side margins finely crenulated by the slightly projecting series of granules. Tail valve almost semicircular, slightly narrower than head valve, the mucro pointed, not swollen, a little posterior when seen from above, posterior slope almost straight, only weakly concave directly behind the mucro; anterior margin rather straight, finely crenulated. Antemucronal area sculptured like central areas, postmucronal area like head valve.

Articulamentum white, not covering the very porous inside margins of the tegmentum. The underfolded part of the tegmentum posteriorly very wide. Apophyses wide apart, on the intermediate valves sharply triangular and rather strongly marked with 2 or 3 wavy growth marks, on the inside meeting the anterior margin at an angle of 100°-110°. Apophyses of tail valve less projecting, trapezoid. Jugal sinus very wide, rather deep, little concave.

Colour of tegmentum white, the exposed parts of valves often covered with a black crust of foreign deposits, sometimes brownish, especially in the grooves separating the rows of granules.

Girdle rather narrow, brownish, covered with elongate, conical, bluntly pointed, somewhat transparent, not imbricating, flat, white scales, 75 µm long, 27 µm broad, ornamented with 6 fine riblets. Towards the margin the scales become much shorter, the outer ones 36 µm long, also 6-ribbed. Scattered among the scales elongate, somewhat curved, calcareous spines are found, gradually tapering to a blunt point, very finely longitudinally striated, 120 µm long, 12 µm thick. Marginal spicules stout, sharply pointed, 100 µm long, 21 µm thick, the visible side with 4-5 narrow riblets. Ventral scales elongate, bluntly rounded at the top, arranged in closely set radial, imbricating rows. At the base the scales are truncated, the distal half marked with 5 faint riblets. Near the inner margin the scales are short, 36 µm long, towards the outer margin they grow gradually longer, the infra-marginal ones measuring 80 µm × 30 µm.

Radula with a short, thinly bladed central tooth, somewhat narrowing towards the base; first lateral narrow, somewhat curved inwardly, with a small blade; major lateral with a bidentate cusp, the outer denticle smaller; spatulate uncinal slightly curved, with a long, narrow cutting edge.

Gills merobranchial, adanal without interspace, 9-10 ctenidia on both sides.

Length 10 mm, breadth 5 mm.

Distribution. A rather common, sublittoral species on both sides of southern South America, from Lat. 42° south to Cape Horn, mostly in a depth of 15-30 m, though Leloup reported it from 250-300 m (Golfo de Ancud, Chile).

Observations. Of the type only the 8 shell plates are left (ZMHU). They agree in all respects with the specimens collected by the Lund University Chile Expedition, described and illustrated here. A study of the types of *L. agesilaus* Dall reveals that they are conspecific with *L. medinae*, as was already supposed by Ferreira (1979b).

L. medinae is closely allied to *L. rugatus* of the North-Eastern Pacific, but differs markedly in the girdle armature and radula.

East Pacific Species

Leptochiton (L.) nexus Carpenter, 1864

(Fig. 36. Map 8)

Type: USNM 16270.
Type locality: California, Catalina Island, 33°23'N 118°24'W, 36-144 m.

Leptochiton nexus Carpenter, 1864a:612 (nom. nud.), 650; Ferreira, 1979b:149, figs 3-6, 8, 35-36 (bibliography and synonymy).
Lepidopleurus (Xiphiozona) heathi Berry, 1919a:5.
Lepidopleurus ambustus Berry, 1907:47 (nom. nud.).
Lepidopleurus (Leptochiton) ambustus Dall, 1919:499.
Lepidopleurus (Leptochiton) lycurgus Dall, 1919:500.

Description. Animal oval, of moderate size, usually not exceeding 20 mm in length, exceptionally up to 25 mm; width: length ratio about 0.59, moderately elevated, not beaked, subcarinated, side slopes convex, colour brown, blotched with minute specks of dark brown or lavender, often with a creamy tan stripe along the jugum; girdle distinctly spiculose, interspersed with many hyaline spines.

Head valve somewhat less than semicircular, posterior margin widely V-shaped. Intermediate valves broadly rectangular, side margins rounded, anterior margin straight to slightly concave, apices inconspicuous, lateral areas not elevated, defined by the sculptural differences only. Tail valve semicircular, mucro median, rather conspicuous, postmucronal slope concave.

Tegmentum uniformly sculptured with minute, subquadrate granules, arranged in radiating series on the head valve, the lateral areas of the inter-

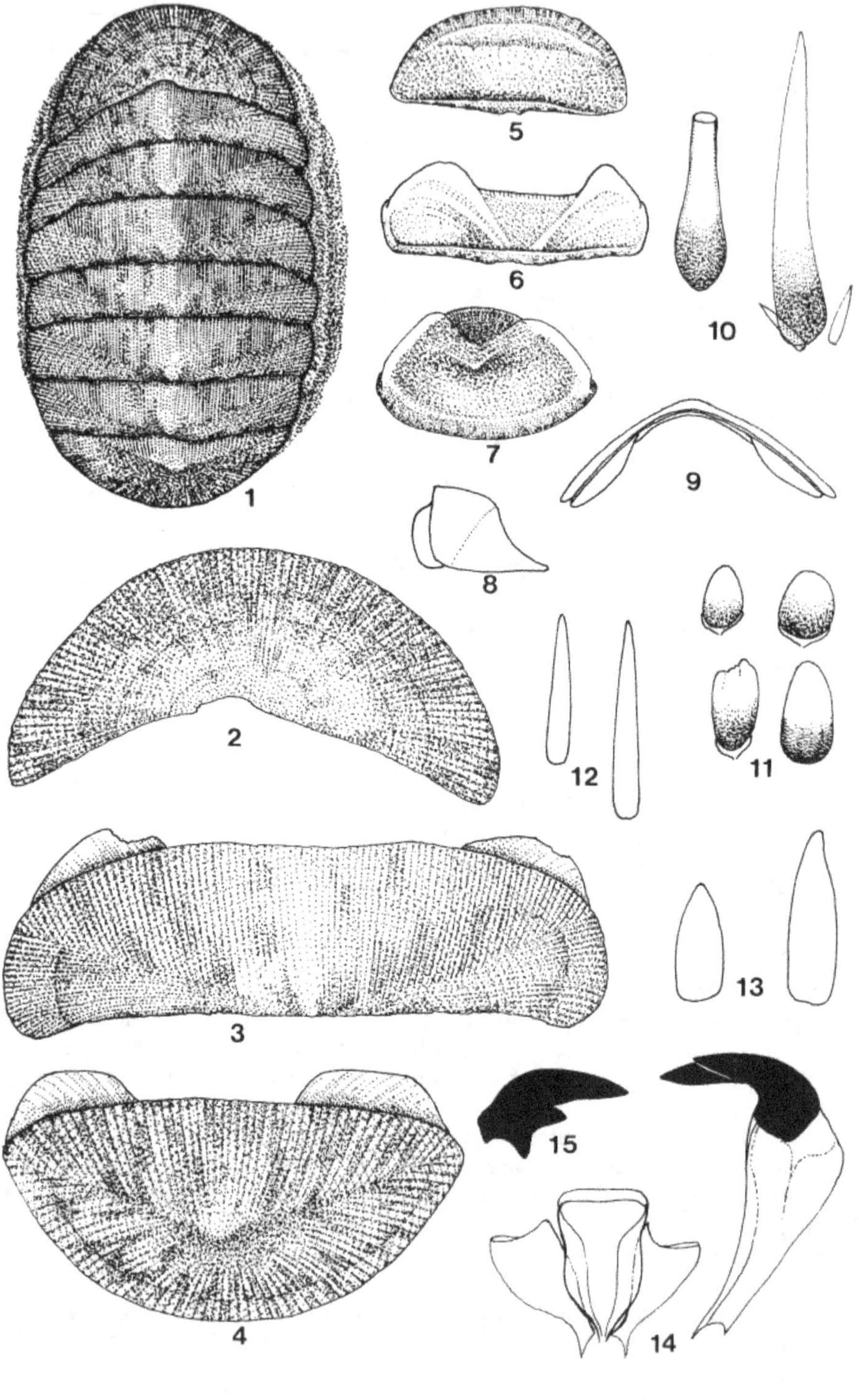
1
2
3
4
5
6
7
8
9
10
11
12
13
14
15

mediate valves and the postmucronal area of the tail valve, in longitudinal rows on the central areas and the antemucronal area, towards the sides the rows converge posteriorly.

Articulamentum greenish brown, apophyses white, narrow, evenly rounded, widely separated, trapezoid in the tail valve.

Girdle moderately wide, coloured like tegmentum, dorsally clothed with small, more or less oval spicules, 48-72 μm long, with many hyaline spines, 220 to over 400 μm long, strewn throughout. There is a marginal fringe of smooth, needle-like spicules, 100-136 μm long, 20 μm thick. Ventral scales elongate, obtusely pointed, truncated at the base, 80-120 μm long, 36 μm wide.

Central radular tooth distally broadening to a narrow blade, first lateral tooth conspicuously notched in the middle of the anterior border, major lateral with a bidentate cusp, the inner denticle much larger than the outer one.

Distribution. *L. nexus* lives intertidally, down to a depth of 144 m and has a continuous distribution from Cohen Island, Alaska, to Punta Abreojos, Baja California. It has also been reported from Bahía de los Angeles in the Gulf of California.

Leptochiton (L.) rugatus (Carpenter *in* Pilsbry, 1892)

(Fig. 37. Map 15)

Syntypes: ANS 35586.
Type locality: Mexico, Baja California, near S. Tomas River, 31°32'N 116°26'W.

Leptochiton internexus var. *rugatus* Dall, 1879:319 (nom. nud.).
Lepidopleurus rugatus Carpenter *in* Pilsbry, 1892:11, pl. 3:figs 67-70.
Leptochiton rugatus; Ferreira, 1979b:146, figs 1-2, 33-34 (bibliography and synonymy).
Leptochiton internexus (Carpenter MS) Dall, 1879:319 (nom. nud.).
Lepidopleurus internexus Carpenter *in* Pilsbry, 1892:12.
Leptochiton cancellatus; Dall, 1879:315, et auct., non *Chiton cancellatus* Sowerby II, 1840.
Lepidopleurus assimilis Thiele, 1909:11, pl. 1:figs 30-39.
Lepidopleurus alascensis Thiele, 1909:13, pl. 1:figs 51-60.

Fig. 36. *Leptochiton (L.) nexus* Carpenter, 1864

1, whole animal, dorsal view, × 9.6; 2, valve I, dorsal view, × 20; 3, valve II, dorsal view, × 20; 4, valve VIII, dorsal view, × 20; 5, valve I, ventral view, × 9.6; 6, valve II, ventral view, × 9.6; 7, valve VIII, ventral view, × 9.6; 8, do, lateral view, × 9.6; 9, valve II, anterior view, × 9.6; 10, large dorsal girdle spicules, × 200; 11, small dorsal spicules, × 200; 12, marginal spicules, × 200; 13, ventral girdle scales, × 200; 14, central, first lateral and major lateral radula teeth, × 200; 15, cusp of major lateral radula tooth, exterior view, × 200.

1, specimen from California, off Redondo, 54-68 m, 1957, E. P. Chace leg., VB 2690a; 2-15, do, K 4893.

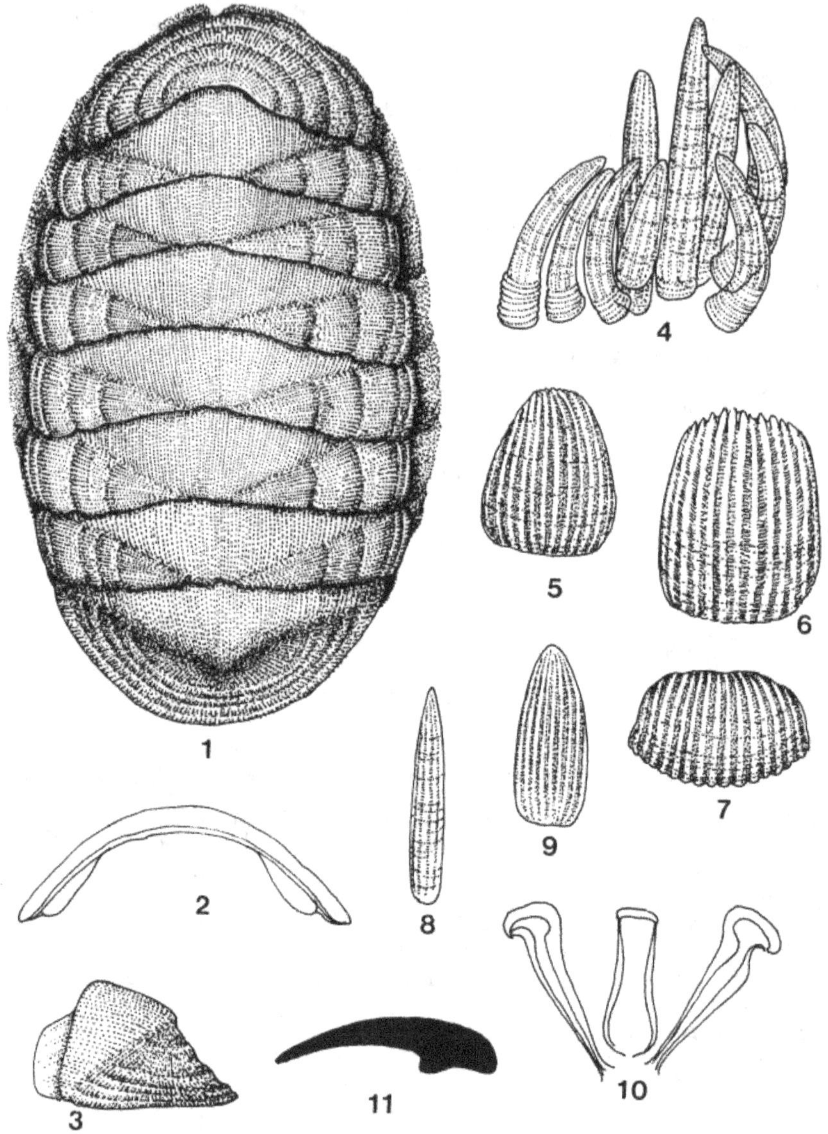

Fig. 37. *Leptochiton (L.) rugatus* (Carpenter *in* Pilsbry, 1892).

1, whole animal, dorsal view, × 9.6; 2, valve IV, anterior view, × 9.6; 3, valve VIII, lateral view, × 9.6; 4, bunch of intersegmental spicules, × 260; 5, dorsal girdle scale, near suture, × 400; 6, sutural scale, × 400; 7, dorsal scale from mid-girdle, × 400; 8, marginal spicule, × 200; 9, ventral scale, × 400; 10, central and first lateral radula teeth, × 400; 11, cusp of major lateral radula tooth, × 400.

1-11, specimen from California, Los Angeles Co., N of Santa Monica, 1947, E. P. Chace leg., K 2843.

Description. Animal rather small, maximum length 15.8 mm, appreciably smaller in its northern range, oval to elongate oval, moderately elevated (dorsal elevation 0.35), back evenly rounded, neither carinated nor beaked, girdle rather narrow.

Head valve somewhat less than semicircular, hind margin regularly notched in the middle. Intermediate valves broadly rectangular, anterior margin slightly convex, posterior margin about straight, apices not indicated, lateral areas more or less elevated. Tail valve semicircular, front margin convex, mucro central to subcentral, somewhat swollen, often slightly pointed, post-mucronal slope variable.

Tegmental sculpture of extremely small granules, forming subgranose riblets, longitudinally disposed on the central areas of the intermediate valves and the antemucronal area of the tail valve, radially oriented on the head valve, the lateral areas, and the post-mucronal area. The end valves and lateral areas clearly show several concentric growth ridges.

Colour of tegmentum and girdle creamy white, sometimes orange or light brownish, frequently blackened by environmental deposits.

Articulamentum whitish, apophyses widely separated, small, triangular, truncated in the tail valve.

Girdle gravelly, dorsally clothed with broad scales, 60 × 36 μm, sculptured with 13-16 riblets, interstices of about equal width, the scales somewhat longer and becoming conical near the sutures, 56 × 44 μm. There are intersegmental bunches of straight or curved, obtusely pointed spicules, up to 144 μm long, all obsoletely longitudinally striated, with several concentric markings at the base. Along the margin identical spicules are found. Ventrally the girdle is paved with imbricating, elongate, obtusely pointed scales, 60 × 24 μm, decidedly sculptured with about 9 longitudinal riblets.

Radula with the central tooth narrowing in front to a small blade; first lateral tooth somewhat aliform, widest in front, major lateral bearing a very long, sharply pointed cusp.

Distribution. *L. rugatus* occurs from the intertidal zone to a depth of 458 m. It is a northern Pacific species found in the Sea of Japan, the Okhotsk Sea, the Bering Sea, and along the western coast of North America from Cohen Island, Alaska, to Magdalena Bay, Mexico. It is also present throughout most of the Gulf of California.

Observations. *L. rugatus* is undoubtedly closely related to the European *L. cancellatus* and *L. sarsi*. It differs from *cancellatus* by its more oval, less elevated shape, by the bunches of intersegmental spicules on the perinotum, and especially in the structure of the radula. *L. sarsi* is still higher elevated and differs much in the armature of the girdle. Moreover, *L. rugatus* grows considerably larger than the others. The antarctic *L. medinae* is also much alike, but for the perinotum and radula.

Leptochiton (L.) incongruus (Dall, 1908)

(Fig. 38. Map 8)

Type: USNM 122969.
Type locality: Gulf of Panama, U.S.S. "Albatross" st. 3354, 07°09.45'N 80°50'W, 589 m.

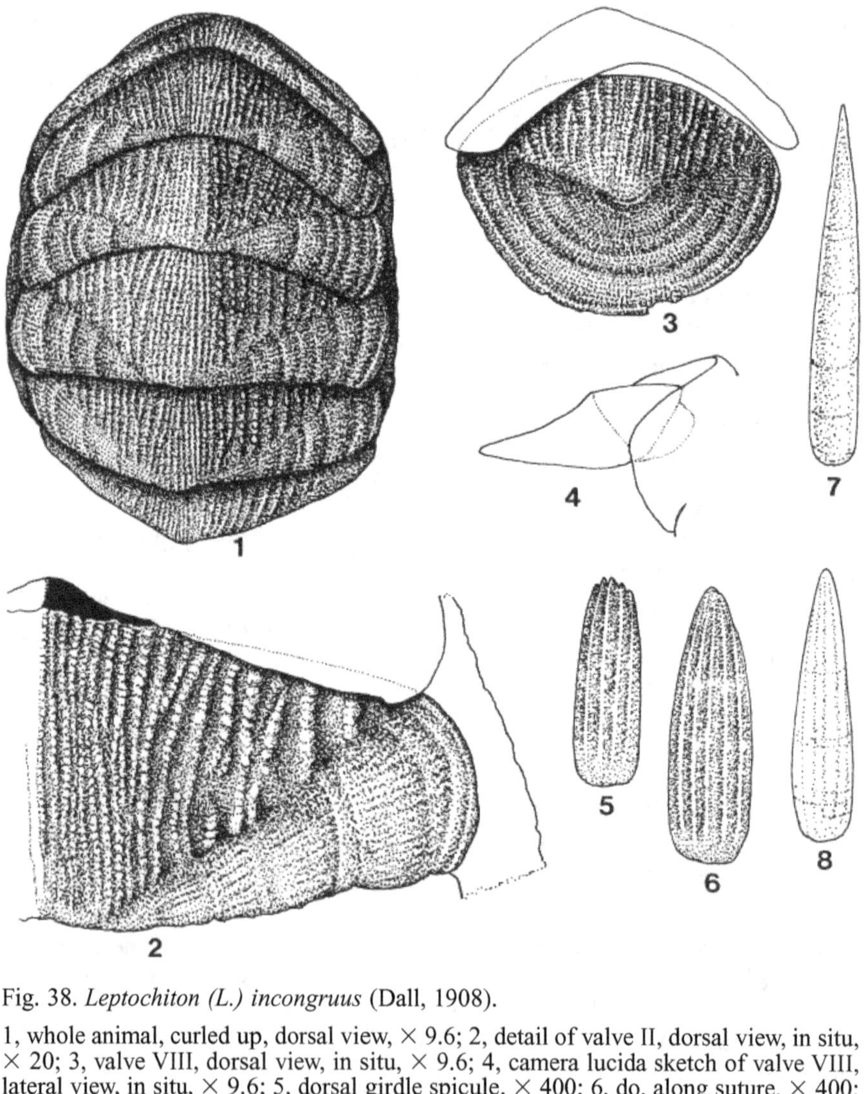

Fig. 38. *Leptochiton (L.) incongruus* (Dall, 1908).

1, whole animal, curled up, dorsal view, × 9.6; 2, detail of valve II, dorsal view, in situ, × 20; 3, valve VIII, dorsal view, in situ, × 9.6; 4, camera lucida sketch of valve VIII, lateral view, in situ, × 9.6; 5, dorsal girdle spicule, × 400; 6, do, along suture, × 400; 7, intersegmental spicule, × 400; 8, ventral girdle spicule, × 400.

1-8, holotype.

Lepidopleurus incongruus Dall, 1908:355; Thorpe *in* Keen, 1971:882.
Leptochiton incongruus; Ferreira, 1979b:157, fig. 29.

Description. Animal rather small, 12-15 mm long, moderately arched, dorsal elevation 0.38, subcarinated, side slopes convex, slightly beaked, lateral areas distinctly defined, decidedly raised; end valves rather flat, mucro central, not prominent, postmucronal slope slightly concave, colour light beige to pale brown, girdle narrow.

Tegmentum peculiarly sculptured: head valve, lateral areas of intermediate valves, and postmucronal area of tail valve very finely granulose, conspicuously concentrically rippled; jugal areas with small, flat-topped granules, arranged in longitudinal rows converging anteriorly; on the pleural areas these rows become well defined ridges with wider, flat interspaces, about 8-10 per side, the ridges sometimes anastomosing, or breaking off abruptly, leaving double-width interspaces the rest of the way.

Girdle densely clothed with bluntly pointed, coarsely ribbed spicules, 64 × 22 µm, with 5 ribs on visible part of spicules; these spicules are larger along the sutures, up to 90-100 µm long, 28-30 µm thick. There are also small intersegmental groups of smooth, pointed spicules, 120 × 17 µm. There is no marginal fringe. Ventral spicules bluntly pointed, weakly ribbed, somewhat slenderer than the dorsal ones, 88 × 20 µm.

Gills forming a very short posterior row.

The holotype, preserved dry, strongly curled, has an estimated length of 12 mm; it measures 5.5 mm in width, 3 mm in height.

Distribution. Apparently an abyssal species. Apart from the holotype only two more specimens (estimated length 15 mm; Scripps coll., CASI 009525) were ever collected, dredged off Salina Cruz, Gulf of Tehuantepec, Mexico, 15°40'N 95°20'W, at a depth of 3541-3612 m.

Leptochiton (L.) albemarlensis Smith & Ferreira, 1977

(Fig. 39. Map 8)

Type: CASG, type collection 58247 (holotype), 58248 (1 paratype).
Type locality: Galápagos Archipelago, Isla Isabella (Albemarle Island), Tagus Cove, 0°16'S 91°22'W, 20 m.

Leptochiton albemarlensis A. G. Smith & Ferreira, 1977:83, figs 1-2; Ferreira, 1979b:157.

Description. Animal small, round-backed, probably only moderately elevated. Over-all decoration of closely spaced, fine granulations on the dorsal side of the valves. Girdle narrow and scaly-spiculose. Colour of both valves and girdle a light cream.

Head valve rounded, somewhat more than semicircular, posterior margin widely V-shaped. Intermediate valves without any trace of carination, side margins evenly rounded, posterior margin straight, apices inconspicuous, lateral areas hardly raised and set off from the central areas. Tail valve semicircular, mucro prominent, rounded, central.

Tegmental sculpture consists of very fine, slightly elongated, little elevated, unconnected granulations quincuncially arranged in close longitudinal as well as diagonal rows. On the lateral areas the quincunx-pattern is formed by curving rows of similar granules. There are 2-3 shallow grooves close and parallel to the side margins, they are somewhat better developed in valves II and VII.

Girdle dorsally covered with closely set, more or less elongated, blunt-pointed, glass-like scales with 4-5 ribs on visible part. Close to the outer margin they measure 40 × 28 μm, along the suture 100 × 36 μm. Interspersed between the scales there are smooth, elongated spicules, up to 128 × 20 μm.

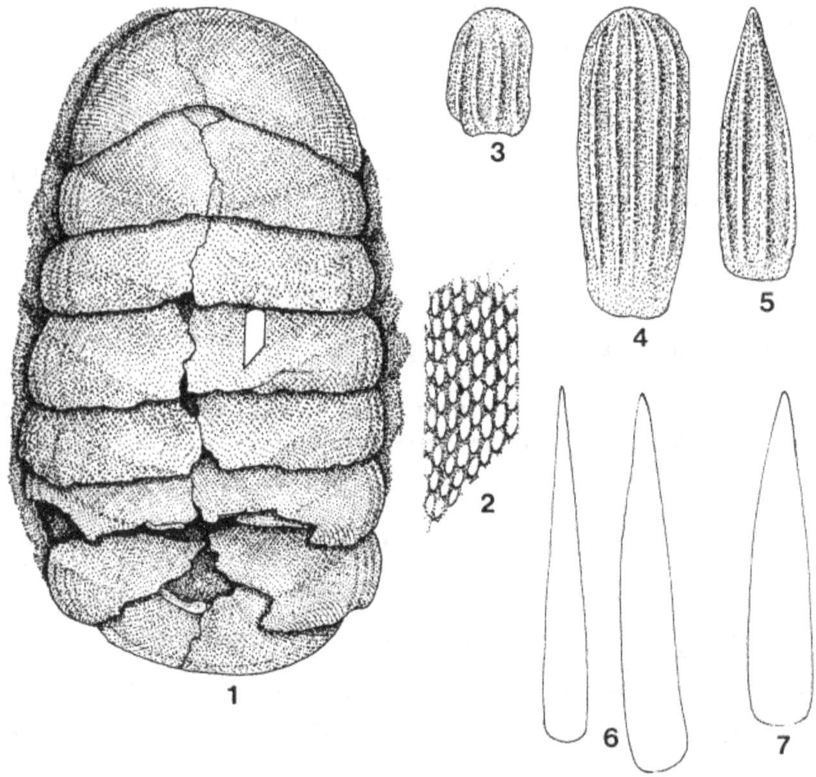

Fig. 39. *Leptochiton (L.) albemarlensis* Smith & Ferreira, 1977.

1, whole animal, dorsal view, × 9.6; 2, granulation of central area (the white patch in fig. 1, valve IV), × 40; 3, dorsal girdle scale, close to outer margin, × 400; 4, do, along the suture, × 400; 5, marginal spicule, × 400; 6, dorsal spicules, × 400; 7, ventral spicule, × 400.

1-7, holotype.

The marginal spicules, 88 × 24 μm, are conical and also coarsely ribbed. Ventrally the girdle is clothed with smooth, elongated spicules, 100 × 24 μm.
Dimensions: length 9.8 mm, width 5.4 mm, height 1.5 mm.

Distribution. *Leptochiton albermarlensis* is only known from the two damaged specimens of the type lot.

Leptochiton (L.) americanus sp. nov.

(Fig. 40. Map 41)

Type: SIO M-559.
Type locality: S of the Gulf of Panamá, 5°58.8'N 81°38.2'W, 1188 m.

Leptochiton rissoi; Ferreira, 1981c:36, figs 1-2 (non *Lepidopleurus rissoi* Nierstrasz, 1905).

Diagnosis. Animal of small to medium size, up to 15 mm long, elongate oval, twice as long as wide, subcarinated, moderately elevated (dorsal elevation 0.35), the sides weakly convex; valves not beaked, lateral areas scarcely raised, mucro of tail valve slightly posterior. Tegmentum granulated in longitudinal, well separated chains on central areas, in more or less radiating rows on end valves and lateral areas. Girdle narrow, covered with striated spicules and almost smooth spines of different sizes. Colour white, often stained with foreign deposits.

Description. Head valve strikingly small, c. one third of a circle, the width 70% of that of the intermediate valves, posterior margin widely V-shaped, with a rounded notch at the apex, anterior slope straight. Tegmentum roughly granulose, the granules well raised, roundish, arranged quincuncially, tending to form radiating rows towards the outer margin, the series broken up by 3-5 concentric growth marks. Intermediate valves rectangular, rather short, 3.5 times as wide as long, the slightly serrated anterior and the posterior margins almost straight, the sides bluntly rounded, apices perceptible but not prominent, never beaked, lateral areas hardly raised. Tegmental sculpture consists of 38 longitudinal rows of irregularly shaped, well raised granules, the interstices as wide as the chains, except for 5 or 6 rows on the jugum, which are finer granulated and closer set; towards the diagonal line the rows are abruptly and strongly converging towards the apex; lateral areas with 12-14 irregularly radiating rows of granules, at the same time quincuncially arranged, crossed by a series of growth marks. Tail valve relatively large, semicircular, anterior margin almost straight, mucro at two thirds of the length, not swollen, posterior slope a little concave; antemucronal area sculptured like the central areas, postmucronal area like the head valve.
Articulamentum white, well developed, especially at the thickened outer margins of the end valves; apophyses rather short, broadly triangular, the inner edges forming an angle of 120° with the posterior margin of the valve,

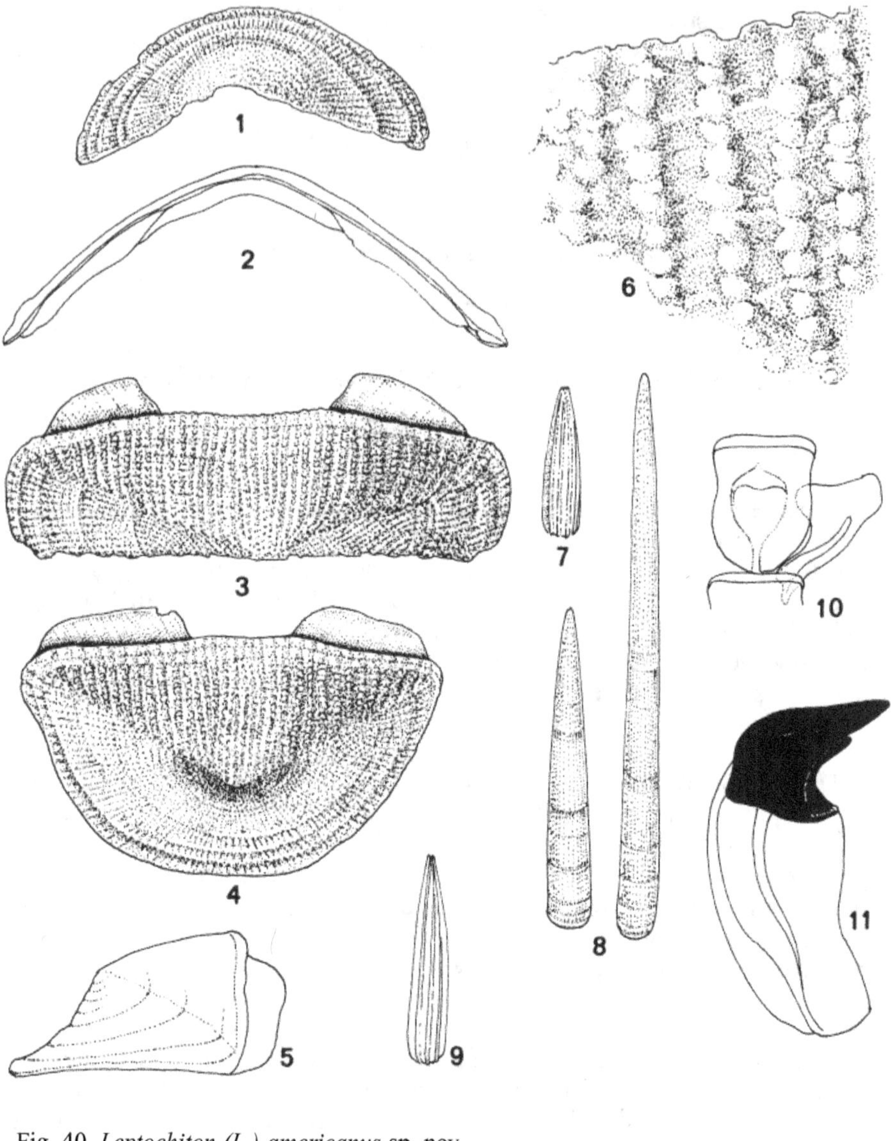

Fig. 40. *Leptochiton (L.) americanus* sp. nov.

1, valve I, dorsal view, × 9.6; 2, intermediate valve, anterior view, × 9.6; 3, do, dorsal view, × 9.6; 4, valve VIII, dorsal view, × 9.6; 5, do, lateral view, × 9.6; 6, part of tegmental sculpture on left pleural area of intermediate valve, × 40; 7, dorsal girdle scale, × 200; 8, dorsal girdle spines, × 200; 9, ventral girdle scale, × 200; 10, central and first lateral radula teeth, × 250; 11, major lateral radula tooth, × 200.

1-6, 10-11, specimen from Chile, off Arica (paratype), 18°30'S 70°34.5'W, 402-311 m, R/V T. Washington, A. Sontar & S. Luke leg., 7.V.1972, SIO M-1392; 7-9, specimen from S of the Gulf of Panamá, 5°58.3'N 81°38.2'W, 1188 m, T. Chase leg., 18.III.1963 (holotype), SIO M-559.

wide apart, leaving a shallow, very wide jugal sinus, convex in valve II, straight to slightly concave in III-VII; apophyses of tail valve relatively wider, trapezoid, the sinus much narrower, straight.

Girdle densely clothed with small, finely ribbed, dorso-ventrally flattened spicules, bluntly pointed, 100 × 24 μm; along the edges of the valves extremely finely striated, almost smooth, slender, white needles are found, varying in length from 200-375 μm, rather sharply pointed, at the base 25-30 μm thick. Ventral side of girdle covered with more strongly ribbed, imbricating scales, arranged in concentric rows, somewhat larger and flatter than those of the dorsal side, 140 × 28 μm.

Central tooth of radula about squarish, almost as broad as long, with a very narrow blade; first laterals without a blade, not reaching the distal edge of the central tooth, rapidly widening anteriorly; major lateral with a flat bicuspid head, the interior denticle much shorter.

Gills merobranchial, adanal without interspace; about 8-12 ctenidia on both sides.

Distribution. Pacific coast of the Americas, from Oregon, 44°39'N, to SW of Iquique, Chile, 21°23'S, in depths of 400-1400 m.

Observations. Under the name of *Leptochiton rissoi* (Nierstrasz, 1905) this species was reported by Ferreira (1981c:36) from five widely separated stations on the American W coast. A close comparison of these specimens, kindly lent to us by courtesy of Dr S. R. Luke of the Scripps Institution of Oceanography, La Jolla, California, and by Dr P. U. Rodda of the California Academy of Sciences, San Francisco, with the type series of *Lepidopleurus rissoi* in the Instituut voor Taxonomische Zoologie, Amsterdam, revealed many differences between the American and the East Indian specimens. *L. americanus* differs from *rissoi* in being smaller, less elevated, the side slopes convex instead of straight, the apices less prominent, in the posterior position of the mucro, which is slightly anterior in *rissoi*, in the granulation of the valves, and especially in the perinotum covering and radular characteristics.

North West Pacific Species

Leptochiton (L.) fuliginatus (Reeve, 1847)

(Fig. 41. Map 33)

Type: BMNH, 51.12.22.261-267 (lectotype + 6 paralectotypes).
Type locality: Korean Archipelago (Voyage of the "Samarang" under Capt. Belcher).

Chiton fuliginatus Reeve, 1847a:24; 1847b:pl. 26:fig. 174.
Lepidopleurus fuliginatus; Pilsbry, 1892:10, pl. 4:fig. 88.

Description. Animal of medium size, elongate oval, moderately elevated (dorsal elevation 0.39), subcarinated, side slopes little convex. Valves rather

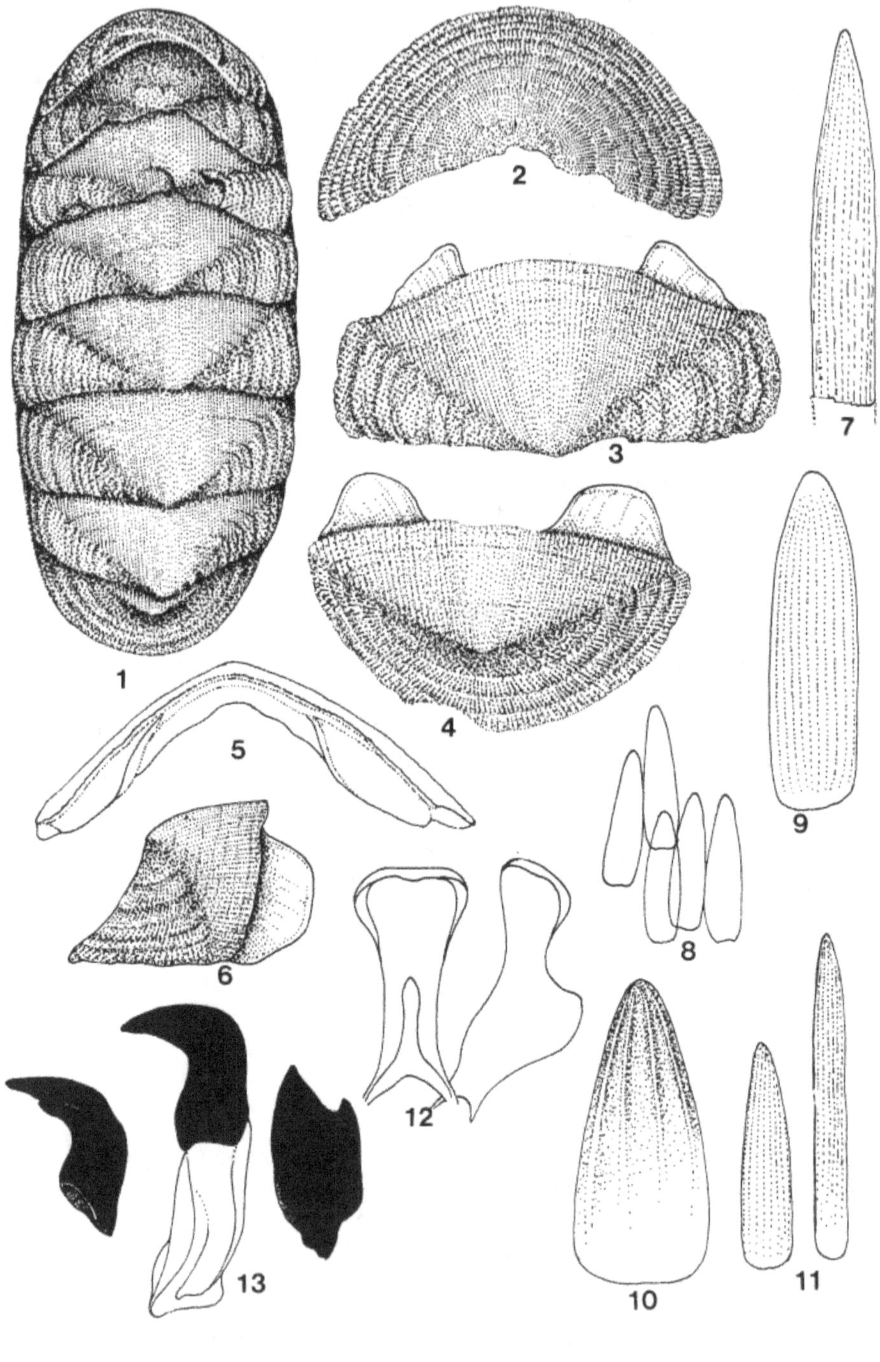
1
2
3
4
5
6
7
8
9
10
11
12
13

solid, not beaked, the apices quite perceptible, lateral areas slightly raised. Head valve, lateral areas, and postmucronal area concentrically wrinkled by growth marks. Girdle narrow, dorsally clothed with scales and spicules.

Tegmentum very finely granulated all over. Head valve semicircular, granules not much raised, roundish, arranged in numerous radiating rows. Anterior margins of intermediate valves somewhat concave in the pleural parts, sligthly convex between the apophyses; posterior margins almost straight, with a slightly projecting apex. Central area with numerous longitudinal rows of round granules, a little converging towards the hind margin, the granules becoming somewhat larger and more raised towards the sides. Lateral areas quite perceptibly raised, with 6-10 concentric growth wrinkles, the granules more or less quincuncially arranged, occasionally giving the impression of forming radiating rows. Tail valve with the mucro a little posterior when seen from above, as the hind slope is rather steep, little concave, the mucro prominent, not projecting. Antemucronal area sculptured like the central areas, postmucronal area like the head valve.

Colour of the tegmentum white, here and there stained with brown or blackish by foreign deposits.

Articulamentum white, rather strongly developed especially towards the outer margins of the end valves. The apophyses small, obliquely triangular with a rounded top in the intermediate valves, broader and trapezoid in the tail valve. Jugal sinus wide, convex.

Girdle clothed with not imbricating, erect, elongate scales, varying in size, gradually tapering to a blunt point, marked with generally four very fine riblets which do not reach the base. The largest about 100 μm long, 45 μm broad. Scattered among them finely grooved spicules are found of various sizes, from 75 μm to 110 μm long (or longer, as most of them are broken). There is a marginal fringe of equal spicules. Ventral part of the girdle densely clothed with imbricating, flat, elongate, almost parallel-sided, bluntly pointed scales, truncate at the base, with c. 10 very fine longitudinal grooves or scratches, the scales becoming larger towards the outer margin, the longest about 110 × 32 μm.

Central tooth of radula rather slender, with a narrow cusp, somewhat pinched in the middle. First laterals with a narrow cusp and an outward projecting wing. Major laterals with a short stalk and a long, black, bifid cusp, the

Fig. 41. *Leptochiton (L.) fuliginatus* (Reeve, 1847).

1, whole animal, girdle folded under, × 4; 2-4, valves I, II, VIII, respectively, dorsal view, × 10; 5, valve II, anterior view, × 10; 6, valve VIII, lateral view, × 10; 7, part of marginal spicule, × 400; 8, ventral girdle scales, × 200; 9, isolated ventral scale, × 400; 10, dorsal girdle scale, × 400; 11, dorsal girdle spicules, × 400; 12, central and first lateral teeth of radula, × 400; 13, major lateral radula tooth, and cusps of major lateral, different views, × 400.

1, lectotype; 2-13, paralectotype.

main denticle strong, sharply triangular, the exterior one small, only little projecting.

Length: 21 mm, width: 8 mm (largest syntype, here designated as the lectotype).

Distribution. Korean Archipelago (Capt. Belcher, *fide* Reeve).

Observations. Sirenko described a *Leptochiton batialis* from Japan (holotype: ZIAS 1856). This we have not seen, but judging from Sirenko's ample description and illustrations it is very closely allied to *fuliginatus*, only differing in minor details, such as a higher dorsal elevation, the granulation of the central areas and, perhaps, the perinotum armature. It may well prove to be a northern deep water subspecies of *fuliginatus*.

Leptochiton (L.) curvatus (Carpenter MS, Pilsbry, 1892)

(Fig. 42. Map 34)

Type: USNM 30754.
Type locality: Okosiri (=Okushiri!), Japan.

Leptochiton (Deshayesiella) curvatus (Carpenter MS) Dall, 1879:314, nom. nud.
Lepidopleurus (Deshayesiella) curvatus (Carpenter MS) Pilsbry, 1892:16, pl. 4:figs 78-81.
Oldroydia percrassa; Jakovleva, 1952 (non Dall, 1894):52, fig. 19, pl. 2:fig. 1.
Deshayesiella extensibilis Sirenko, 1973b:660, fig. 1 (type: ZIAS).

Description. Animal of moderate size, elongate oval, the shell plates rather solid, rather flat (dorsal elevation 0.34), the back evenly rounded, not carinated; intermediate valves slightly beaked, lateral areas distinctly raised.

Head valve a little more than semicircular, tegmentum sculptured with roundish pustules, becoming larger towards the outer margin, randomly arranged, apparently forming concentric lines, caused by many growth lines forming rather coarse ridges.

Sides of intermediate valves somewhat thrown forward, the posterior margin widely V-shaped, the apex slightly projecting, rather sharp, not swollen. Anterior margin almost straight, except in valve II, which is decidedly longer than the others, somewhat projecting anteriorly. Central areas sculptured with 7-10 longitudinal, somewhat curved chains of pustules, rather widely separated on the pleural parts; the rather wide jugal tract finely granulose, the oval granules arranged in hardly raised, close set longitudinal rows, somewhat diverging from the apex. Lateral areas raised, strongly concentrically ribbed by coarse growth marks, ornamented with concentric rows of pustules. Tail valve large, semicircular, depressed, the mucro little projecting, somewhat anterior, posterior slope almost straight, convex towards the outer margin. Antemucronal area sculptured like the central areas, post-mucronal area like the head valve.

Colour dirty white. Articulamentum white, well developed. The under-

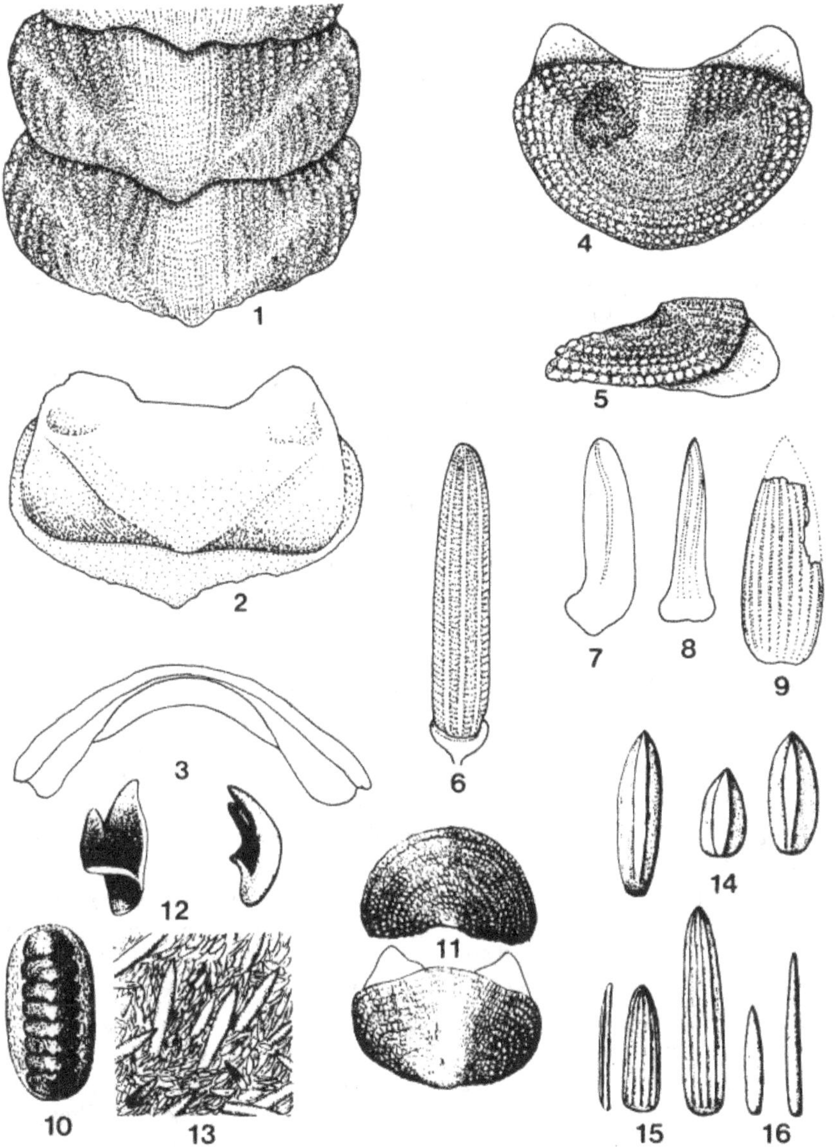

Fig. 42. *Leptochiton (L.) curvatus* (Carpenter *in* Pilsbry, 1892).

1, valve IV and V, dorsal view, × 9.6; 2, valve VI, ventral view, × 9.6; 3, do, anterior view, × 9.6; 4, valve VIII, dorsal view, × 9.6; 5, do, lateral view, × 9.6; 6, dorsal girdle spicule, × 400; 7-9, ventral girdle scales, × 400; 10, whole animal, dorsal view, × 1.16; 11, valves I and II, dorsal view, × 5.6; 12, cusps of major lateral radula teeth, × 80; 13, part of dorsal side of girdle, × 32; 14, ventral girdle scales, × 190; 15, dorsal girdle scales, × 190; 16, dorsal spicules, × 32.

1-9, holotype; 10-16, after Sirenko, 1973 (*D. extensibilis*).

folded part of the tegmentum beneath the apex extremely wide. Apophyses small, triangular in all valves. Jugal sinus wide, shallow, straight.

Girdle moderately wide, white, dorsally covered with elongate, bluntly pointed, longitudinally ribbed scales, 90-150 µm long, 28-32 µm broad. Among them many long, smooth, straight spicules of various sizes are found, up to 440 × 60 µm. Marginal spicules like the dorsal ones. Ventral scales flat, imbricating, elongate egg-shaped, longitudinally finely ribbed.

Length 14 mm, breadth 6.5 mm (holotype).

Distribution. Japan Sea: Okushiri (Adams); Bay of Posjet, Vostok, and Popov (Sirenko). Okhotsk Sea: off Kunashir Id. (Sirenko), 1-20 m.

Observations. The holotype of *curvatus* is a dried, disarticulated specimen, deprived of most of its girdle armature. We could only detect a few broken scales and spicules. Nevertheless we do not hesitate to synonymize *Deshayesiella extensibilis* Sirenko, 1973 with *curvatus*, as we do not find any differences to justify specific separation, except for the ventral girdle scales, which are typified with one wide longitudinal rib according to Sirenko. As the girdle of the type of *curvatus* is in such a bad shape, we better rely on Sirenko's description and drawings. That author also described the radula: the major lateral bearing a bidentate cusp, the inner denticle being smaller than the outer one.

Oldroydia bidentata Is. Taki, 1938, is much the same, so we were apt to consider it another synonym of *curvatus*. There are, however, enough differences to withold us from such a step: *O. bidentata* is much larger, the unique type (which was lost during World War II) measuring 45 × 23 mm, whereas the largest of 107 specimens of "extensibilis" measures 23 × 10 mm. *O. bidentata* was procured in a depth of 100-140 m, *D. extensibilis* is a sublittoral species, occurring in depths of 1-20 m. In *bidentata* the girdle encroaches at the sutures, which is not the case in the present species. Most of all: the valves II-VIII of *bidentata* are strongly forwardly projecting at the jugum.

L. curvatus was made the type of the subgenus *Deshayesiella* by Dall (ex Carpenter MS), but the distinguishing features of that subgenus (spiculose girdle, sculptural differentiation of jugal and pleural areas, elongate body, curved valves) do not justify its separation from *Leptochiton*, as all of them are, in one way or another, found in many *Leptochiton* species. That's why we consider *Deshayesiella* as a synonym of *Leptochiton*.

Leptochiton (L.) hakodatensis (Thiele, 1909)

(Fig. 43. Map 33)

Type: ZMHU.
Type locality: Japan, Hakodate.

Chiton (Leptochiton) concinnus Gould, 1859:164 (type in USNM, b 2077, Hako-

dadi, Japan, laminarian zone); 1862:117; Carpenter, 1864a:586. (*Non: Chiton concinnus* Sowerby, 1840).
Leptochiton concinnus; Dall, 1879:316, 318.
Lepidopleurus concinnus; Pilsbry, 1892:11; Thiele, 1909:10; Is. Taki & Iw. Taki, 1929c:161, 163.
Lepidopleurus hakodatensis Thiele, 1909:10, pl. 1:figs 11-20; Is. Taki, 1938a:326, pl. 14:fig. 1, pl. 16: figs 1-4, 6-8, pl. 17:figs 6-8; Jakovleva, 1952:54, fig. 15, pl. 1:fig. 3; Is. Taki, 1962:31; Iw. Taki, 1964b:408; Sirenko, 1976a:88.

Description. Animal oval, length 1.7 × breadth, moderately elevated (dorsal elevation 0.40), valves thin, back evenly rounded, not carinated; intermediate valves of equal width, not beaked,' lateral areas not raised, hardly perceptible; girdle moderately wide in the middle, narrow at the extremities.

Head valve semicircular, with 90-100 very fine radiating rows of roundish granules, the interstices very narrow, near the margin a few very faint concentric lines of growth. Intermediate valves rather short, posterior margin almost straight, umbo inconspicuous, not projecting, anterior margin slightly concave in the jugal part, regularly convex towards the sides. Central areas with closely set, parallel, longitudinal chains of roundish, connected, flat, little elevated granules, about 40-50 rows at both sides of the median axis, the jugal series somewhat weaker pronounced. Lateral areas not raised, though sligthly swollen, granular all over, the granules arranged in quincunx, tending to form radial rows towards the sides. Posterior valve semicircular, a little narrower than the head valve; mucro central, moderately produced, bluntly pointed. Antemucronal area regularly arched, sculptured like central areas; postmucronal area like head valve; posterior slope concave directly behind the mucro, anterior margin straight.

Articulamentum weakly developed, white, the apophyses narrow, triangular with rounded top, wide apart in the intermediate valves, trapezoid in tail valve. Jugal sinus wide, almost straight, or slightly concave.

Colour light yellowish, almost white, the interstices of the rows of granules light brownish, girdle light yellowish brown.

Girdle dorsally densely covered with not imbricating, flat, egg-shaped scales, marked with 4-6 fine, narrow riblets which do not reach the base, 80 μm long, 45 μm wide. Among them occur straight, or slightly bent, smooth, sharply pointed spicules, 125 μm long, 20 μm thick. Ventral side covered with closely set, radiating rows of small, imbricating scales, obtusely pointed, widest in the middle, with about 3 very weak riblets near the top, 50 μm long, 30 μm wide; towards the outer margin the scales become longer and more sharply pointed, the inframarginal ones like flattened spines, also with about 3 well pronounced riblets in the distal half, up to 120 μm long. Marginal spicules small, smooth, needle-like, 60-80 μm long.

Central tooth of radula distally much widened, with a narrow blade. Major lateral with a bidentate cusp, the denticles bluntly pointed, the outer one a little shorter and weaker than the inner one.

Gills merobranchial, adanal, with 12 to 14 ctenidia on both sides. Length 9 mm, breadth 5 mm.

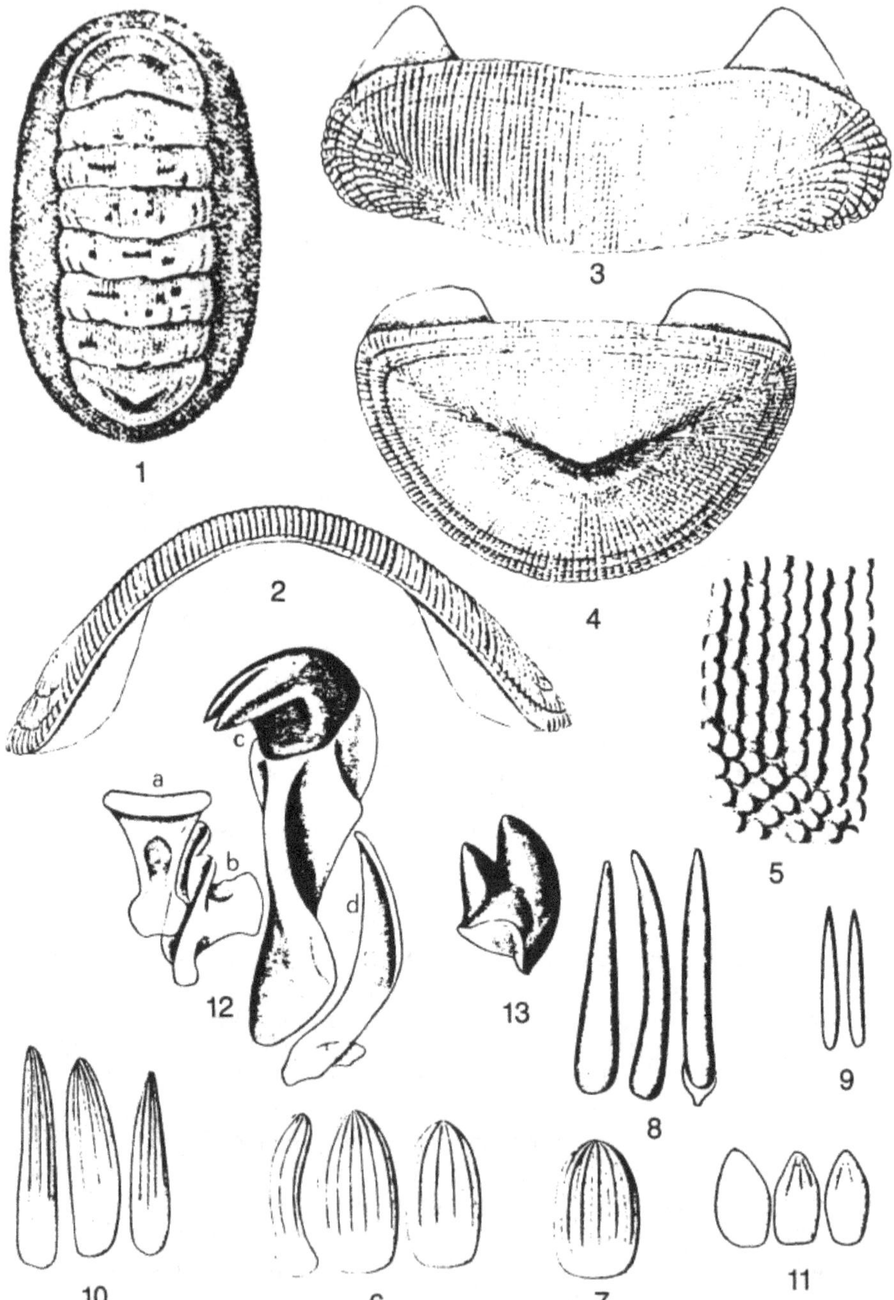

Distribution. Japan Sea: Gulf of Amur, entire coast of Peter the Great Bay, Vostok Bay, Moneron Id, Hakodate, Mutsu Bay. East China Sea: Nagasaki. Pacific Ocean: off Yokohama. Bathymetrical range: 0-80 m.

Observations. The species was excellently described and illustrated by Is. Taki, 1938. His drawings are reproduced here. The only specimen we could study was sent to the junior author by Dr B. Sirenko. It was collected in Moneron Id, Japan Sea, in a depth of 35-40 m. The specimen, preserved in alcohol and a little curled up, measures 7.4 × 5.4 mm; it fits the descriptions of Thiele and Taki in all respects.

The type of *Chiton (Leptochiton) concinnus* Gould, 1859, is a dried, partly curled-up specimen, valves VII and VIII loose, reddish brown, the anterior and lateral parts of the valves yellowish, the girdle pale buff. Close comparison either with Thiele's description and figures, or with the single specimen of L. *hakodatensis* at our disposal, reveals their inseparability. So Thiele's supposition (1909:10) was correct, and he did well in renaming it, as Gould's name was preoccupied by *Chiton concinnus* Sowerby, 1840, a probable synonym of *Ischnochiton pusio* (Sowerby, 1832).

Leptochiton (L.) diomedeae Berry, 1917

(Fig. 44. Map 34)

Holotype: USNM 215625 (S. S. B. 95).
Type locality: Japan, off Shiono Misaki Light, U.S.S. "Albatross" sta. 4967, 446-462 m (=light house in Shiono misaki, Wakayama Pref.).

Leptochiton diomedeae Berry, 1917b:1, pl. 1:figs 1-3, pl. 2; Ferreira, 1979b:163.
Lepidopleurus (Deshayesiella) diomedeae; Is. Taki & Iw. Taki, 1929c: 162; Is. Taki, 1962:32; Iw. Taki, 1964b:408.

Description. Animal of moderate size, little elevated (dorsal elevation 0.30), elongate oval, almost twice as long as wide, side slopes straight, the valves carinated, decidedly beaked.

Head valve semicircular, posterior margin widely V-shaped, anterior slope concave in the apical region, flattening towards the outer margin, tegmentum finely granulated all over, the granules somewhat conical, oval, arranged in quincunx. Intermediate valves wide and short, the anterior

Fig. 43. *Leptochiton (L.) hakodatensis* (Thiele, 1909).

1, whole animal, dorsal view, × 6.3; 2, valve IV, anterior view, × 19; 3, do, dorsal view, × 19; 4, valve VIII, dorsal view, × 15; 5, tegmental sculpture, × 65; 6-7, dorsal girdle scales, × 250; 8, dorsal spines, × 250; 9, marginal spicules, × 250; 10, infra-marginal scales, × 250; 11, ventral scales, × 250; 12, elements of radula: a central, b first lateral, c major lateral, d minor lateral radula tooth, × 300; 13, cusp of major lateral tooth, × 250.

All after Is. Taki, 1938.

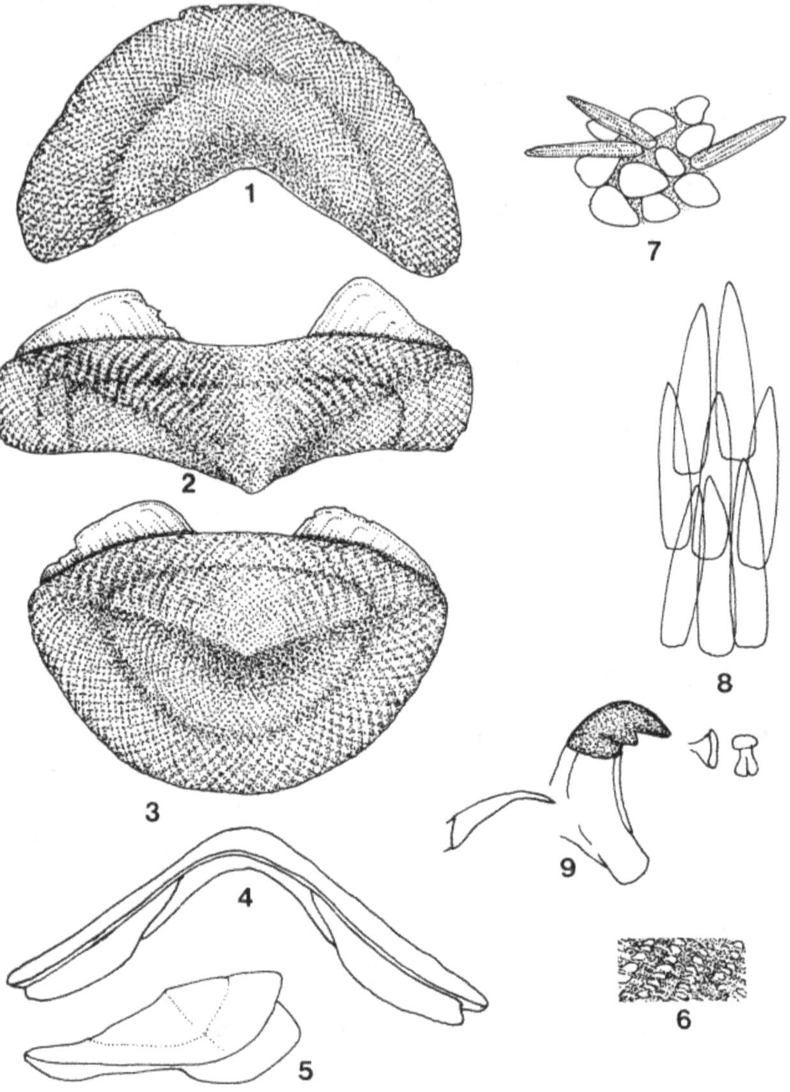

Fig. 44. *Leptochiton (L.) diomedeae* Berry, 1917.

1, valve I, dorsal view, × 9.6; valve III, dorsal view, × 9.6; 3, valve VIII, dorsal view, × 9.6; 4, valve III, anterior view, × 9.6; 5, valve VIII, lateral view, × 9.6; 6, tegmental sculpture of central area, × 20; 7, armature of girdle, dorsal side, × 200; 8, ventral girdle spicules, × 320; 9, isolated radula elements, × 130.

1-9, holotype; 7-9, after Berry, 1917.

margin somewhat concave at the jugal area, convex at the pleurae, except for valve II, which is decidedly longer, the anterior margin regularly convex. Posterior margin in all valves concave at both sides of the sharply pointed, projecting apex. Lateral areas slightly raised, separated from central area by a faint fold. Central areas sculptured with oblique, sinuate series of oval granules, rather widely separated, especially towards the side margins, the interstices pitted, as the granules of neighbouring series are conncected by fine threads, crossing the series perpendicularly. Lateral areas sculptured like head valve. Tail valve a trifle narrower than head valve, semicircular, rather flat, mucro sligthly anterior, well marked but not prominent, posterior slope very little concave directly behind the mucro, flattening towards the outer margin. Antemucronal area sculptured like central areas, post-mucronal area like head valve.

Articulamentum white, strongly calloused at outer margin of end valves. Apophyses of intermediate valves broadly triangular, jugal sinus less than 1/3 of valve width, rather deep, somewhat convex in valve II, concave in valves III-VII. Tail valve with short and wide, regularly rounded apophyses and shallow, almost straight jugal sinus.

Colour of tegmentum light yellowish.

Girdle densely covered with short, bluntly conical, smooth spicules, 20-30 µm long, 15-20 µm wide, interspersed with longer, faintly striated spines, 60-80 µm long, 10-12 µm thick. Equal spines are found here and there at the margin. Ventrally the girdle is closely beset with somewhat flattened, smooth, torpedo-shaped spicules-scales, distally tapering to a blunt point, the base truncated, 65 µm long, 14 µm wide.

Central tooth of radula short, mushroom-like in outline, with a narrow blade; major laterals with a strong, bidentate cusp.

Gills merobranchial; 12 ctenidia on each side.

Length 15 mm, breadth 8 mm.

Distribution. Japan: Wakayama Pref., Shiono Misaki, 446-462 m (Berry, type locality). Apparently a deep water species.

Leptochiton (L.) hirasei (Is. Taki & Iw. Taki, 1929)

(Fig. 45. Map 33)

Type: lost during World War II.
Type locality: Japan, Prov. Kii: near Yuzaki-Ôjima, in Setokamayama-mura.

Lepidopleurus hirasei Is. Taki & Iw. Taki, 1929c: 157, figs 44-58; Bergenhayn, 1933:4.

Description. Animal small, elongate oval, twice as long as broad, rather flat (dorsal elevation 0.32), smoothly arcuate, not carinated; intermediate valves of equal width, not beaked, lateral areas hardly perceptible.

Head valve almost semicircular, anterior margin minutely undulated,

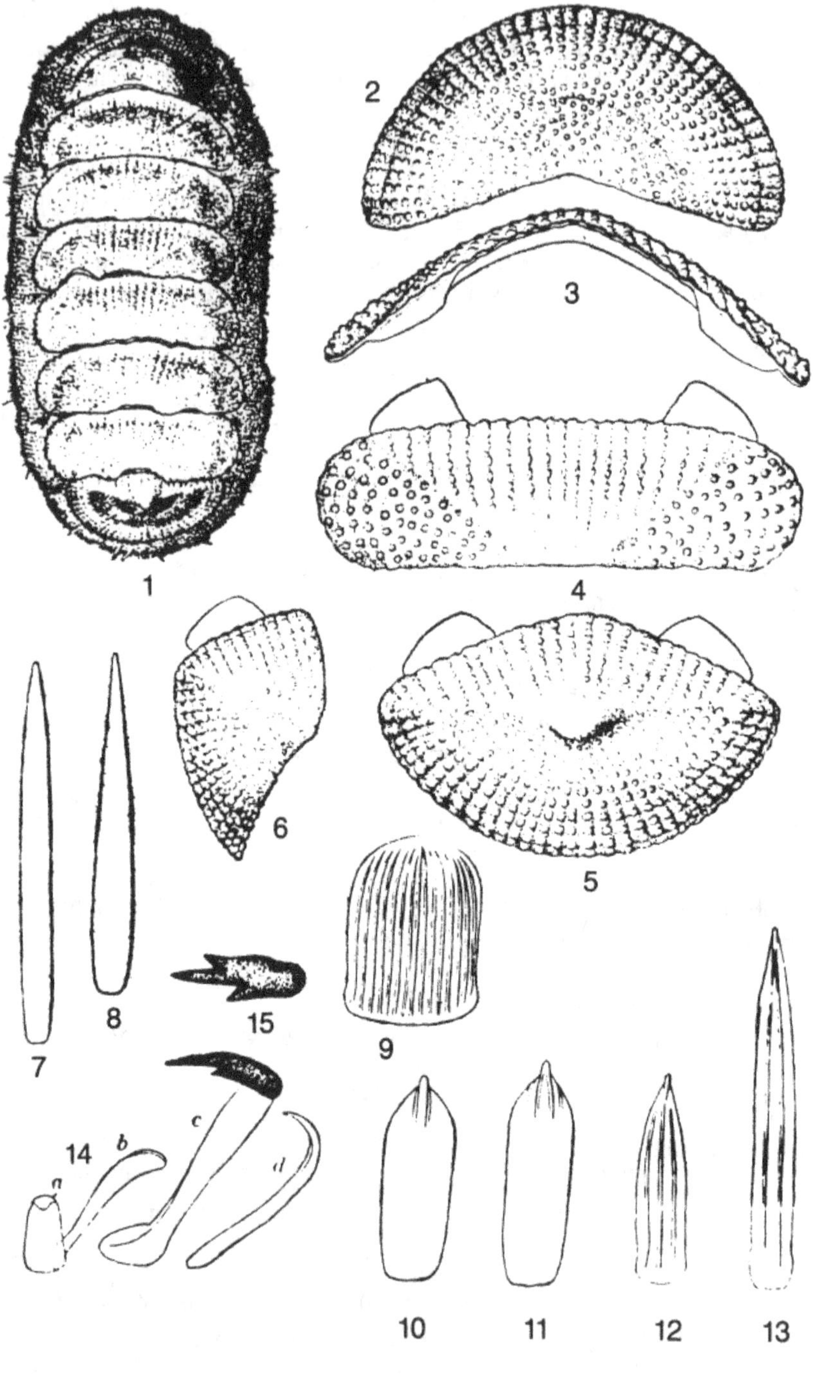
1
2
3
4
5
6
7
8
9
10
11
12
13
14
a
b
c
d
15

posterior angle very obtuse, side slope straight. Tegmentum sculptured with about 50 radial rows of inconspicuous flat pustules, traversed by about 15 concentric incremental lines. Intermediate valves very broad and short, anterior margin conspicuously waved, posterior margin almost straight, lateral margins well arcuate, jugum indiscriminate. Central area ornamented with 12-14 longitudinal or slightly oblique rows of minute, elliptic granules on each side. Lateral areas hardly raised, very faintly convex, with six or more radial rows of low but conspicuous granules, at the same time arranged concentrically. Tail valve with the mucro a little anterior, bluntly pointed; antemucronal area slightly convex, rather indistinctly sculptured like the central areas. Posterior slope moderately concave; postmucronal area sculptured like head valve with about 50 radial rows of pustules, but less distinct.

Articulamentum rather smooth, sligthly pitted beneath the pustules on the tegmentum; callus not thickened; apophyses thin, low, narrow, subtriangular, acute in front, slope towards jugal sinus rather steep, outer inclination smooth, more or less arcuate, separated by a broad, smoothly-edged sinus. Marginal region of all valves somewhat thickened.

Colour light brown between the rows of granules; median part of end valves with a concentric brownish band; a similiar band on intermediate valves, though very faint; margins of valves greyish, inside white.

Girdle densely clothed with imbricating, minute, oblong scales, 60 × 40 mm, with 15-17 fine striae on dorsal side. Scattered among them strong, needle-like, smooth spines are found, 100 μm long, 12 μm thick. Sutures between the valves armed with a few longer spines, up to 130 × 30 μm, directed somewhat backward. Marginal spicules variable in shape, from flat, sharply pointed ones, 55 × 18 μm, to long, dagger-like ones, 100 × 12 μm, all of them with 2-6 striae. Among them occur several calcareous needles, 200 × 25 μm, sticking out of the margin. Ventral side covered with imbricating, flat, elongate scales with acute points, distally with 2-4 faint striae, 70 × 20 μm.

Radula very small, all elements extremely slender and elongated. Central tooth elongate-oblong, cusp simple, slightly curved backward. First laterals oar-shaped, with a smoothly pointed, curved, outwardly extending, simple cusp, tapering towards the base. Major laterals well developed, with a sharply tridentate cusp.

Fig. 45. *Leptochiton (L.) hirasei* (Is. Taki & Iw. Taki, 1929).

1, whole animal, dorsal view, × 16; 2, valve I, dorsal view, × 38; 3, intermediate valve, anterior view, × 38; 4, do, dorsal view, × 38; 5, valve VIII, dorsal view, × 38; 6, do, lateral view, × 38; 7-8, dorsal girdle spines, × 400; 9, dorsal scale, × 400; 10-11, ventral scales, × 400; 12-13, marginal spines, × 400; 14, radula elements: a central, b first lateral, c major lateral, d minor lateral radula tooth, × 400; 15, cusp of major lateral, × 400.

All after Taki & Taki, 1929.

Gills merobranchial, adanal; 5 ctenidia on both sides.
Length 4.5 mm, breadth 2.2 mm (largest specimen = holotype)
(altered after Taki & Taki, 1929)

Distribution. Japan: Prov. Kii, near Yuzaki-Ôjima in Seto-kamayamamura; Ôshima, near Kushimoto, littoral. Bergenhayn reported it from Sagami Wan, Okinose, 300 m.

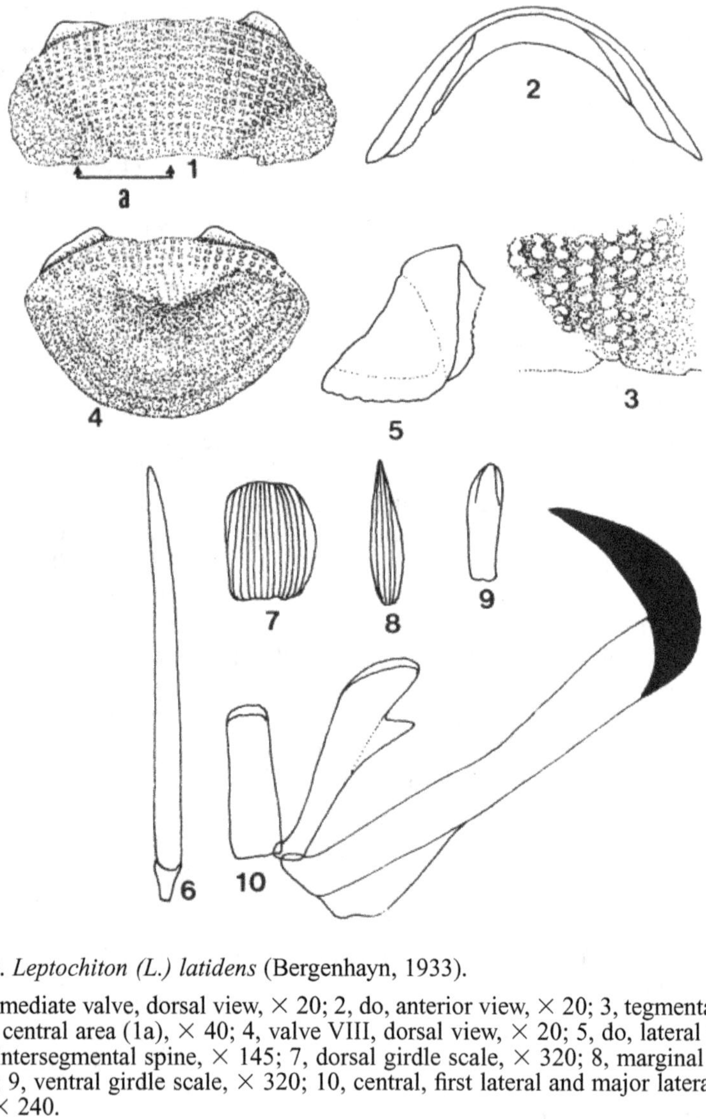

Fig. 46. *Leptochiton (L.) latidens* (Bergenhayn, 1933).

1, intermediate valve, dorsal view, × 20; 2, do, anterior view, × 20; 3, tegmental sculpture of central area (1a), × 40; 4, valve VIII, dorsal view, × 20; 5, do, lateral view, × 20; 6, intersegmental spine, × 145; 7, dorsal girdle scale, × 320; 8, marginal spicule, × 320; 9, ventral girdle scale, × 320; 10, central, first lateral and major lateral radula teeth, × 240.

1-10, holotype; 6-10, after Bergenhayn, 1933.

Leptochiton (L.) latidens (Bergenhayn, 1933)

(Fig. 46. Map 34)

Type: UUZM 131 (holotype only).
Type locality: Japan: Aburazuba, Sagami Misaki Marine Biological Station, 2-3 m.

Lepidopleurus latidens Bergenhayn, 1933:8, pl. 2:figs 21-23, text-fig. 2; Is. Taki, 1962:32; Iw. Taki, 1964b:408.

Description. Animal small, oval, valves thin, rather elevated (dorsal elevation 0.45), back evenly rounded, not carinated nor beaked.

Head valve large, little more than semicircular, tegmentum sculptured with small, roundish, somewhat convex granules arranged in quincunx, with 4 or 5 concentric growth lines; posterior margin very widely V-shaped. Intermediate valves hardly more than twice as wide as long, anterior margin regularly rounded, posterior margin almost straight; apex hardly perceptible, not projecting, lateral areas distinctly raised. Central areas sculptured with 36 longitudinal rows of flat granules of different forms and sizes, generally oval, sometimes connected; on the jugum the granules are weakly developed, the rows fairly closely set, towards the outer margins the granules are higher elevated, the rows wider apart, somewhat converging posteriorly, the interstices as wide as the granules. Lateral areas raised, with round, little pronounced granules in quincunx. Tail valve about 5/6 the width of the head valve, mucro anterior of the centre, posterior slope rather steep, little concave; antemucronal area short, sculptured like the central areas, postmucronal area like head valve, with a few concentric growth marks.

Articulamentum white; apophyses short, wide apart, triangular with rounded top, broader on tail valve. Jugal sinus wide, shallow, straight or little convex.

Colour of tegmentum dirty white.

Girdle rather narrow, densely clothed with squarish, little curved scales, not imbricating, marked with 13-15 fine grooves, 48 µm long, 40 µm wide. On the perinotum tongues between the valves long, smooth, straight or slightly curved, calcareous spines in chitinous cups are found, 360 µm long. Marginal spicules sharply pointed, with longitudinal furrows, 60 µm long, 12 µm thick. Ventral side of girdle covered with radiating, imbricating rows of flat scales, about rectangular in outline, distally bluntly pointed, with two weak grooves near the top, the base truncate, 45 µm long, 15 µm wide.

Central tooth of radula rectangular, 3 times as long as wide, with a narrow blade; first lateral elongate, narrow, with a small blade and a wing-like projecting exteriorly; cusp of major lateral with one long, sharply pointed denticle; outer marginal plate 8 times wider than long.

Gills merobranchial, adanal, with 9 ctenidia on left, 10 on the right side.

Length ca 5 mm, width ca 3 mm.

Distribution. Only known from the type locality.

Observations. The unique type is a disarticulated specimen, deprived of its radula, kept in alcohol. A glass jar contains the tail valve and two intermediate valves. Bergenhayn states that the granules on the head valve are arranged in radiating series, but close examination of his photograph (pl. 2:fig. 21) reveals that they are arranged more or less quincuncially, which seems logical.

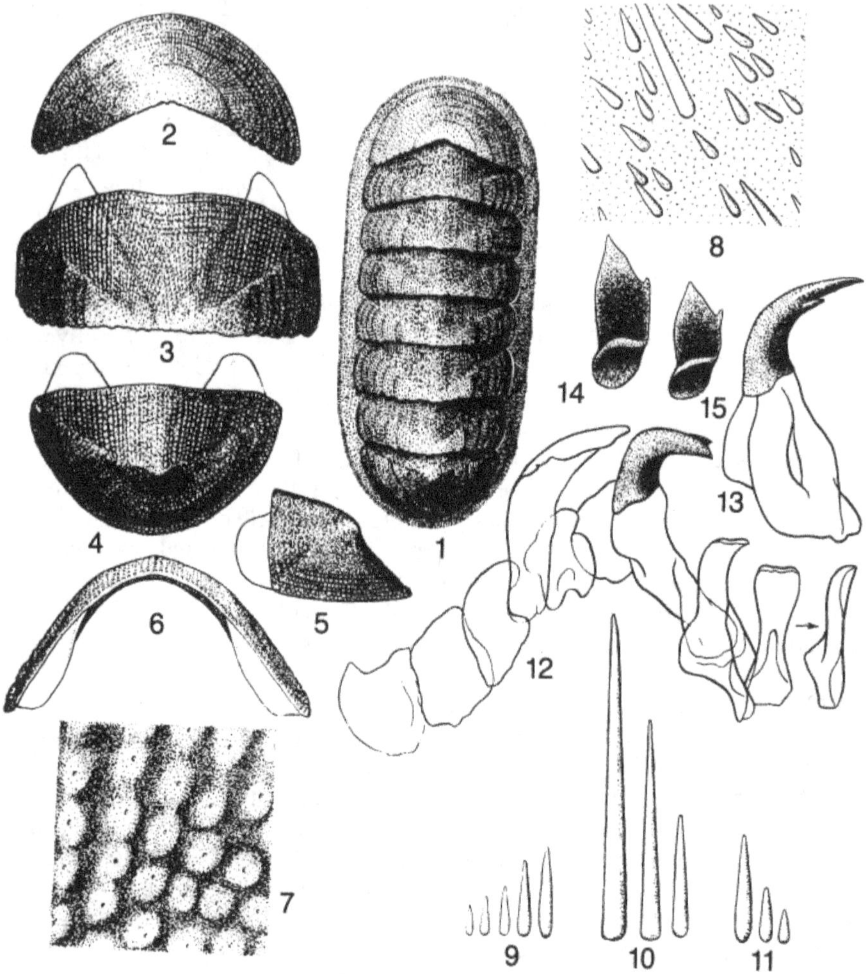

Fig.47. *Leptochiton (L.) batialis* Sireňko, 1979.

1, whole animal, dorsal view, × 2.4; 2, valve I, dorsal view, × 10; 3, valve V, dorsal view, × 10; 4, valve VIII, dorsal view, × 10; 5, do, lateral view, × 10; 6, valve V, anterior view, × 10; 7, sculpture of tegmentum near diagonal line, × 90; 8, part of dorsal side of girdle, × 105; 9, dorsal girdle spicules, × 105; 10, dorsal girdle spines, × 105; 11, ventral girdle spicules, × 105; 12, half radula row, × 105; 13, isolated major lateral radula tooth, × 105; 14-15, cusp of major lateral tooth, × 105.

1-14, holotype; 15, paratype; all after Sirenko, 1979.

The name refers to the outer marginal radula plate, which, however, was not figured by the author.

The descriptions of the girdle elements and radula have been borrowed from Bergenhayn's original description.

Leptochiton (L.) batialis Sirenko, 1979

(Fig. 47. Map 33)

Holotype: ZIAS 1856.
Type locality: Pacific coast of northern Honshū (Japan), off Kuji, E. S. "Vityaz" sta. 6671, 2500 m.

Leptochiton batialis Sirenko, 1979:118, figs 1-12.

Description. Animal of medium size, elongate, shell rounded, yellowish white with dark umber, partly black, deposits of iron salts; girdle very narrow, dirty yellowish brown.

Head valve 1.1 times as wide as tail valve. Intermediate valves regularly arched, the apices somewhat projecting, anterior margins laterally somewhat obliquely cut off, lateral areas slightly raised, mucro of tail valve central, not projecting, posterior slope slightly concave.

Sculpture of the tegmentum uniformly granular. The granules become gradually larger towards the side margins, varying from 50-80 μm in diameter. On the head valve they are arranged in radial series, on the lateral areas and post-mucronal area the arrangement is irregular. On the central areas the granules are arranged in longitudinal rows, the posterior ends somewhat diverging in the jugal part. Growth marks clearly defined as coarse riblets.

Apophyses small, triangular with rounded top on intermediate valves, trapezoid on tail valve.

Girdle narrow, at fifth valve 0.8 mm wide, armed with numerous spicules and needles. Spicules semi-transparent, distally pointed, up to 125 μm long; needles very long, up to 0.5 mm, scattered over the dorsal surface, also forming a marginal fringe. Ventral spicules like those of dorsal side, but more closely set.

Radula 10 mm long; central tooth long and slender, major lateral not large, with a bifid cusp, the exterior denticle very small.

There are 12-14 pairs of gills, extending from the anterior margin of valve VII to the anal papilla.

Length 25 mm, breadth 10 mm (holotype); length 18-26 mm (six paratypes).

Distribution. Pacific coast of northern Honshū (Japan), off Kuji, 2500 m (holotype and 6 paratypes); off the coast of the isle of Shikotan, 1450-1500 m.

Observations. The above description is a brief translation of Sirenko's original description. *L. batialis* is very closely related to *L. fuliginatus* (Reeve), if not conspecific with it. The minor differences between the two (*batialis* is more highly arched, and the granulation of the central areas seems to vary a little) may well fall within the range of intraspecific variation.

Indo-Pacific Species

Leptochiton (L.) rissoi (Nierstrasz, 1905)

(Fig. 48. Map 24)

Lectotype: ITZ (design. A. J. Ferreira, 1979b).
Type locality: Celebes Sea, N of Manado, "Siboga" sta. 126, 3°27.1'N 125°18.7'E, 2053 m.

Lepidopleurus rissoi Nierstrasz, 1905:6, figs 5, 52-55; Ferreira, 1979b:163, figs 30-32.
Non: *Leptochiton rissoi;* Ferreira, 1981c:36, figs 1-2 (=*L. americanus* sp. nov.)

Description. Animal very large for the genus, up to 25 × 13 mm, elongate oval, rather elevated (dorsal elevation 0.41), carinated, side slopes straight, valves beaked, girdle spiculose.

Head valve semicircular, posterior margin V-shaped, tegmentum sculptured with fine, quincuncially arranged granules and several concentric growth ridges. Intermediate valves broadly rectangular, side margins more or less rounded, posterior margin decidedly concave at both sides of the apex, lateral areas slightly raised, clearly indicated, sculptured with close set granules, arranged in quincunx, and several prominent, concentric growth ridges; central areas sculptured with longitudinal rows of granules, about 28 per side, very fine and close set on the jugum, getting larger, more widely spaced and slightly posteriorly converging towards the side margins. Tail valve somewhat more than semicircular, anterior margin a little convex, mucro central, slightly pointed, posterior slope strongly concave, antemucronal area sculptured like central area of intermediate valves, postmucronal area like head valve.

Colour of tegmentum whitish, with abundant orange-brown to black deposits, girdle yellowish in the smaller specimens, brownish in the larger ones.

Articulamentum white, apophyses small, broadly triangular, short and rounded in the tail valve.

Girdle densely clothed with elongate, pointed, weakly longitudinally striate spicules, ca 146 × 33 µm in mid-girdle, much larger, 300 × 53 µm, along the sutures. No marginal spicules present. Ventral spicules elongate, pointed, weakly longitudinally striate, up to 140 × 36 µm.

Central radula tooth elongate, sinuated at both sides, well cusped with arcuate edge, first lateral tooth narrow, bearing an elongate, exteriorly curved blade, cusp of major lateral with one strong, sharp, curved denticle.

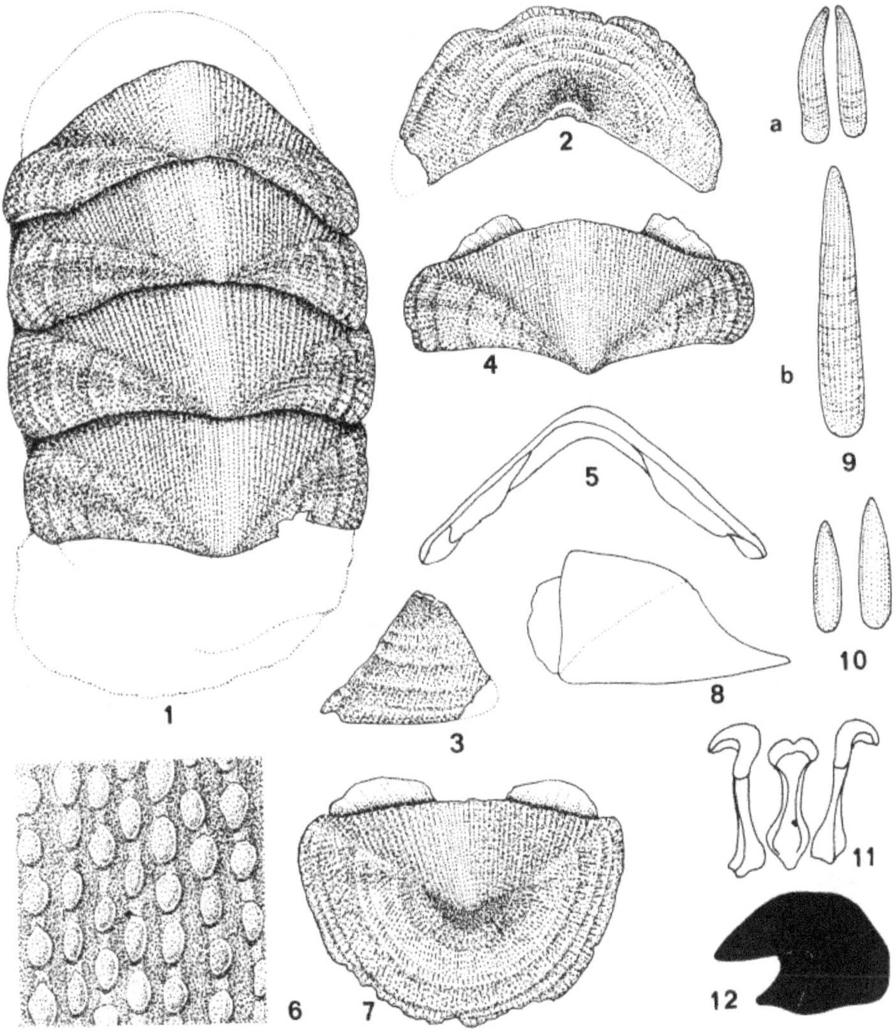

Fig. 48. *Leptochiton (L.) rissoi* (Nierstrasz, 1905).

1, whole animal, slightly curled up, partly disarticulated, valves III-VI in situ, dorsal view, × 4.8; 2, valve I, dorsal view, × 4.8; 3, do, lateral view, × 4.8; 4, valve II, dorsal view, × 4.8; 5, do, camera lucida sketch, anterior view, × 4.8; 6, detail of sculpture on left pleural area of valve II, × 40; 7, valve VIII, dorsal view, × 4.8; 8, do, camera lucida sketch, lateral view, × 4.8; 9, dorsal girdle spicules, a from mid-girdle, b along the sutures, × 120; 10, ventral girdle spicules, × 120; 11, central and first lateral radula teeth, × 200; 12, cusp of major lateral tooth, × 200.

1-12, lectotype.

Gills merobranchial, adanal without interspace, with up to 14 ctenidia on both sides.

The length of the ten known specimens varies from 5.5 to 25 mm, the breadth from 3.5 to 13 mm.

Distribution. The syntypic series was obtained in six "Siboga" stations, the localities ranging from 3°27.1'N to 10°39'S, and from 119°8.5' to

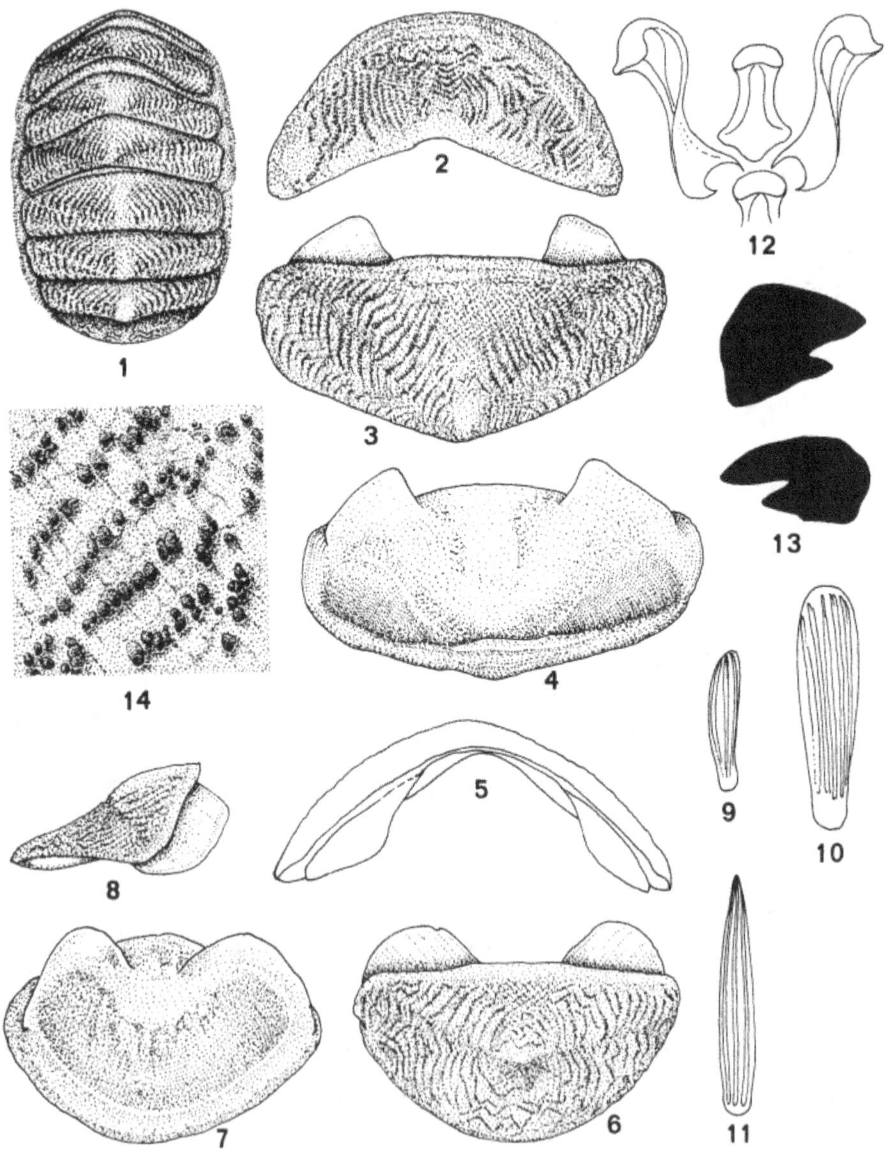

131°0.5'E. The bathymetric range extends from 216 to 2053 m.

A very large specimen (approximately 40 × 22 mm) was procured by the French "Corindon" – Makassar Expedition, 1980, in Makassar Strait, sta. CH 280, 01°59'S 119°10'E, 715-800 m (MNHN).

Observations. Ferreira (1918c) reported *L. rissoi* from the W coast of the Americas, from Oregon to Iquique, Chile, 400-1400 m, but the specimens were wrongly identified. They are described in the present Monograph as *L. americanus* sp. nov.

Leptochiton (L.) lineatus (Nierstrasz, 1905)

(Fig. 49. Maps 22, 44)

Lectotype: ITZ (design. A. J. Ferreira, 1979).
Type locality: Savu Sea, "Siboga" sta. 297, 10°39'S 123°40'E, 520 m.

Lepidopleurus lineatus Nierstrasz, 1905:8, figs 4, 48-51; Ferreira, 1979b:163, figs 23-24.
Lepidopleurus belknapioides Leloup, 1981a:317, fig. 1, pl. 1:figs 1-3; Kaas, 1982b: 87.
Leptochiton lineatus; Kaas, 1982b:87.

Description. Animal of moderate size, up to 17 mm long, elongate oval, rather elevated (dorsal elevation 0.40), subcarinated, side slopes convex, valves solid, though not to be compared to the valve sculpture of *Lepidopleurus*, not beaked.

Head valve somewhat less than semicircular, widely V-shaped behind, sculptured with minute, elevated granulae, arranged in a most peculiar pattern, in more or less radiating series on the central part, in concentric zigzag series near the outer margin. Intermediate valves broadly rectangular, side margins rounded, posterior margin straight, apices inconspicuous, lateral areas weakly raised but well marked, slightly depressed behind the diagonal ridge, pleural areas sculptured with wavy, strongly posteriorly diverging series of minute granulae, interstices weakly latticed, the series more or less concentrically continuing onto the lateral areas, jugal area finely reticulate by series of extremely fine granules in quincunx. Tail valve semi-

Fig.49. *Leptochiton (L.) lineatus* (Nierstrasz, 1905).

1, whole animal, slightly curled up, dorsal view, × 4.8; 2, valve I, dorsal view, × 9.6; 3, valve II, dorsal view, × 9.6; 4, do, ventral view, × 9.6; 5, do, anterior view, camera lucida sketch, × 9.6; 6, valve VIII, dorsal view, × 9.6; 7, do, ventral view, × 9.6; 8, do, lateral view, × 9.6; 9, dorsal girdle spicule, × 400; 10, do, sutural spicule, × 400; 11, ventral girdle spicule, × 400; 12, central and first lateral radula teeth, × 200; 13, cusps of major lateral radula teeth, × 200; 14, detail of sculpture on left pleural area of valve II, × 40.

1-14, paralectotypes, Indonesia, Makassar Strait, 0°34.6'N, 119°8.5'E, 1301 m, "Siboga" sta. 88.

circular, anterior margin straight, mucro subcentral, not prominent, back slope almost straight or a little concave, antemucronal area sculptured like pleural areas of intermediate valves, postmucronal area like head valve.

Colour of tegmentum golden yellow, granules and perinotum somewhat more yellowish brown.

Articulamentum white, strongly developed, apophyses small, triangular, rounded in valve VIII, separated by a straight sinus, about 1/3 the width of the valve.

Girdle narrow, dorsally very densely beset with small, club-shaped, finely longitudinally striated spicules, placed side by side, the "heads" sticking out, 45 × 10 µm. Along the margins of the valves, there is a row of larger spicules of the same shape, 92 × 22 µm. There is no marginal fringe. Ventrally the girdle is densely covered with more or less imbricating, spindle-shaped, longitudinally grooved, sharply pointed spicules, 80 × 12 µm.

Central radular tooth tapering to a small, rounded blade, first lateral tooth helmet-shaped, major lateral with a short bifid cusp, the main denticle strong, sharply triangular, the exterior one much smaller.

Gills increasing in number with the size of the animal, 12-18 ctenidia on both sides.

The length of the ten known specimens varies from 7 to 17 mm, the breadth from 4.5 to 9 mm, six of them measure 12 × 7.5-8 mm.

Distribution. The type series was obtained in four "Siboga" stations, the localities ranging from 0°34.6'N to 10°39'S, and from 119° to 123°40'E. The bathymetric range extends from 450 to 1301 m.

The French "Corindon" – Makassar Expedition, 1980, procured a specimen in Makassar Strait, sta. CH 209, 00°07'S 117°53'E, 490 m. Leloup (1981a) reported it under the new name *belknapioides* from sta. 44 of the French "Musorstom" 1 – Philippines Expedition, 1976, in the N part of the Verde Island Passage, between the Lubang Islands and Luzon, 592-610 m. Kaas (1982b) recorded it from 10 stations of the "Musorstom" 2 – Philippines Expedition, 1980, in the same area, 299-682 m.

Leptochiton (L.) planus (Nierstrasz, 1905)

(Fig. 50. Map 22)

Type: ITZ (holotype).
Type locality: Celebes Sea, N of Manado, "Siboga" sta. 126,3°27.1'N 125°18.7'E, 2053 m.

Lepidopleurus planus Nierstrasz, 1905:9, figs 6, 56-59.

Description. Animal of rather small size, originally 9 × 5.5 mm, now 7.6 × 5 mm (sligthly curled), elongate oval, exceptionally flat (dorsal elevation c. 0.13), subcarinated, side slopes little concave, valves rather heavy, little or not beaked, girdle very wide for the genus.

Head valve (now badly broken) semicircular, posterior margin somewhat wavy. Intermediate valves broadly rectangular to kidney-shaped, side margins rounded, posterior margin slightly beaked in valves II and III, about straight in the others, apices inconspicuous, lateral areas not raised, not

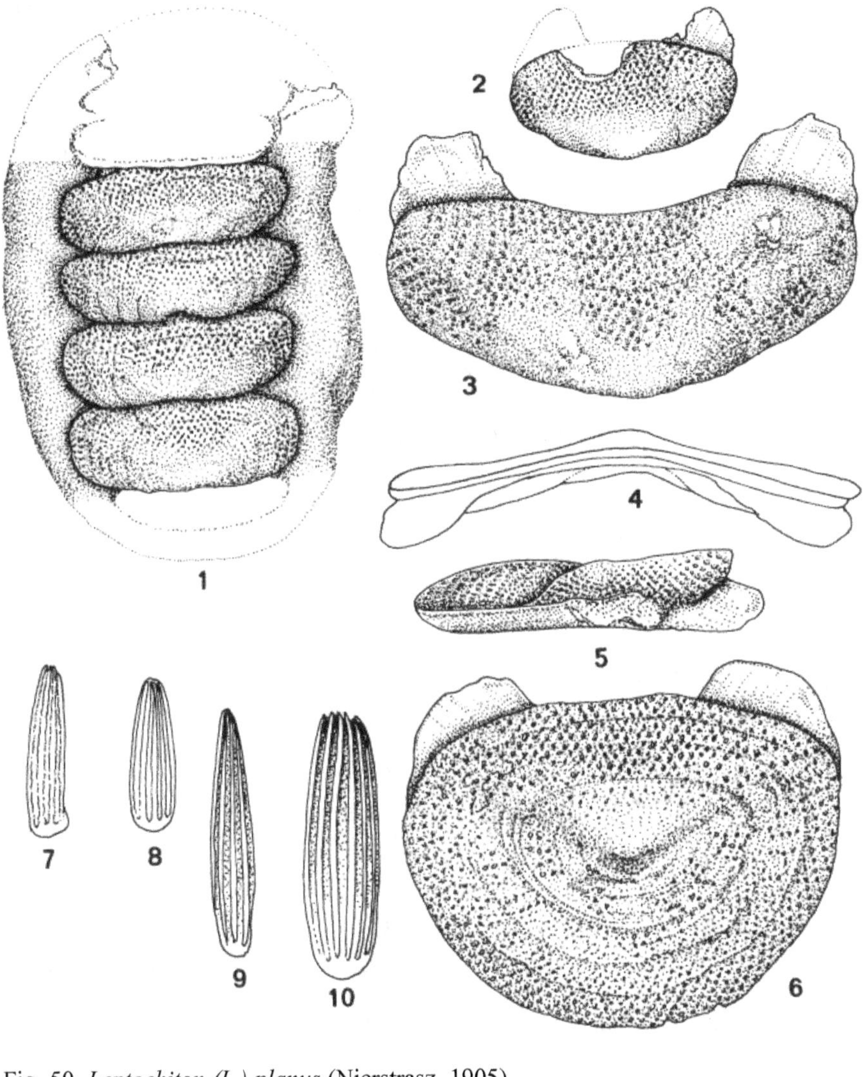

Fig. 50. *Leptochiton (L.) planus* (Nierstrasz, 1905).

1, whole animal, partly disarticulated, dorsal view, × 9.6; 2, valve II, def., dorsal view, × 9.6; 3, valve III, dorsal view, × 20; 4, do, anterior view, camera lucida sketch, × 20; 5, valve VIII, lateral view, × 20; 6, do, dorsal view, × 20; 7, ventral girdle spicule, × 400; 8, marginal spicule, × 400; 9-10, dorsal girdle spicules, × 400.

1-10, holotype.

indicated, separated from the central area by a very weak, diagonal depression. Tail valve more than semicircular, very long with regard to the width, 1:b = ca 0.73, flat, more or less saucer-shaped as the outer margin is sligthly curled upward, mucro about central, small, rather inconspicuous, postmucronal slope slightly concave.

Tegmentum finely and regularly granulated all over, the granules small, roundish, only little raised, well separated, arranged in quincunx, often worn off along the outer margins.

Colour of tegmentum yellowish, granules brownish, contrasting very much with the background.

Articulamentum thick, white, valves II-VII with rather long, widely separated, evenly rounded apophyses, "thrown forward", III-VII with a concave, broad sinus, apophyses of valve VIII more or less trapezoid.

Girdle wide and fleshy, colour brown, dorsally densely clothed with small, ribbed spicules, mostly with a blunt top, a few are narrower and more sharply pointed, 80-90 μm long, 16-26 μm broad. There is a marginal fringe of about three rows of short, truncated, conical spicules, more weakly ribbed, 50 × 16 μm. Ventral side of the girdle very sparsely beset with small, still weaker ribbed, truncated spicules, 50 × 12 μm.

The radula was unfortunately taken out of the unique specimen and apparently got lost. It was never illustrated.

Gills merobranchial; Nierstrasz (1905:9) counted 19 ctenidia on the right side.

Distribution. Only known from the type locality.

Leptochiton (L.) niasicus (Thiele, 1906)

(Fig. 51. Map 24)

Type: ZMHU.
Type locality: Indonesia, Nias Strait. "Tiefsee Exp." sta. 194,0°15.2'N 98°8.8'E, 614 m; sta. 199, 0°15.5'N 98°4'E, 470 m; sta. 203, 1°47.1'N 96°58.7'E, 660 m.

Lepidopleurus niasicus Thiele, 1906:327, pl. 29:figs 1-5.

Description. Animal oval, rather small, moderately elevated, subcarinated, side slopes almost straight, end valves semicircular, relatively large.

Fig. 51. *Leptochiton (L.) niasicus* (Thiele, 1906).

1, whole specimen, dorsal view, × 4.4; 2, another specimen, lateral view, × 3; 3, valve I, dorsal view, × 22; 4, intermediate valve, dorsal view, × 22; 5, valve VIII, dorsal view, × 22; 6, intermediate valve, anterior view, × 22; 7, do, ventral view, × 9.6; 8, valve VIII, ventral view, × 9.6; 9, dorsal girdle scales, × 350; 10, marginal spicules, × 350; 11, ventral scales, × 350; 12, dorsal spicule, × 350; 13, valve VIII, lateral view, × 22; 14, part of a radula row, × 350.

1-14, syntypes, 1-2, 9-12, 14, after Thiele, 1909.

Mucro of tail valve somewhat anterior, not prominent, posterior slope straight. Intermediate valves short, bluntly beaked, hind margin slightly concave at both sides of the apex, lateral areas not raised, hardly perceptible.

Tegmentum finely granulated all over; granules somewhat oval, toward

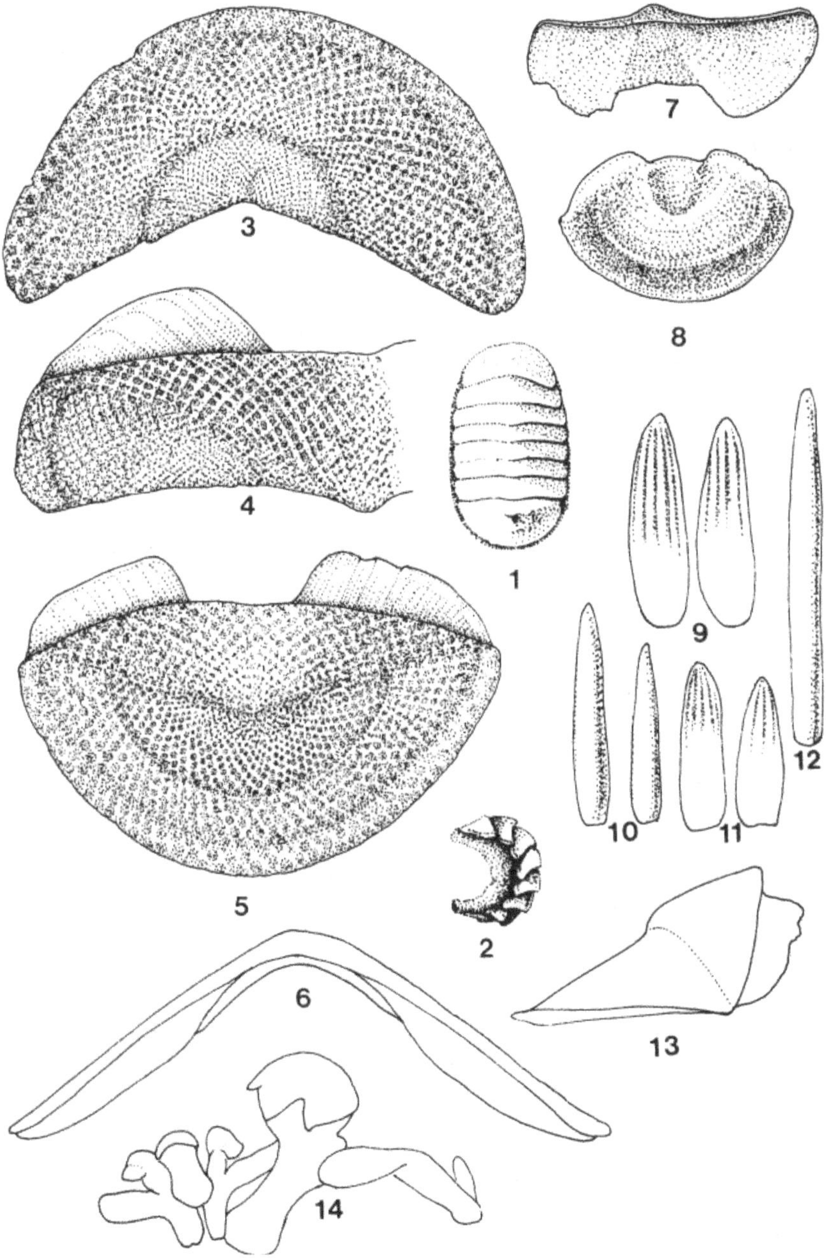

the margins round, not much raised, yellowish brown, quincuncially arranged, the interstices white, the granules closer set and less elevated on the lateral areas and end valves.

Articulamentum white, solid, decidedly thickened at the outer edges of the end valves. Apophyses rounded triangular, trapezoid in the tail valve; jugal sinus wide, shallow, slightly concave.

Girdle narrow, dorsally closely set with narrow, elongate, bluntly pointed scales, 75 μm long, 21 μm wide, distally marked with 3 or 4 longitudinal ridges. Scattered among them a few long, round, almost smooth, bluntly pointed spicules are found, 130 μm long. Ventral scales like the dorsal ones, but smaller, thinner, and less distinctly grooved, 60 μm long, 18 μm wide. Marginal spicules like the dorsal ones, but smaller, 65-80 μm long, 12 μm thick.

Central tooth of the radula pinched in the middle, with a well developed cusp; first laterals with a small outward bent cusp and a peculiar long and narrow outer wing, extending between the major lateral teeth. Cusp of major lateral with two strong, equal, triangular teeth and a small interior one.

Gills merobranchial, 10 ctenidia at both sides.

Length 7 mm, breadth 4 mm.

Distribution. Only known from the type locality.

Leptochiton (L.) andamanicus (Smith, 1906)

(Fig. 52. Map 38)

Type: BMNH 1906.10.12.86 (lectotype, designated herein); 1906.10.12.87-89 (three paralectotypes).
Type locality: Bay of Bengal, off N Sentinel Id, Andaman Is, 432 m.

Lepidopleurus andamanicus E. A. Smith, 1906:251; Annandale & Stewart, 1909:pl. 23:figs 4, 4a; Winckworth, 1940:19.

Description. Animal elongate oval, up to 12 mm long, 7 mm broad (lectotype, slightly curled), moderately elevated (dorsal elevation 0.42), carinated, side slopes almost straight to very slightly convex, the valves beaked, tegmentum very finely shagreened, lateral areas hardly elevated, decidedly marked by diagonal lines.

Head valve semicircular, posterior margin widely V-shaped; intermediate valves broadly rectangular, anterior margin strongly convex in valve II, straight to slightly concave in II-VIII. Lateral areas hardly raised but decidedly marked by diagonal lines. Tegmentum very finely shagreened by curved rows of tiny, round punctures, arranged in quincunx as well on the central as on the lateral areas, resulting in a reticulate appearance when seen under high magnification. The valves are decidedly beaked, the posterior margin slightly concave at both sides of the roundly pointed apices. Valve VIII semicircular, the mucro central, little pronounced, postmucronal slope slightly concave.

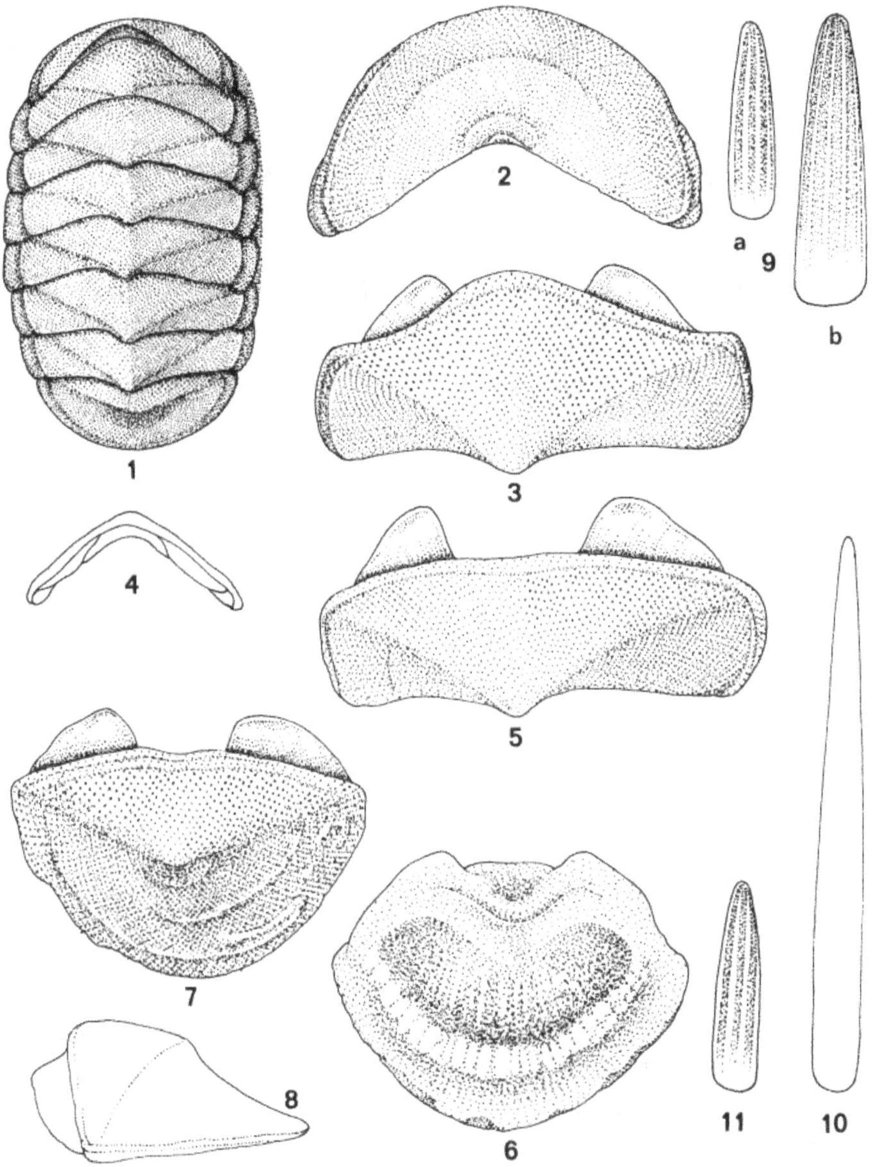

Fig. 52. *Leptochiton (L.) andamanicus* (Smith, 1906).

1, whole animal, dorsal view, × 4.5; 2, valve I, dorsal view, × 9; 3, valve II, dorsal view, × 9; 4, do, anterior view, × 4.5; 5, valve VII, dorsal view, × 9; 6, valve VIII, ventral view, × 9; 7, do, dorsal view, × 9; 8, do, lateral view, × 9; 9, dorsal girdle spicules, a from mid-girdle, b along the valves' margins, × 375; 10, intersegmental needle, × 375; 11, ventral girdle spicule, × 375.

1, lectotype; 2-11, paralectotype.

Articulamentum white, solid, the apophyses broadly triangular, regularly rounded, the jugal sinus little less than one third of the width of the valves, widest and strongly convex in valve II, slightly concave in subsequent valves.

In the lectotype there is a deep, sharp, concentric growth mark in all valves, close to the outer margin, not present in one of the paralectotypes.

Colour of tegmentum dirty white or light yellowish, the girdle a shade darker.

Girdle dorsally as well as ventrally densely covered with straight, bluntly pointed spicules, 68 × 16 µm, longitudinally ribbed with ca 4 riblets on the visible part. Sutural spicules larger, up to 96 × 24 µm, equally ribbed. There are a few straight, smooth intersegmental needles, 184 × 16 µm. Marginal fringe absent.

Distribution. Andaman Islands: off N Sentinel Id, 432 m (type, E. A. Smith).

Observations. E. A. Smith compared the present species with *L. arcticus*, with which it has nothing to do. *L. andamanicus* is easily recognizable by its extremely finely pitted tegmentum and spiculose – not scaly – girdle. The Winckworth collection in BMNH contains a specimen from Port Blair which is wrongly labelled *'L. andamanicus'*. It represents a specimen of a new species of the Sg. *Parachiton*, and is described elsewhere in this Monograph as *L. (P.) ronaldi* sp. nov. (BMNH, 1952.11.17.320).

Leptochiton (L.) vitjazi (Sirenko, 1977)

(Fig. 53. Map 43)

Holotype: ZIAS 1856, (J. A. Birshtein).
Type locality: Solomon Sea, Bougainville trough, E. S. "Vityaz" sta. 3655, 6920-7657 m.

Lepidopleurus vitjazi Sirenko, 1977:1108, figs 1-13.

Description. Animal small, elongate, the shell somewhat angular, white, girdle very narrow, dirty yellowish brown.

Head valve slightly wider than the tail valve. Intermediate valves broadly rectangular, apex only little projecting, broken off in most valves. Diagonal lines weakly marked. Mucro of tail valve behind the centre, slightly projecting.

Sculpture of the tegmentum uniformly granular. On the head valve, the lateral areas of the intermediate valves and the postmucronal area of the tail valve, the small granules, ca 50 µm in diameter, are arranged without distinct order, in some places much crowded, forming short, curved series, in other places more or less regularly disposed. On the central areas the arrangement is about identical, more regular and more crowded towards the jugum, inclined to form waving lines near the side margins. Lines of growth weakly indicated.

Apophyses small, the shape changes from angular with a rounded top (II) to trapezoid (VIII).

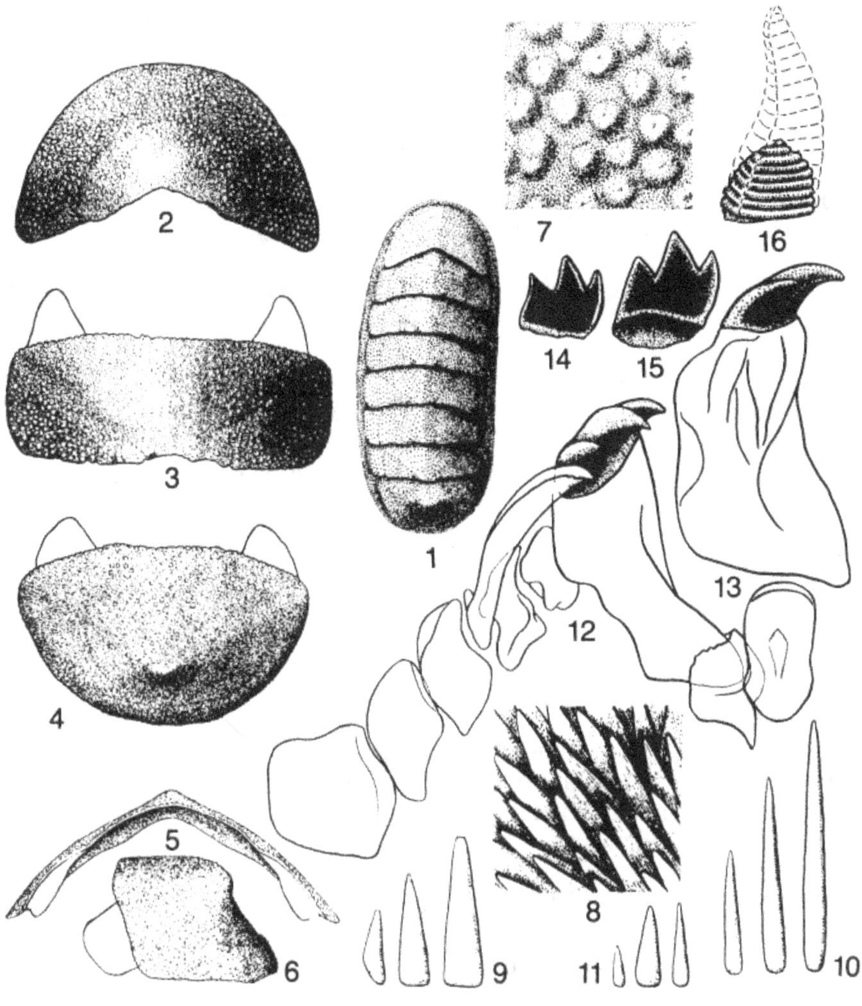

Fig. 53. *Leptochiton (L.) vitjazi* (Sirenko, 1977).

1, whole animal, dorsal view, × 3; 2, valve I, dorsal view, × 9; 3, valve V, dorsal view, × 9; 4, valve VIII, dorsal view, × 9; 5, valve V, anterior view, × 9; 6, valve VIII, lateral view, × 9; 7, sculpture of the tegmentum in the region of the diagonal line, × 89; 8, part of the girdle, dorsal view, × 170; 9, dorsal girdle spicules, × 170; 10, dorsal girdle spines and marginal spines, × 170; 11, ventral girdle spicules, × 170; 12, half row of the radula, × 170; 13, isolated major lateral radula tooth, × 170; 14-15, cusp of major lateral tooth, × 170; 16, gill, × 26 (the dotted lines indicate the size of a gill *L. japonicus*, lenght of the animal 10 mm).

2-11, 14, 16, holotype; 1, 12-13, 15, paratype; all after Sirenko, 1977.

Girdle clothed with spicules and longer spines. Spicules transparent, pointed, in the middle of the girdle about 85 μm long, near the valves up to 115 μm; spines, 100-215 μm long, chiefly at the sutures, sparsely on the rest of the girdle. The marginal needles are alike. Ventral spicules much the same as those on the dorsal side, but shorter, up to 60 μm long.

Radula 4 mm long, central tooth rounded at the base, major lateral tooth with a mighty cusp, with three sharp denticles, the central one largest.

Gills 4 at both sides, rather short, c. 400 μm long and wide, each with 9-10 lamellae.

Length 11 mm, breadth 5.3 mm, height 3 mm (holotype); length 14 and 12 mm, breadth 6.3 and 5 mm respectively (2 paratypes).

Distribution. Only known from the ultra-byssal Bougainville trough (Solomon Sea) in a depth of 6920-7657 m.

Observations. *L. benthedi* (Leloup, 1981) from the Glorieuses Islands, N entrance of the Mozambique Channel, is very much alike.

Leptochiton (L.) foresti (Leloup, 1981)

(Fig. 54. Map 44)

Type: MNHN.
Type locality: Philippines: Verde Island Passage, "Musorstom" 1, 1976, Exp., sta. 25, 14°02.7'N 120°20.3'E, 191-200 m.

Lepidopleura foresti Leloup, 1981a:321 (sic!), fig. 2, pl. 2:figs 7-10.
Leptochiton foresti; Kaas, 1982b:88.

Description. Animal elongate oval, rather small, up to 14 × 7.5 mm, little elevated (dorsal elevation 0.26), carinated, the valves slightly beaked, tegmentum regularly granulated, lateral areas not raised; girdle narrow, densely spiculose. Colour white.

Valve I semicircular, finely granulated, the granules roundish to oval, arranged in radiating rows, about 100-120 by many intercalations, and in somewhat wavy concentric rows as well. Posterior margin widely V-shaped, with a small rounded notch at the apex, anterior slope straight. A few weakly pronounced growth ridges always perceptible. Intermediate valves short, about three times as wide as long, anterior margin slightly rounded to almost straight, finely serrated, sides bluntly rounded, posterior margin concave at both sides of the slightly beaked apex; back carinated, side slopes straight, tending to become a little concave towards the side margins. Central area sculptured with ca 24 longitudinal chains of well raised, roundish granules on either side of the jugum, the interstices almost as wide; towards the sides the chains are rather strongly converging towards the apex; on the jugum the granules are less pronounced and the chains closer. Lateral areas hardly indicated, not raised, differentiated by the sculpture only. There are 12-14 (by intercalation) radiating rows of granules, less raised than on the central area, interstices very narrow; generally there are 2-4 weakly pronounced growth ridges. Tail valve semicircular, rather flat, the mucro central, hardly raised, posterior slope very weakly concave or almost straight; antemucronal area sculptured like the central areas, postmucronal area like the head valve.

Articulamentum well developed, forming a flat, thickened outer ridge along the margins of the end valves. Apophyses relatively long, the inner margin straight, forming an angle of 120° with the anterior margin of the valves, regularly narrowing towards the sides, wide apart, dorsally finely striated; jugal sinus very wide, deep, flat.

Girdle white, densely clothed with straight, imbricating, longitudinally ribbed, bluntly pointed spicules, directed outward, about 80 × 20 μm. There is a fringe of much longer, only very finely striated, slightly curved and sharply pointed supra-marginal spicules, 144 × 18 μm. Ventral spicules small, smooth, imbricating, regularly tapering to a blunt point, 48 × 13 μm.

Radula with a rather wide central tooth, the blade deeply bilobed, narrowing towards the base, first laterals about triangular, with a small notch on

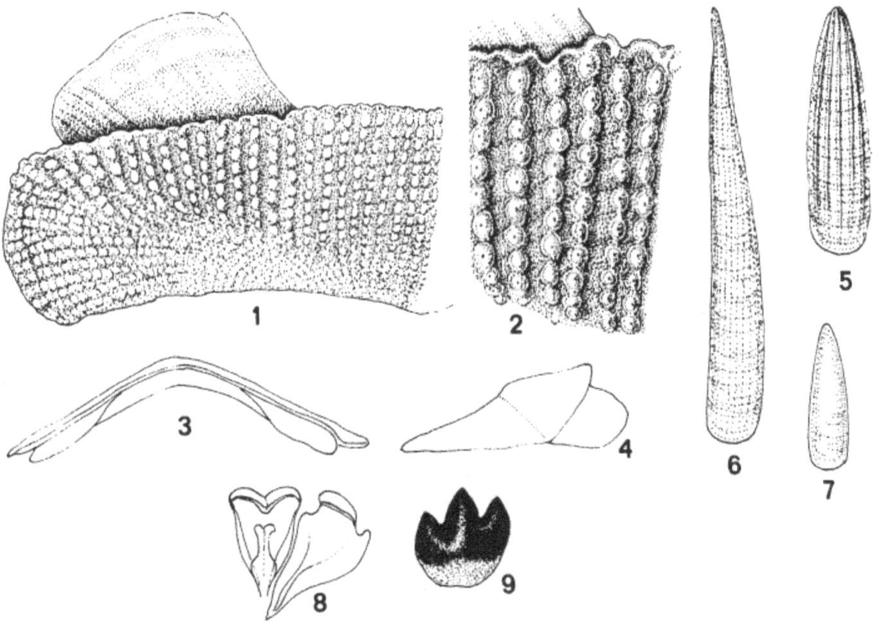

Fig. 54. *Leptochiton (L.) foresti* (Leloup, 1981).

1, valve IV, dorsal view of left half, × 20; 2 detail of sculpture of same valve, central area, × 40; 3, valve IV, anterior view, × 9.6; 4, valve VIII, lateral view, × 9.6; 5, dorsal girdle spicule, × 400; 6, supra-marginal girdle spicule, × 400; 7, ventral spicule, × 400; 8, central and first lateral radula teeth, × 200; 9, cusp of major lateral radula tooth, × 200.

1-9, specimen from Philippines, N. of Verde Island Passage, between Lubang Ids and Luzon, Musorstom 1 Exp., 1976, sta. 34, 14°01'N 120°15.8'E, 191 m, MNHN.

both sides of the narrow blade; major laterals with a tridentate cusp, the teeth of equal width, the central one slightly longer than the others.

Gills merobranchial, abanal; about 16-18 ctenidia on both sides.

Distribution. Only known from the large type set, collected between 13°49.4' to 14°02.7'N and 120°04.8' to 120°28'E, in depths varying from 150-484 m. The "Musorstom" 2, 1980, Expedition procured many more specimens in the same area.

Observations. Leloup compared this species with *L. pergranatus* (Dall, 1889) (not Kaas, 1972!) and *L. africanus* (Nierstrasz, 1905), both totally different species. In fact *L. foresti* is more closely related to *L. hakodatensis* (Thiele, 1909), which, however, differs in its bidentate major lateral radula teeth, more scaly girdle, and round-backed valves.

Leptochiton (L.) juvenis (Leloup, 1981)

(Fig. 55. Map 44)

Type: MNHN (holotype + one paratype).
Type locality: Philippines, Verde Island Passage, 13°59.5'N 120°15.6'E, 179-194 m ("Musorstom" 1 Exp., 1976, sta. 62).

Lepidopleurus juvenis Leloup, 1981a:321, fig. 3, pl. l:figs 4-7.
Leptochiton juvenis; Kaas, 1982b:88.

Description. Animal elongate oval, rather small, up to 12 mm long, 6 mm broad, subcarinated to carinated, little elevated (dorsal elevation 0.27-0.30), valves not beaked, lateral areas hardly raised, extremely finely granulated all over, girdle narrow, covered with elongate scales and spicules and long inter-segmental spines. Colour white.

Valve I rather large, semicircular, posterior margin widely V-shaped, with a rounded notch at the apex, anterior slope straight, tegmentum extremely finely granulated in quincunx as well as in radiating series; there are always 3-6 concentric growth lines. Intermediate valves rectangular, II a little longer than the others, which are three times as broad as long, the apices perceptible but never beaked, lateral areas hardly or not raised, side slopes slightly convex, posterior margin almost straight, anterior margin straight to slightly concave, side margins somewhat truncated. Central areas sculptured with c. 60 longitudinal rows of clearly cut, connected, roundish granules, interstices very narrow, pitted, the granules of one row alternating with those of the neighbouring series, the rows abruptly ending at the diagonal lines; lateral areas sculptured with much smaller granules, less raised, quincuncially arranged; always with a few growth lines. Valve VIII semicircular, the mucro a little anterior, small, little raised, posterior slope a little concave directly behind the mucro; anterior margin somewhat sinuate; antemucronal area sculptured like the central areas, postmucronal area like the head valve.

Articulamentum white, well developed, solid; the apophyses rather long, almost half the length of the valve, narrow, bluntly triangular, wide apart; jugal sinus as wide as the apophyses together, deep, straight; in the tail valve the apophyses are more or less trapezoid, the jugal sinus much narrower.

Girdle narrow, clothed with elongate egg-shaped scales, distally sculptured with 5-6 sharp riblets, 48 × 22 µm; among them, randomly dispersed, small, needle-like spicules occur, 44 × 5 µm, along the margins of the valves

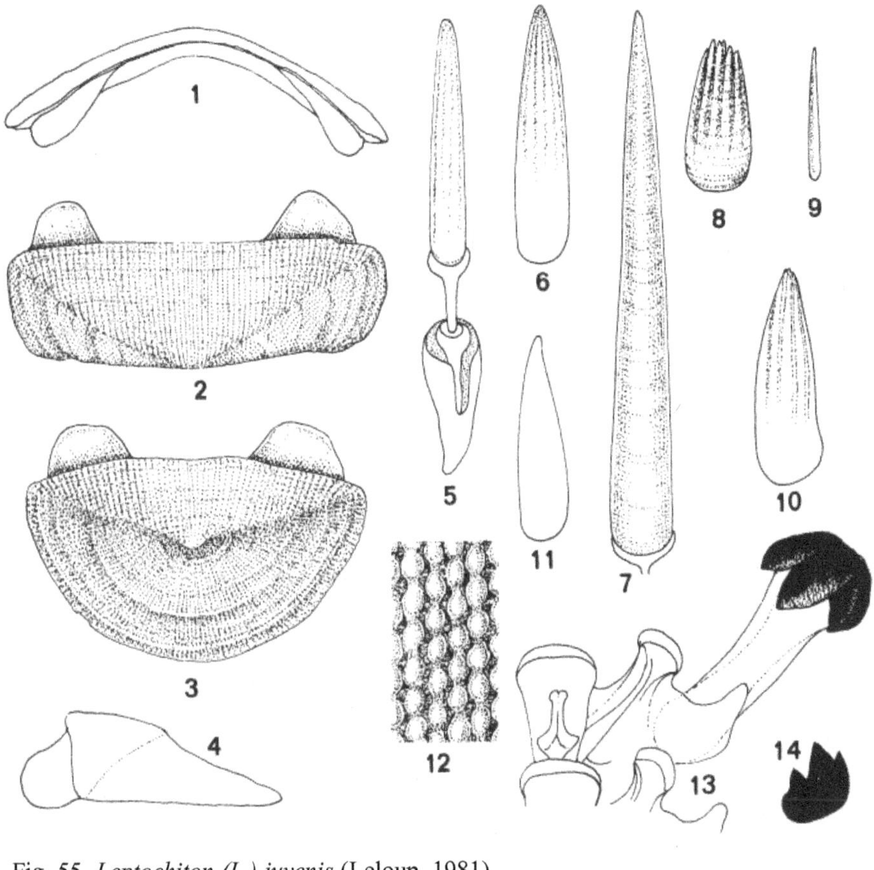

Fig. 55. *Leptochiton (L.) juvenis* (Leloup, 1981).

1, valve IV, anterior view, × 9.6; 2, do, dorsal view, × 9.6; 3, valve VIII, dorsal view, × 9.6; 4, do, lateral view, × 9.6; 5, "Ringschaftnadel" sensu Thiele, on perinotum tongue between valves, × 400; 6, marginal spicule, × 400; 7, intersegmental spine, × 400; 8, dorsal girdle scale, × 400; 9, dorsal spicule, × 400; 10, ventral scale, near outer margin, × 400; 11, do, near inner margin, × 400; 12, detail of tegmental sculpture on central area of intermediate valve, × 40; 13, central, first and second (major) lateral radula teeth, × 200; 14, cusp of major lateral tooth, × 200.

1-4, 6-13, specimen from the Philippines, N of Verde Island Passage, between Lubang Ids and Luzon, 14°00'N 120°19'E, 188-198 m, Musorstom 2 Exp., 1980, sta. CP1, P. Bouchet coll., MNHN; 5, 14, specimen from same area (paratype), 13°59.5'N 120°15.6'E, 179-194 m, Musorstom 1 Exp., 1976, sta. 62, MNHN.

larger spicules-scales are found, weakly ribbed on the distal half, 72 × 24 μm; the perinotum tongues between the valves bear a small group of "Ringschaftnadeln" (sensu Thiele), weakly striated, 82 × 12 μm, and 3-4 long, almost smooth spines, up to 176 × 20 μm; marginal spicules finely striated, 96 × 18 μm; ventral side of the girdle clothed with rows of imbricating, flat, elongate, distally pointed scales, 68 × 16 μm.

Central tooth of the radula elongate, distally slightly widening, with a narrow blade, first laterals with a narrow blade and a long outward, forwardly directed wing, major laterals with a tridentate cusp, the central denticle slightly longer.

Distribution. The "Musorstom" Philippines 1 (1976) and 2 (1980) Expeditions procured 10 more specimens from four stations in the same restricted area where the juvenile types, described by Leloup, were hauled. Most of them are adults, caught between 13°41' and 14°00N and between 120°15.6' and 120°50'E, Verde Island Passage, between the Lubang Islands and Luzon, Philippines.

Leptochiton (L.) philippinus (Leloup, 1981)

(Fig. 56. Map 44)

Type: MNHN.
Type locality: Philippines, Verde Island Passage, 13°15.5'N 120°28'E, 484-448 m ("Musorstom" 1 Exp., 1976, sta. 43).

Lepidopleurus philippinus Leloup, 1981a:322, fig. 4, pl. 2:figs 1-3.
Lepidopleurus porosus Leloup, 1981a:322, figs 5-6, pl. 2:figs 4-6; Kaas, 1982b:89.
Leptochiton philippinus; Kaas, 1982b:89.

Description. Animal elongate oval, of moderate size, up to 26 mm long, 12 mm broad, carinated, moderately to highly elevated (dorsal elevation from 0.39 to 0.50), valves slightly beaked, lateral areas little raised, tegmentum very finely and faintly punctate in quincunx; girdle narrow, clothed with elongate, striated spicules. Colour white, the shells mostly covered with black or dark brown foreign deposits.

Head valve semicircular, posterior margin widely V-shaped, anterior slope straight, tegmentum regularly quincuncially punctated, the punctures shallow, roundish, towards the anterior margins of the intermediate valves and towards the outer margins of all valves these punctures are sometimes replaced by hardly raised pustules; a series of deeply engraved concentric growth lines is always present in all valves. Intermediate valves rectangular, anterior margin convex in the jugal part, posterior margin straight, except for the little projecting apex, side margins almost rectangularly cut off; jugum rather sharply carinated, side slopes somewhat convex, lateral areas slightly raised, the whole surface punctated as in the head valve. Tail valve semicircular, the mucro central to a little posterior, not swollen, anterior margin almost straight; antemucronal area somewhat vaulted, posterior slope weakly concave.

Articulamentum white, well developed, thickened towards the outer margins, with a broad, flat outer ridge in the end valves. Apophyses broadly

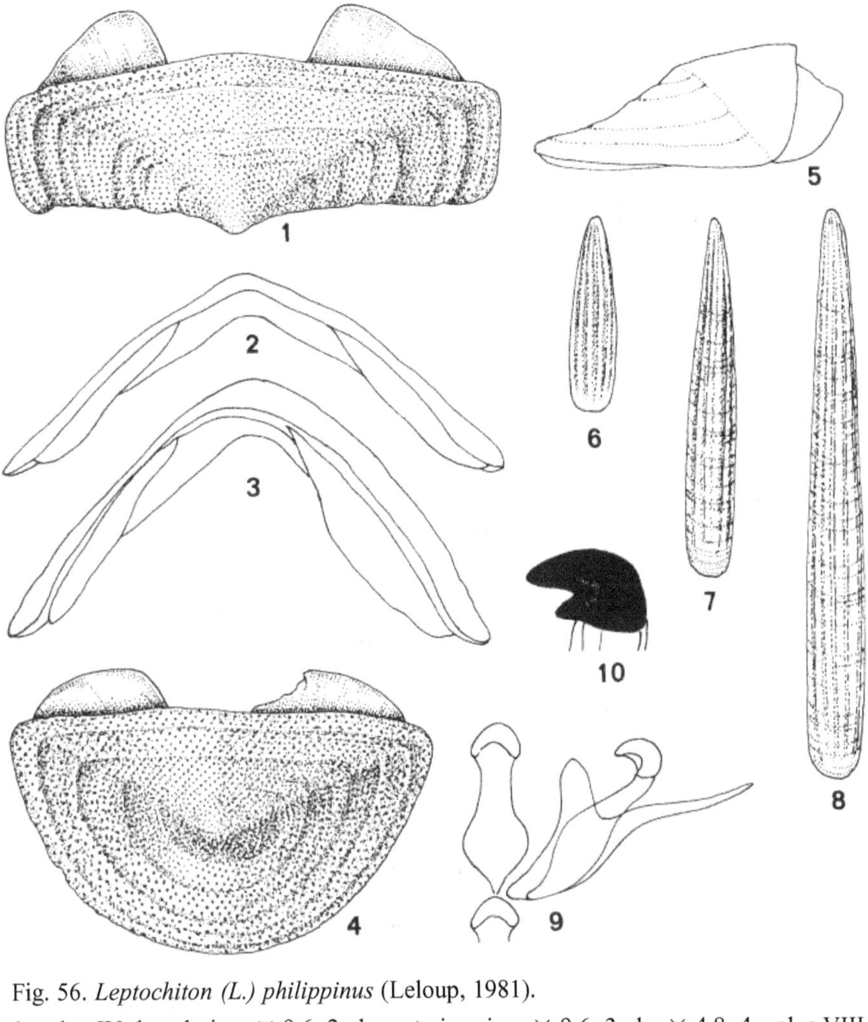

Fig. 56. *Leptochiton (L.) philippinus* (Leloup, 1981).

1, valve IV, dorsal view, × 9.6; 2, do, anterior view, × 9.6; 3, do, × 4.8; 4, valve VIII, dorsal view, × 9.6; 5, do, lateral view, × 9.6; 6, ventral girdle spicule, × 400; 7, dorsal spicule, × 400; 8, supra-marginal spicule, × 400; 9, central and first lateral radula teeth, × 400; 10, cusp of major lateral radula tooth, × 200.

1-2, 4-10, specimen from Philippines, N of Verde Island Passage, between Lubang Ids and Luzon, Musorstom 2 Exp., 1980, sta. CP38, 12°53'N 122°27'E, 1650-1660 m, P Bouchet coll., MNHN; 3, specimen from same area, Musorstom 1 Exp., 1976, sta. 47, 13°40.7'N 120°30'E, 757-685 m, holotype of *Lepidopleurus porosus* Leloup, 1981, MNHN.

triangular, the inner margins forming an angle of 110° with the anterior margin of the valve, the sinus as wide as the apophyses. In valve VIII they are relatively shorter and wider, more trapezoid, the sinus much narrower.

Girdle narrow, dorsally densely clothed with bluntly pointed, elongate spicules, 120 × 16 μm, with about 5 strong riblets on the visible part, and much longer, equally sculptured supra-marginal spicules, 188 × 20 μm. Ventral spicules also ribbed, lanceolate, 64 × 16 μm.

Radula with a rather long central tooth, widest in the proximal half, pinched in the middle, with a small, strongly rounded blade; first lateral teeth narrow, sinuate, with a small, curved, interiorly directed blade, rising from a bilabiate calyx, the outer lip very long and narrow; major lateral teeth with a bidentate cusp, the inner denticle much smaller.

Distribution. Philippines; known from ten stations of the "Musorstom" Expeditions 1 (1976) and 2 (1980), in a restricted area in the Verde Island Passage, between the Lubang Islands and Luzon, from 12°53' to 14°09'N and from 120°02' to 122°36'E, in depths varying from 177 to 1660 m.

Observations. *L. philippinus* bears a close resemblance with *L. andamanicus* (Smith). In fact the two are so similar that we first were apt to believe them to be conspecific. However, the numerous growth marks present in *philippinus* (none or only one present in the type set of *andamanicus*) leave us in doubt, and as long as the radula of *andamanicus* is unknown, we hesitate to take a decision.

The unique type of *L. porosus* Leloup ("Musorstom" 1, 1976, Exp.: sta. 47, in the same area, 757-685 m), a disarticulated specimen, of which the radula and girdle are mounted in balsam (MNHN), proves to be a very large (26 × 12 × 6 mm), highly elevated (see Fig. 56:3), aberrant form of *philippinus*. In sculpture, perinotum covering and radula *porosus* is in all respects identical with *philippinus*. It appears that the type of *"porosus"* was injured in a juvenile stage, after which growth became irregular, the broken right half of valve VI being repaired by an abnormal growth of valve VII.

Leptochiton (L.) benthedi (Leloup, 1981)

(Fig. 57. Map 45)

Type: MNHN.
Type locality: Indian Ocean, SE of the Glorieuses Is, 3700 m, "Benthedi" sta. 90-CH, 11°44'S 47°30'E, 1977.

Lepidopleurus benthedi Leloup, 1981b:2, figs 2-3, pl. 1:fig. 5.

Description. Animal small, up to 9.5 mm long, 4 mm broad, elongate oval, moderately elevated, subcarinated, the side slopes almost straight. Valves thin and brittle, evenly granulose in a quincuncial pattern. Lateral areas of intermediate valves hardly or not indicated, not perceptibly raised, apices inconspicuous. Colour of tegmentum white to light greyish.

Valve I semicircular, posterior margin widely V-shaped, evenly and closely granulated in quincunx; intermediate valves of about equal width, broadly rectangular, posterior and anterior margins almost straight, the sides only slightly rounded. Lateral areas hardly indicated, not raised; tegmentum sculptured with roundish to oval, close set and well raised granules in a quincuncial pattern, continuous on central and lateral areas. Valve VIII semicircular, the mucro little pronounced, slightly posterior when seen from above, postmucronal slope concave, tegmentum sculptured like the intermediate valves, postmucronal and antemucronal areas hardly separated.

Articulamentum weakly developed, the apophyses short, bluntly triangular, the inner edge oblique. Jugal sinus rather wide, flat to slightly convex.

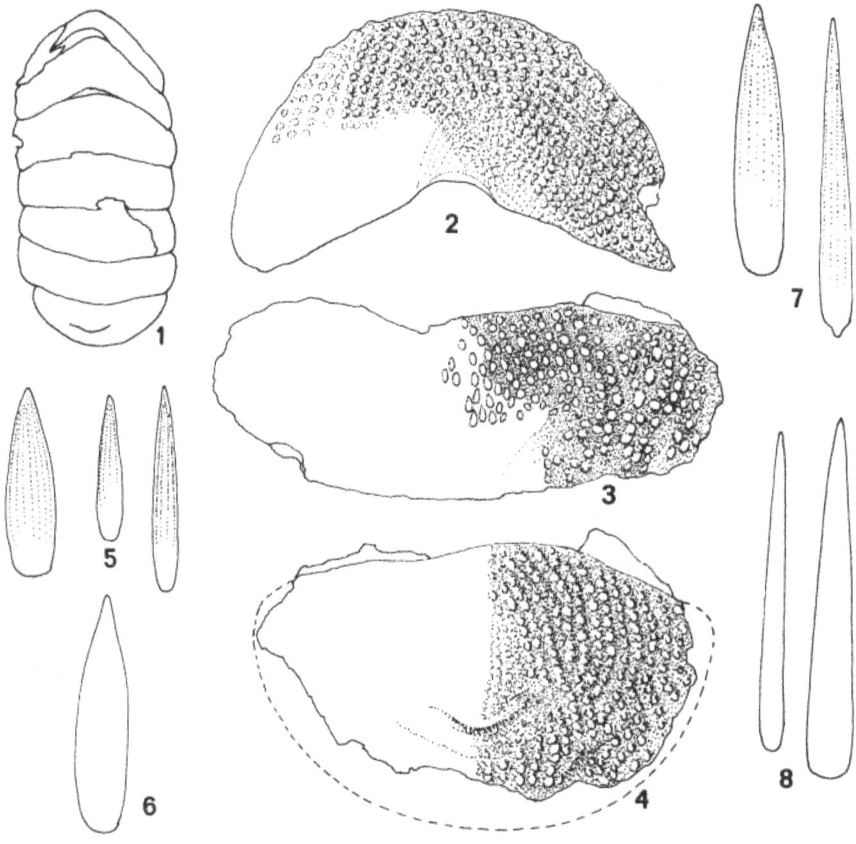

Fig. 57. *Leptochiton (L.) benthedi* (Leloup, 1981).

1, whole animal, dorsal view, camera lucida sketch, after a photograph in Leloup, 1981, × 7.2; 2-4, valves I, III, VIII respectively, dorsal view, × 20; 5, dorsal girdle spicules, × 280; 6, ventral spicule, × 280; 7, dorsal spicules along the valves' margins, × 280; 8, marginal spicules, × 280.

1-8, paratypes.

Girdle narrow, densely covered with conical, abruptly pointed, finely longitudinally striated spicules, directed towards the outer margin, 80-90 µm long, 12-24 µm thick, those along the sutures of the valves somewhat larger, almost smooth, up to 136 × 24 µm. Ventral spicules smooth, shaped like the dorsal ones, 100 × 24 µm. There is a marginal fringe of straight, smooth needle-like spicules, up to 160 µm long, 12-24 µm thick.

Length 9.5 mm, breadth 4.0 mm (largest paratype); breadth 3 mm (holotype, strongly curled up, now disarticulated).

Distribution. Only known from two "Benthedi"-stations, 87-CH and 90-CH, both SE of the Glorieuses Is at the northern entrance of the Mozambique Channel, in a depth of 3716 and 3700 m respectively.

Observations. Leloup compared *L. benthedi* with *L. lineatus* and *L. setigera* of Nierstrasz (1905), but it is quite unlike those species. It is easily recognizable by the granulation of the valves, which forms curved or wavy lines in a quincuncial pattern, exactly as in *L. vitjazi*, with which it may be identical.

Leptochiton (L.) nierstraszi (Leloup, 1981)

(Fig. 58. Map 45)

Type: IRSN IG 26.374.
Type locality: Malagasia: reef off Tuléar (M. Peyrot-Clausade coll.).

Lepidopleurus (Pilsbryella) nierstraszi Leloup, 1981d:27, fig. 13, pl. 2:fig. 1.

Description. Animal very small, up to 6 × 3.5 mm, oval, rather elevated (dorsal elevation 0.38), back rounded to subcarinated, valves not beaked, lateral areas slightly raised, tegmentum sculptured with highly raised, roundish pustules arranged in quincunx. Girdle narrow, covered with small, imbricating, ribbed scales and spicules of different sizes. Colour white.

Head valve semicircular, anterior margin reversed V-shaped, with a rounded notch at the apex, anterior slope concave; tegmentum sculptured with rather large, strongly pronounced, roundish pustules in quincunx. Intermediate valves broadly rectangular, three times as wide as long, anterior margin convex in valve II, straight to weakly concave in III-VII, posterior margin almost straight, the apices well marked but not projecting. Central areas divided into a broadly V-shaped jugal area, bordered by the inner margins of the apophyses and the apex, and the pleural areas, only separated by a different pattern of sculpture: on the jugal area consisting of well separated, round, hollow pustules, neatly arranged in quincunx; on the pleural areas the pustules are slightly smaller, arranged in 12-14 more or less juxtaposed longitudinal rows, slightly diverging posteriorly, separated by rather wide interstices. Lateral areas perceptibly raised, the pustules smaller, in quincunx, those along the posterior and side margins very long, dentating the

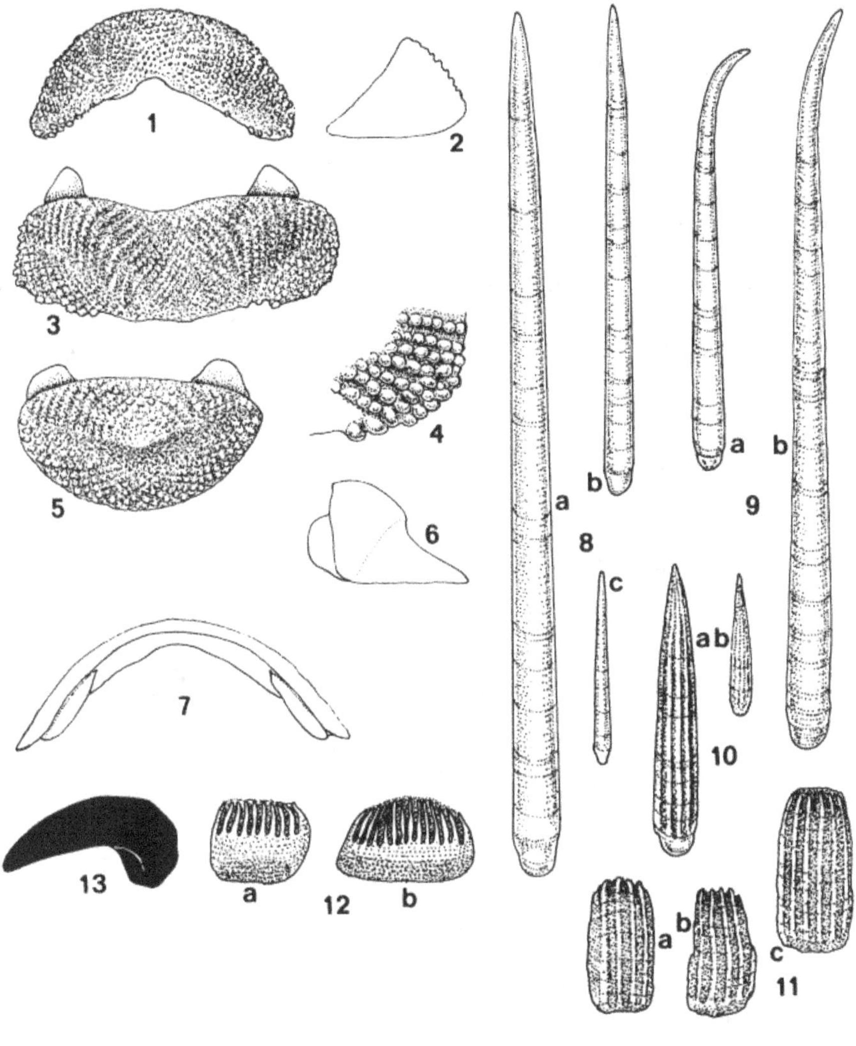

Fig. 58 *Leptochiton (L.) nierstraszi* (Leloup, 1981).

1, valve I, dorsal view, × 20; 2, do, lateral view, × 20; 3, valve V, dorsal view, × 20; 4, sculpture of lateral area, × 40; 5, valve VIII, dorsal view, × 20; 6, do, lateral view, × 20; 7, valve VII, anterior view, × 20; 8a-c, dorsal girdle spicules of different sizes, × 400; 9a-b, intersegmental spicules, × 400; l0a-b, marginal spicules, × 400; 11 a-c, ventral scales, × 400, a-b near inner margin, c near outer margin; 12a-b, dorsal scales, × 400, a near outer margin, b from mid-girdle; 13, cusp of major lateral radula tooth, × 800.

1-13, paratype from Malagasia, Tuléar, Songoritelo reef (M. Peyrot-Clausade coll., sta. 34.7).

sutures. Tail valve transversely oval, the length about 0.6 of the breadth, the mucro central to somewhat anterior; anterior margin convex, antemucronal area swollen, sculptured like intermediate valves, postmucronal area like the head valve; posterior slope concave.

Articulamentum thin, white; apophyses small, triangular, wide apart, jugal sinus shallow, wide, convex in valves II and VIII, straight to slightly concave in III-VII.

Girdle narrow, densely covered with bent, imbricating scales, much wider than long, up to 50 × 25 μm, dorsally sculptured with 10-12 short ribs, the basal part smooth. Among the scales randomly dispersed smooth, straight, slender needles are found, varying in length from 60 to 280 μm. On the perinotum tongues between the valves small groups of smooth needles are present, the top slightly bent, 140-250 μm long. Ventral side of girdle covered with imbricating, flat, elongate scales, distally truncated, sculptured with 4-6 rather strong riblets, narrower than the interstices, varying in size from 44 × 22 μm near the inner margin, to 54 × 26 μm at the outer edge. There is a marginal fringe of ribbed spicules, up to 100 μm long.

Major lateral radula teeth with a unicuspid head, the denticle 30 μm long, slightly curved.

Distribution. Malagasia (Madagascar), reefs off Tuléar; Réunion, reef off St. Pierre.

Observations. Leloup compares this species with *L. simplex* (Nierstrasz, 1905), synonymized with *L. alveolus* by Ferreira (1979b: 152), which lacks sutural needles, and with *L. latidens* (Bergenhayn, 1933), which is also totally different. The author ranked it under the Sg. *Pilsbryella* Nierstrasz, 1905, which, however, is characterized by the possession of long chitinous intersegmental hairs, not present in *L. nierstraszi*.

Leptochiton (L.) cancelloides Kaas, 1982

(Fig. 59. Map 44)

Type: MNHN (holotype + 1 paratype).
Type locality: Philippines, Verde Island Passage, between the Lubang Ids and Luzon, 13°32'N 121°07'E ("Musorstom" 2, 1980, Expedition, Sta. DR33, P. Bouchet coll.).

Leptochiton (L.) cancelloides Kaas, 1982b:89, figs 1-13.

Diagnosis. Animal small, up to 6 × 3 mm, elongate oval, rather elevated (dorsal elevation 0.45), back evenly rounded, side slopes convex; lateral areas of intermediate valves little elevated; tegmentum granulated all over, in longitudinal series on central areas of intermediate valves and on antemucronal area of tail valve, quincuncially on head valve, on lateral areas of intermediate valves, and on postmucronal area of tail valve; mucro of tail valve about central, posterior slope deeply concave; girdle narrow, spiculose; major lateral radula tooth with a unicuspid head. Colour white.

Description. Head valve semicircular, posterior margin widely V-shaped, anterior slope straight, tegmentum granulose, granules roundish, arranged in

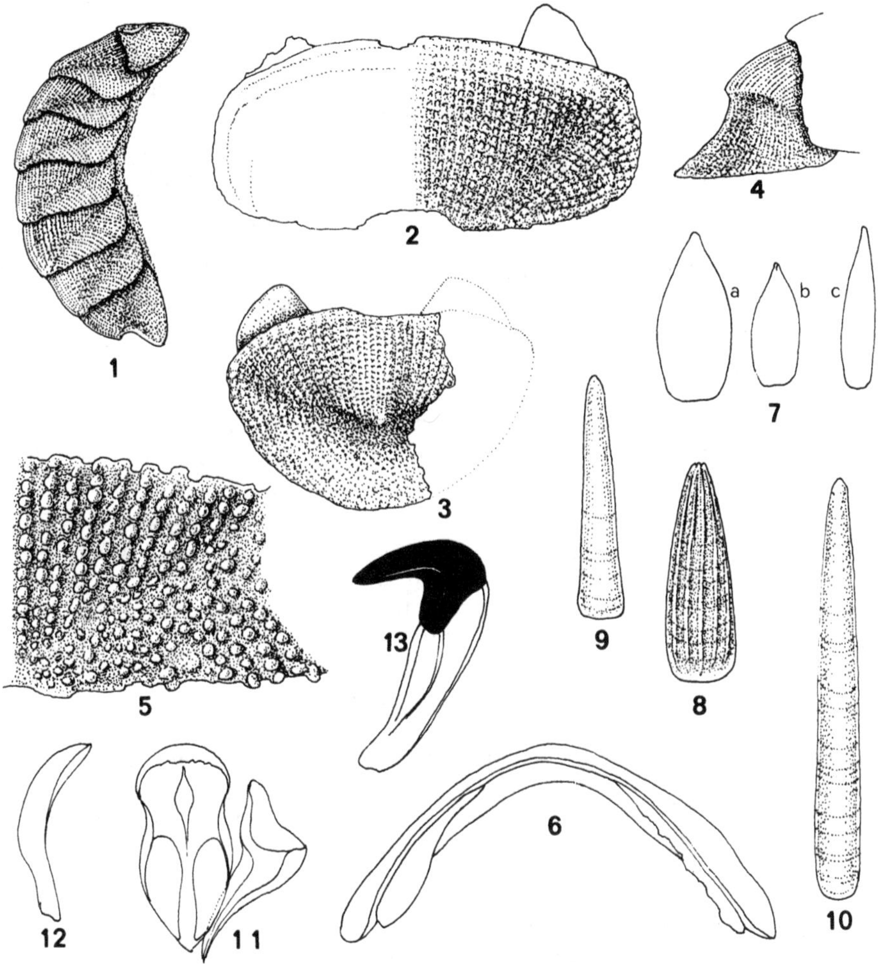

Fig. 59. *Leptochiton (L.) cancelloides* Kaas, 1982.

1, whole animal, lateral view, × 9.6; 2, valve II, dorsal view, × 20; 3, valve VIII, dorsal view, × 20; 4, do, lateral view, × 20; 5, tegmental sculpture on right half of valve III, × 40; 6, valve III, anterior view, × 20; 7, ventral girdle scales, × 400; a, c, near outer margin, ventral and lateral view respectively, b near inner margin; 8, dorsal girdle spicule, × 400; 9, marginal spicule, × 400; 10, intersegmental spicule, × 400; 11, central and first lateral radula teeth, × 400; 12, spatulate uncinal tooth, × 400; 13, major lateral tooth, × 200.

1, 4, 7-10, holotype; 2-3, 5-6, 11-13, paratype, after Kaas, 1982.

quincunx, at the same time in radial rows, with many intercalations; intermediate valves little more than twice as wide as long, anterior margin convex in valve II, concave in the jugal part of II-VIII, posterior margin straight, apex hardly indicated, mostly worn away, lateral areas little raised, central areas with c. 50 longitudinal rows of roundish, well raised pustules, the interstices almost as wide, on the lateral areas the granules are arranged in curved series, about perpendicular to the diagonal lines and at the same time parallel to these; tail valve semicircular, the mucro central to slightly posterior when viewed from above, somewhat swollen, inclining, the posterior slope deeply concave directly behind it; antemucronal area highly vaulted, with 40 or more longitudinal series of granules, anterior margin convex towards the sides, concave in the middle, postmucronal area sculptured like the head valve.

Articulamentum well developed, white; apophyses wide apart and narrowly triangular in the intermediate valves, with a wide, slightly concave sinus (convex in valve II), wider and more trapezoid in valve VIII, the sinus narrower, concave.

Girdle dorsally clothed with outwardly directed, bluntly pointed, longitudinally ribbed spicules, 72 × 24 μm, with 4-5 riblets on the visible half; there is a marginal fringe of smooth, bluntly pointed spicules, 80 × 18 μm; on the perinotum tongues small groups of longer, smooth spicules are found, 140 × 16 μm. Ventral side of the girdle paved with concentric rows of flat, imbricating, lanceolate scales, 40 × 16 μm near the inner margin, becoming gradually larger towards the outer margin, the largest 56 × 25 μm.

Radula with a short and wide central tooth, pinched in the anterior half, with a narrow, rounded blade; first laterals short, triangular, without a blade; major laterals with a strong, unidentate cusp.

Distribution. Only known from the type locality.

Observations. This new species bears a superficial resemblance to the European *L. cancellatus* (Sowerby), from which it differs in sculpture and, especially, in the characteristics of the perinotum elements and the radula.

Fig. 60. *Leptochiton (L.) liratus* (H. Adams & Angas, 1864).

1, whole animal, dorsal view, × 9; 2, valves I, II, in situ, dorsal view, × 19; 3, sculpture on tegmentum of valve IV, right half, × 38; 4, valve VIII, ventral view, × 9; 5, do, dorsal view, × 9; 6, do, lateral view, × 9; 7, valve IV, dorsal view, × 9; 8, do, ventral view, × 9; 9, do, anterior view, × 9; 10, dorsal girdle scales, a near suture, b from mid-girdle, × 300; 11, ventral scales, a from mid-girdle, b infra-marginal, × 190; 12, marginal spicule, × 190; 13, dorsal spicule, × 190; 14, central and first lateral radula teeth, × 375; 15, cusp of major lateral tooth, × 375.

1-15, specimen from rocks near Marino, Gulf St Vincent, S Australia, B. J. Weeding leg., K 2737.

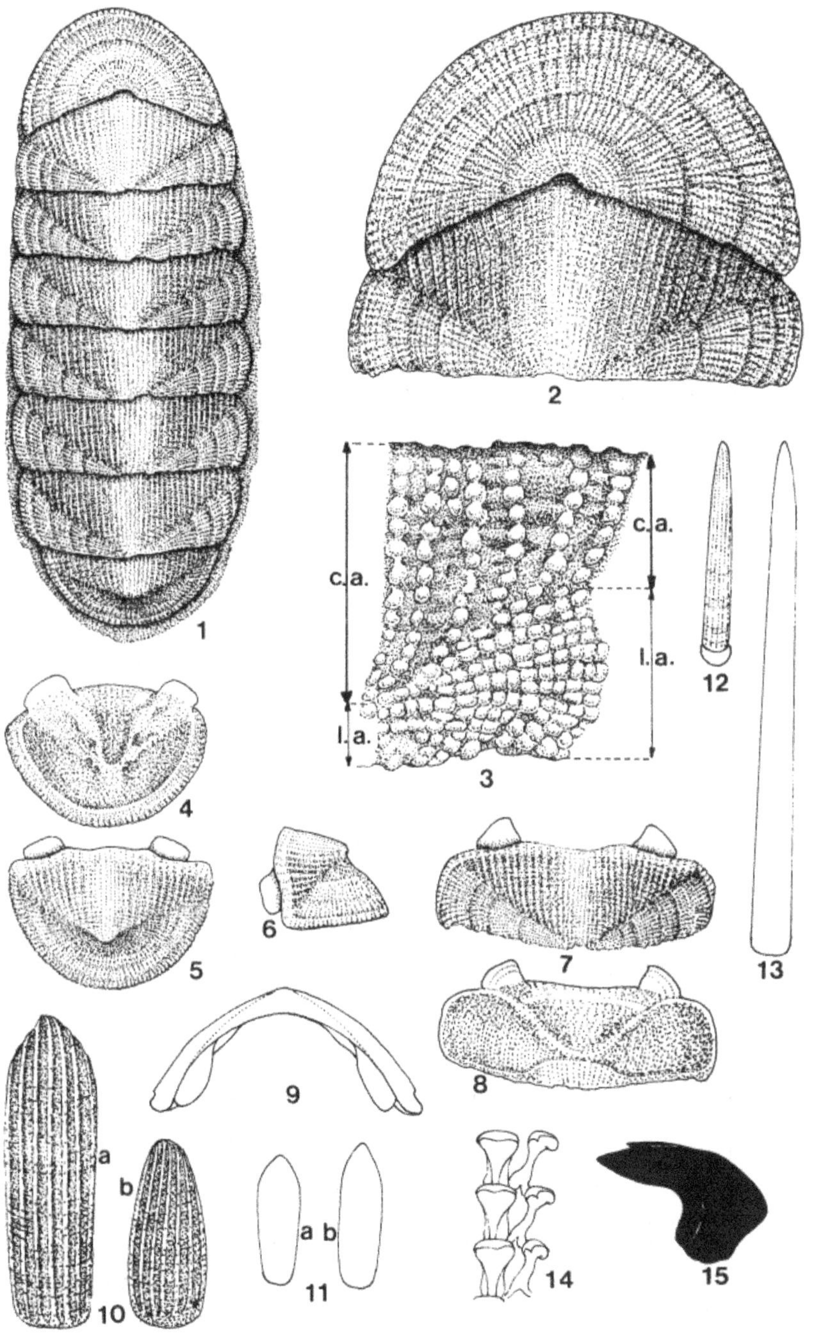
1
2
c. a.
c. a.
l. a.
l. a.
3
12
13
4
5
6
7
8
9
a
b
10
a b
11
14
15

Australian Species

Leptochiton (L.) liratus (H. Adams & Angas, 1864)

(Fig. 60. Map 20)

Neotype: SAMA D 13735.
Type locality: South Australia, Yorke Peninsula, Hardwicke Bay.

Lepidopleurus liratus H. Adams & Angas, 1864:192; Ashby, 1921a:155; 1923a: 218, pl. 16:fig. 1; 1928b:175.
Lepidopleurus inquinatus; Sykes, 1896:86, pl. 6:fig. 4; Bednall, 1897:141, et auct. Non *Chiton inquinatus* Reeve, 1847.
Lepidopleurus profundus May, 1923:pl. 14:fig. 2 (type: SAMA, May-collection); Ashby, 1927a:94. Non *Lepidopleurus profundus* Ashby, 1923a:221, pl. 16:figs 2, 2a [=*Leptochiton collusor* (Iredale & Hull, 1925)].
Terenochiton liratus; Iredale & Hull, 1925:342, pl. 39:fig. 4; Cotton & Godfrey, 1940:477, fig. 458; Macpherson & Gabriel, 1962:3; Cotton, 1964:17, fig. 5.
Parachiton profundus; Iredale & Hull, 1925:346, pl. 39:fig. 21.

Description. Animal rather small, up to 10 mm long, elongate oval, moderately elevated (dorsal elevation 0.43), subcarinated, valves not beaked, girdle narrow.

Head valve semicircular, posterior margin widely V-shaped, notched in the middle. Intermediate valves broadly rectangular, anterior margin jugally slightly excavated, side margins somewhat rounded, posterior margin straight, apices inconspicuous, side slopes evenly rounded, lateral areas decidedly raised. Tail valve somewhat more than semicircular, mucro at the posterior 1/3 of the valve, posterior slope about straight, only a little excavated behind the mucro.

Tegmentum uniformly granulose; the granules arranged in radiating rows on the end valves and the lateral areas of the intermediate valves, coalescing to longitudinal chains on the central areas, the chains wide apart in the pleural parts, the interstices latticed. There are c. 65 radiating rows of granules on the head valve, c. 12 on the lateral areas (by splitting), c. 36 on the central areas. There is a concentric ridge formation all around the shell, especially on the lateral areas, with 4-6 strong growth marks.

Colour of tegmentum yellowish to light ochraceous, with small, irregularly disposed white spots.

Articulamentum pinkish white, apophyses very small, triangular, wide apart.

Girdle dorsally covered with only slightly bent, long scales, 82 × 35 μm, with about 7 fine ribs, narrower than the interstices, the scales much larger along the sutures, 140 × 42 μm, and many, randomly disposed, long, smooth, pointed spicules, up to 368 μm long, 29 μm wide. There is a marginal fringe of longitudinally striated, slender spicules, 160 × 20 μm, embedded in short chitinous cups. Ventrally the girdle is paved with elongate, bluntly pointed, imbricating scales, 92 × 25 μm, the marginal ones somewhat larger, 105 × 32 μm.

Central tooth of the radula pinched in the middle, with a well developed cusp, first lateral tooth with a small outward bent cusp, major lateral with a strong bidentate cusp, the exterior denticle smaller.

Distribution. This littoral species has been found from Western Port, Victoria, west to Nuyts Archipelago, South Australia; near Fremantle and Normalup, Western Australia, and along the southeast coast of Tasmania.

Observations. Thanks to Miss Alison Green, Curator of Invertebrate Zoology at the Tasmanian Museum and Art Gallery, we received two specimens of *Lepidopleurus profundus* May, 1923 (TMAG E-3035) on loan. The specimens belonged to the collection of W. L. May, collected in 1910 in D'Entrecasteaux Channel, SE Tasmania, in a depth of 18 m. One of the specimens is strongly curled up, the other flat but broken along the median axis, 10.2 × 4.8 mm. A careful examination of both specimens and comparison with specimens of *L. liratus* leads us to the conclusion that no differences of any importance exist, neither in shape and sculpture of the valves, nor in the armature of the girdle. Consequently *L. profundus* May must be regarded as a synonym of *L. liratus*.

Leptochiton (L.) matthewsianus (Bednall, 1906)

(Fig. 61. Map 21)

Neotype: SAMA D13734.
Type locality: South Australia, Gulf St Vincent, Marino.

Lepidopleurus matthewsi Tate & May, 1901:412 (nom. nud.).
Lepidopleurus matthewsianus Bednall *in* Bednall & Matthews, 1906:92, pl. 9:figs 1, la-f; Ashby, 1923a:218, pl. 16:figs 5, 5a; Ashby, 1927a:93; Ashby & Cotton, 1930:58.
Lepidopleurus niger Torr, 1911:105, pl. 225:figs 5a-f (type: SAMA D11686); Ashby, 1923a:220, pl. 16:fig. 4.
Terenochiton erratus Hull, 1923a:159, pl. 24:figs 6-9 (type: AM); Iredale & Hull, 1925:341, pl. 39:figs 7-10.
Terenochiton matthewsianus; May, 1923:pl. 14:fig. 3; Iredale &Hull, 1925:341, pl. 39:fig. 3; Cotton, 1964:18, figs 6-7.
Terenochiton niger; Iredale & Hull, 1925:341; Cotton & Weeding, 1939:182; Cotton, 1964:18, fig. 8.

Description. Animal small, maximum 10 mm long, generally c. 6 mm long, the width about half the length, elongate oval, almost parallel sided, moderately elevated (dorsal elevation 0.34), round backed, with only the slightest indication of a carina, valves not beaked, girdle narrow. Colour of tegmentum pale buff, pinkisk white, or a drab grey.

Head valve semicircular, posterior margin widely V-shaped, notched in the middle. Intermediate valves broadly rectangular, fore and hind margins straight, apices inconspicuous, side slopes evenly convex, lateral areas hardly or not raised. Tail valve semicircular, anterior margin decidedly convex, mucro at posterior third of valve, pointed, not pronounced, posterior slope about straight with only a slight concavity directly behind the mucro.

Tegmentum uniformly sculptured with round, well separated granules, very little raised above the tegmental surface. The granules are arranged in radiating rows on the end valves and the lateral areas of the intermediate valves, in longitudinal rows, slightly posteriorly converging towards the sides, on the central areas. There are 65-70 anastomosing rows on the head valve, 8-10, sometimes more by splitting, on the lateral areas, c. 38 on the central areas.

Articulamentum whitish, apophyses very small, triangular, wide apart.

Girdle dorsally covered with closely packed, slightly curved, erect scales, only the 'heads' showing as 'half moons', 44 × 32 μm, the sutural one larger, 66 × 36 μm, with 10-12 strong ribs, the ribs as wide as the pitted interstices. There are intersegmental groups of 3-5 smooth, slender spicules, up to 160 μm long. Marginal fringe composed of longitudinally grooved, slender spicules, up to 140 μm long, with 5-6 grooves on the visible half of them. Ventrally the girdle is clothed with imbricating, obtusely pointed scales, 60 × 22 μm along the inner margin, 80 × 32 μm along the outer margin.

Gills merobranchial, adanal with interspace, 9-10 ctenidia on each side.

Central tooth of radula more or less rectangular, with a narrow blade, first lateral tooth rather slender, of the same size as the central tooth, major lateral strongly cusped, with one curved, sharply pointed denticle, spatulate uncinal tooth with a long, inwardly curved blade.

Distribution. *L. matthewsianus* is a subtidal species which has been found in many places along the southern coast of Australia, from Victoria to Albany, Western Australia; also along the northern coast of Tasmania.

Observations. *Lepidopleurus niger* Torr, 1911 does not differ from *matthewsianus*. After disarticulation of a dried, 5.8 mm long specimen of *'niger'* from Cape Jervis, Gulf St Vincent, S. Australia (B. J. Weeding leg., K 2528) the shape and granulation of the valves, the armature of the girdle, and the radula appeared to be exactly alike those of *matthewsianus*. The only difference is the black pigmentation of the animal, which is still obvious in dried specimens, whereas live *matthewsianus* is reddish (yellowish brown when dried).

We believe the black form to be a mutation of *matthewsianus,* either with black grains in the blood, or with black pigmented dermal tissue, thus of no importance from a taxonomical point-of-view.

Leptochiton (L.) columnarius (Hedley & May, 1908)

(Fig 62. Map 23)

Type: AM C-29060.
Type locality: Tasmania, off Cape Pillar, 183 m.

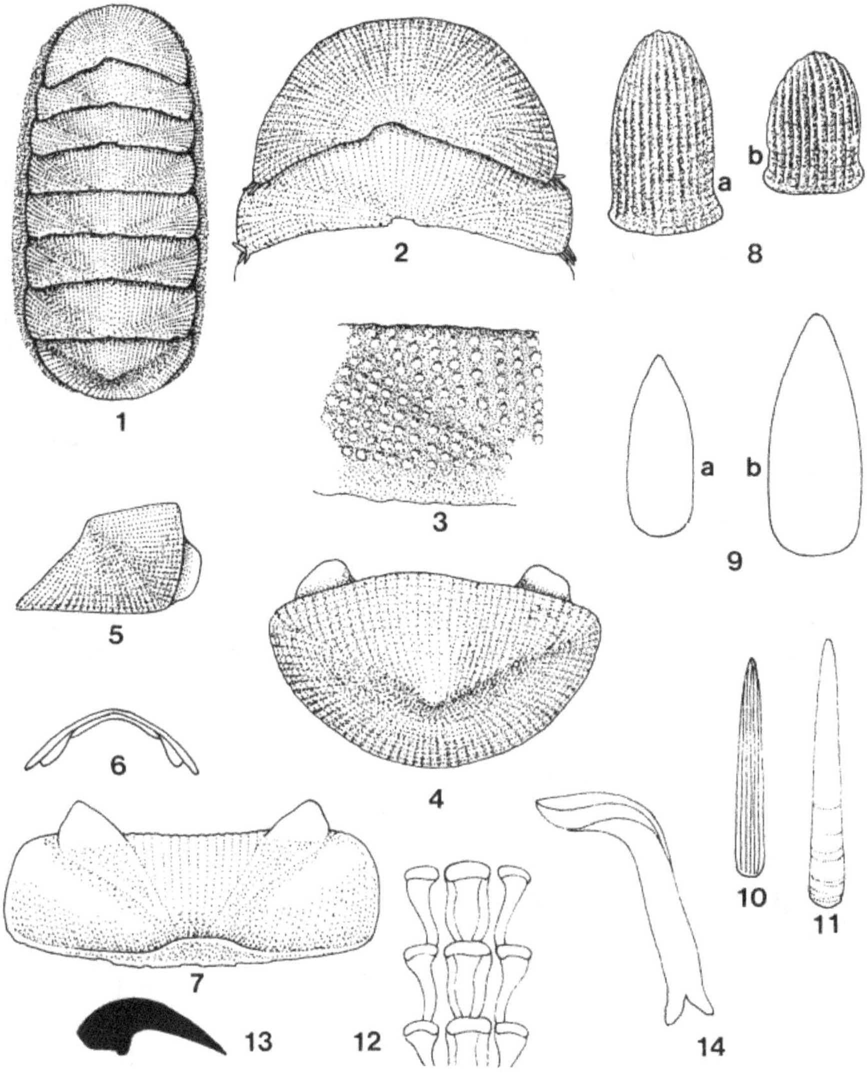

Fig. 61. *Leptochiton (L.) matthewsianus* (Bednall, 1906).

1, whole specimen, dorsal view, × 9.6; 2, valves I, II, in situ, dorsal view, × 20; 3, sculpture on tegmentum of valve IV, left half, × 40; 4, valve VIII, dorsal view, × 20; 5, do lateral view, × 20; 6, valve IV, anterior view, × 9.6; 7, do ventral view, × 20; 8, dorsal girdle scales: a sutural, b from mid-girdle, × 400; 9, ventral girdle scales: a from mid-girdle, b infra-marginal, × 400; 10, marginal spicule, × 200; 11, intersegmental spicule, × 200; 12, central and first lateral radula teeth, × 400; 13, cusp of major lateral tooth, × 400; 14, spatulate uncinal tooth, × 400.

1-14, specimen (5.4 × 2.75 mm) from Gulf St Vincent, S Australia, B. J. Weeding leg., K 2916.

Lepidopleurus columnarius Hedley & May, 1908:123, pl. 29:figs 27-28; Gatliff & Gabriel, 1917:24; Ashby, 1923a:219, pl. 16:figs 3, 3a-b; Kaas, 1977:85.
Lepidopleurus pelagicus Torr, 1912b:165, pl. 5:figs 2a-f; Ashby, 1923a:220, pl. 16: fig. 3b.
Parachiton columnarius; Iredale & Hull, 1925:345, pl. 39:figs 11-12; Cotton, 1964: fig. 14.
Parachiton pelagicus; Cotton & Weeding, 1939:183; Cotton, 1964:22, fig. 13.
Non: *Lepidopleurus columnarius;* Ashby, 1929b:372, pl. 32:fig. 9 [=*Leptochiton subantarcticus* (Iredale & Hull, 1930)].

Description. Animal rather small, length 8 mm, breadth 3-4 mm, elongate, highly to very highly arched (dorsal elevation varies from 0.58-0.86; relative breadth varies accordingly), more or less carinated, valves hardly or not beaked, girdle narrow. Colour of tegmentum waxen, pinkish or pale sulphur-yellow, the girdle having a slightly deeper shade.

Head valve semicircular. Intermediate valves roughly rectangular, somewhat obliquely truncated at the sides, anterior margin slightly concave in the centre, posterior margin almost straight, apices inconspicuous, side slopes slightly convex, lateral areas somewhat raised. Tail valve relatively small, the length 2/3 of the breadth, mucro at the posterior 1/4 of the valve, little elevated, posterior slope about straight, save for a small excavation directly behind the mucro.

Tegmentum uniformly, finely granulose. Central areas with c 64 longitudinal chains of coalescing granules, anastomosing at both sides of the jugum, posteriorly converging on the pleurae, interstices narrower than the chains. On the head valve, the lateral areas of the intermediate valves, and the post-mucronal area of the tail valve, the granules are as large as on the central areas, but less pronounced, without interspaces, quincuncially arranged, at the same time forming radiating rows (12-14 on the lateral areas).

Articulamentum whitish, apophyses small, triangular, wide apart.

Girdle dorsally clothed with elongate oval scales, with 6-7 ribs, interspersed with smooth, slender spicules (all broken in holotype), the largest (broken) one measuring 120 × 16 µm. There is a fringe of stout, bluntly pointed, longitudinally ribbed spicules, 100 × 24 µm. Ventral scales elongate, 4 × as long as wide, bluntly pointed, narrowing towards the base, 100 × 24 µm.

Distribution. *L. columnarius* has been dredged in deep water near the south-eastern coasts of Victoria and South Australia, in Bass Strait, and along the northern and eastern coast of Tasmania.

Observations. We were able to study a sample of 15 intermediate valves and one tail valve, dredged in Gulf St Vincent, S. Australia (? according to the late Rev. B. J. Weeding, who provided the material from the collection of W. G. Torr). A small strip of paper, accompanying this material, reads in pencil: "sta. 9. 300 fms", so we assume that the valves once belonged to the "several valves ... dredged from 300 fathoms off the south-east coast of South Australia" (Torr, 1912:166).

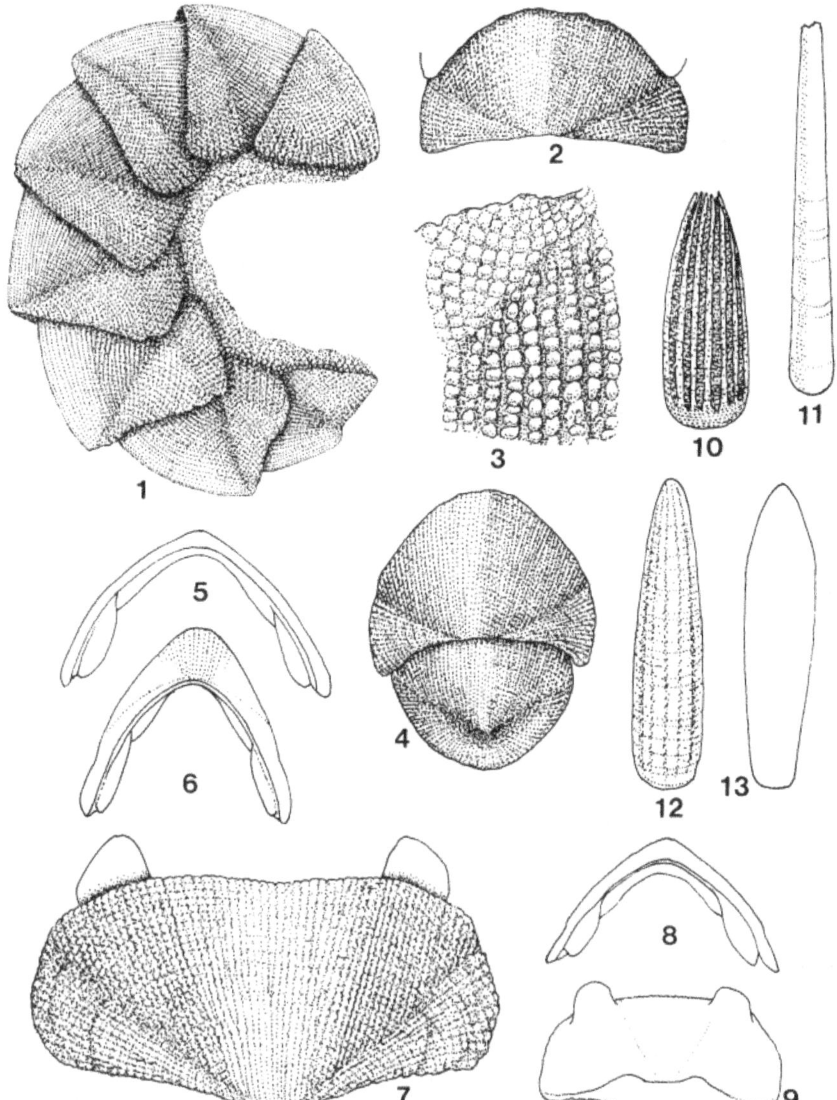

Fig. 62. *Leptochiton (L.) columnarius* (Hedley & May, 1908).

1, whole specimen, lateral view, × 9.6; 2, valve IV, in situ, dorsal view, × 9.6; 3, detail of tegmental sculpture on valve IV, × 40; 4, valves VII and VIII, in situ, dorsal view, × 9.6; 5, valve IV, anterior view, × 9.6; 6, intermediate valve, anterior view, × 12; 7, another loose valve, dorsal view, × 20; 8, do, anterior view, × 10; 9, do, ventral view, × 10; 10, dorsal girdle scale, × 400; 11, dorsal spicule, × 400; 12, marginal spicule, × 400; 13, ventral scale, × 400.

1-5, 10-13, holotype; 6, Gulf St Vincent, S Australia, 548 m, W. G. Torr leg., one of 16 loose valves, labelled "*Lepidopleurus pelagicus* Torr"; 7-9, another loose valve, do, K 2826.

The single tail valve is exactly alike that of the holotype of *columnarius*, except for being a trifle smaller and the posterior slope being a little concave. The position of the mucro, and the number of granulose lirulae on the antemucronal area, however, are exactly in accordance with those of *columnarius*. The dorsal elevation of the intermediate valves varies from 0.54 - 0.86 (in the holotype of *columnarius* 0.58), the number of granulose lirulae on the central area varies from 62 - 68 (in *columnarius* 64).

In our opinion Cotton & Weeding (1939:183) were misled by Torr's questionable illustrations when they restored Torr's name, which had been synonymized (correctly, in our opinion) with *columnarius* by Gatliff & Gabriel (1917:24). Iredale & Hull (1925:345) suppose that Torr's shell might prove to be a subspecies of *columnarius*.

We prefer to regard *columnarius* and *pelagicus* as conspecific, until more material from the Flindersian as well as from the Peronian region may turn up and disprove this opinion.

Leptochiton (L.) badius (Hedley & Hull, 1909)

(Fig. 63. Map 23)

Type: AM C-30459.
Type locality: New South Wales, Port Jackson, Balmoral and Shark Island.

Lepidopleurus badius Hedley & Hull, 1909:260, pl. 73:figs 1-2; Ashby, 1918c:80; Ashby 1924a:314; Ashby, 1927a:93.
Terenochiton badius; Iredale & Hull, 1925:340, pl. 39:figs 1-2; Macpherson & Gabriel, 1962:3, fig. 8.
Lepidopleurus glauerti Ashby, 1929c:50, fig. 7 (Type: WAM 12876); Wells, 1977: 34.
Terenochiton iscus Cotton & Weeding, 1939:182, pl. 7:fig. 1 (Type: SAMA D 1232); Cotton & Godfrey, 1940:478, fig. 459; Cotton, 1964:19, fig. 9.
Terenochiton glauerti; Cotton, 1964:20, fig. 10.

Description. Animal small, up to 6 × 3.5 mm, elongate oval, almost parallel-sided, moderately elevated (dorsal elevation 0.33), subcarinated, valves thin, not beaked, girdle narrow, scaly with interspersed needles. Colour normally white, mostly yellowish brown by foreign deposits.

Head valve almost as broad as the intermediate valves, semicircular, posterior margin widely V-shaped, notched in the middle. Intermediate valves broadly rectangular, side margins evenly rounded, hind margin straight, not umbonated, side slopes slightly convex, lateral areas little raised, hardly differentiated from the central areas. Tail valve as wide as the head valve, more or less oval, mucro antemedian, pointed but not prominent, postmucronal slope straight.

Tegmentum uniformly granulose, the roundish, only little elevated granules arranged in an irregular quincuncial pattern on the head valve, the lateral areas of the intermediate valves, and the postmucronal area of the tail valve, in 35-40 longitudinal, sometimes anastomosing rows on the central areas, the rows slightly curved on the pleurae and posteriorly converging near the outer sides.

Articulamentum white, weakly developed, apophyses widely separated by a more or less flat sinus, convex in valves II and VIII.

Girdle dorsally clothed with imbricating, strongly curved, round-topped scales, measuring 40 μm at the base, length 25 μm, sutural scales somewhat longer, up to 40 μm, all scales ornamented with 15-16 fine riblets, the inter-

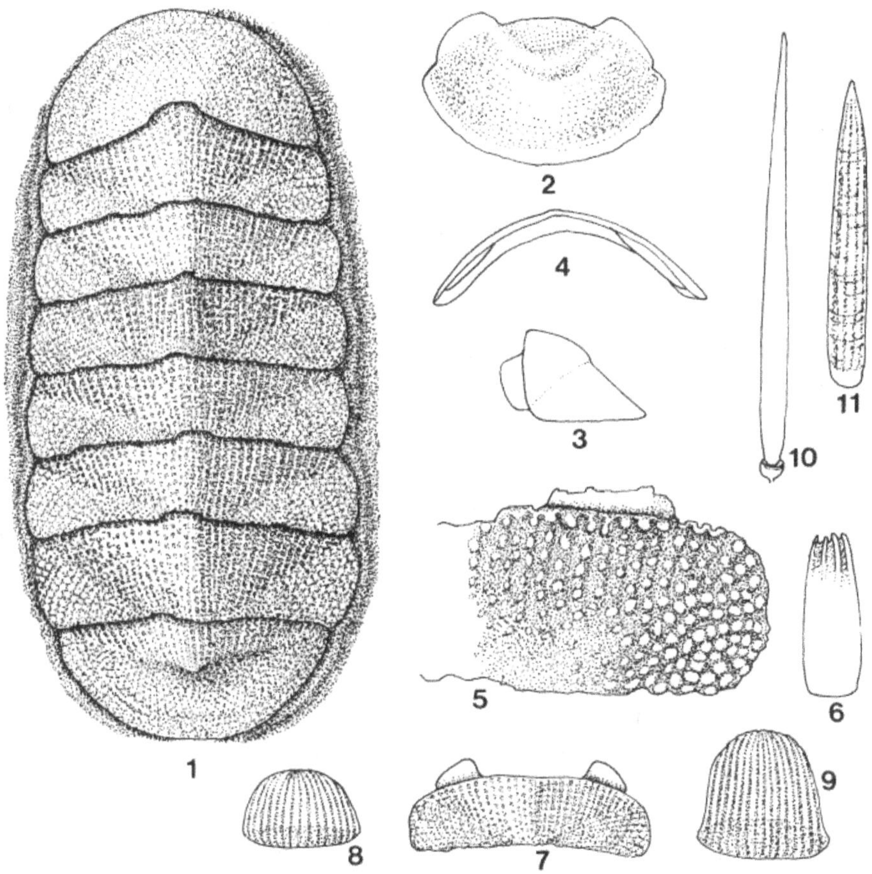

Fig. 63. *Leptochiton (L.) badius* (Hedley & Hull, 1909).

1, whole animal, dorsal view, × 20; 2, valve VIII, ventral view, × 20; 3, do, lateral view, × 20; 4, valve VII, anterior view, × 20; 5, do, dorsal view of right half, × 40; 6, ventral girdle scale, × 400; 7, valve V, dorsal view, × 20; 8, dorsal scale from mid-girdle, × 400; 9, do, along the suture, × 400; 10, dorsal needle, × 400; 11, marginal spine, × 400.

1-6, specimens from Long Reef, N.S.W., Australia, T. Iredale leg., K 2259; 7-11, specimens from Arno Bay, Spencer Gulf, S Australia, B. J. Weeding leg., labelled *"Terenochiton iscus* Cotton & Weeding, 1939", K 2260.

stices as wide as the riblets. In good specimens sutural tufts can be observed of several smooth, straight needles, varying in length from 144 to 164 μm, embedded in minute chitinous cups. There is a marginal fringe of straight, longitudinally coarsely ribbed spicules, 100 × 15 μm. In many specimens, the dorsal needles and the marginal spicules are partly or completely broken off. Ventrally the girdle is covered with elongate scales, about 52 × 20 μm, sculptured with 4-5 strong ribs near the truncated top.

Dimensions: 6 × 3.5 mm (Hedley & Hull, 1909); from 3 × 1.5 to 5 × 2.5 mm (Ashby, 1924); 3.4 × 2.5 mm, 4.8 × 2.5 mm (figured specimen, K 2259).

Distribution. *L. badius* lives in shallow water along the coasts of Australia, from Point Cartwright, Queensland, to Spencer Gulf, S. Australia, and also near Robbins Island, NW. Tasmania, W. to Rottnest Id., W. Australia.

Observations. Towards the western range of its distribution, the species becomes even smaller, up to 5 × 3 mm, and there are a slightly smaller number of longitudinal rows on the central areas, with somewhat less pustules to a row, while the ventral scales are more weakly sculptured. This form was considered a different species, *Terenochiton iscus,* by Cotton & Weeding (1939: 182). After a thorough study of several specimens we conclude that *T. iscus* is conspecific with, *L. badius* and can only be considered an extreme form of this.

Thanks to Dr Fred E. Wells, Head of the Department of Malacology of the Western Australian Museum, a paratype of *Lepidopleurus glauerti* Ashby, 1929, (WAM N-12875), could be studied. It is a strongly curled-up specimen from Rottnest Island, W. Australia, preserved in alcohol, estimated length 4.5 mm. Apparently the specimen had been dried before it was put into alcohol, as the soft parts of the animal are quite shrunk. As a result almost all the shell plates are broken. Nevertheless we are certain that it is merely a local form of *badius,* only differing in being a trifle more elevated, and in having the granules slightly more raised, which may well fall within the range of specific variation.

Leptochiton (L.) liratellus (Iredale & Hull, 1925)

(Fig. 64. Map 20)

Type: probably lost.
Type locality: South Queensland, Port Cartwright.

Terenochiton liratellus Iredale & Hull, 1925:342, pl. 39:fig. 5, MacKay, 1930:289, 291, 293.

Description. "Like a miniature *T. liratus;* shell small, elongate oval, moderately elevated, subcarinated. Colour pink.

Anterior valve finely closely rayed with about fifty pustulose ribs. Median

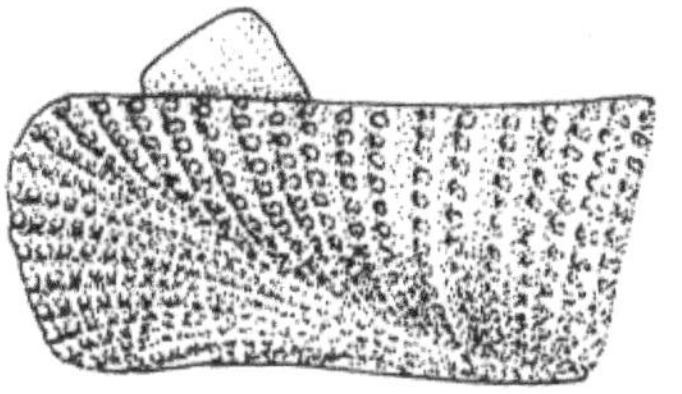

Fig. 64. *Leptochiton (L.) liratellus* (Iredale & Hull, 1925).

Left half of an intermediate valve, approximately × 30. After Iredale & Hull, 1925.

valves: lateral areas well differentiated and rayed similarly to the anterior valve; central areas strongly sculptured with fifteen longitudinal rows of pustules at each side, more distant and widely separated towards the margin, massed on jugum, but still distinct. Posterior valve with mucro ante-median; post-mucronal area concave, sculptured as anterior valve, rays distinct; ante-mucronal area like central areas.

Interior: pinkish white.

Girdle: covered with minute scales.

Dimensions: 5.5 × 3 mm."

Distribution. Port Cartwright in S Queensland, under stones on seaward side of rocky escarpment, below low water mark (Hull); Caloundra, Coolum, and Keppel Bay, Queensland (MacKay).

Observations. According to Iredale & Hull the type should be stored in the Queensland Museum. We were informed by Dr John Stanisic, curator of mollucsc at this institution, that there is no record of the type of *L. liratellus* in the museum register. "The surprising thing is that there are a number of other Iredale & Hull types present in our collection. These are all registered and were donated by Bassett-Hull at various times early this century" (J. S. in litt., 02.10.1981).

That's why we could do no better than to reproduce the original, description and figures, by which we hope the species may be recognizable.

MacKay (1930:290) found small sutural tufts of 3-4 long, curved, pointed, glassy spicules, which "are in no way to be confused with the spiculose edge of the girdle, which is quite distinct".

Leptochiton (L.) sperandus (Iredale & Hull, 1925)

(Fig. 65. Map 20)

Type: AM C126815.
Type locality: New South Wales, off Ulladulla, 126 m.

Terenochiton sperandus Iredale & Hull, 1925:342, pl. 39:fig. 6.

Description. Animal small, 4.5 × 2.8 mm, elongate oval, almost parallel-sided, moderately elevated (dorsal elevation 0.31), subcarinated, valves not beaked, girdle scaly-spiculose.

Head valve almost as wide as the intermediate valves, semicircular, posterior margin very widely V-shaped, notched in the middle. Intermediate valves broadly rectangular, side margins more or less rounded, posterior margin straight, apices inconspicuous, side slopes slightly convex, lateral areas raised. Tail valve somewhat smaller than the others, mucro elevated, antemedian, postmucronal slope concave.

Tegmentum uniformly sculptured with rather coarse, flat-topped, cylindrical granules, the granules arranged in c. 55 radiating, often splitting, rays on the head valve, 8 on the lateral areas of the intermediate valves, interrupted by a few irregular growth marks, and 30 on the postmucronal area of the tail valve. On the central areas there are 26-28 longitudinal rows of granules, the 8-10 jugal ones narrow and close together, the pleural ones wide apart.

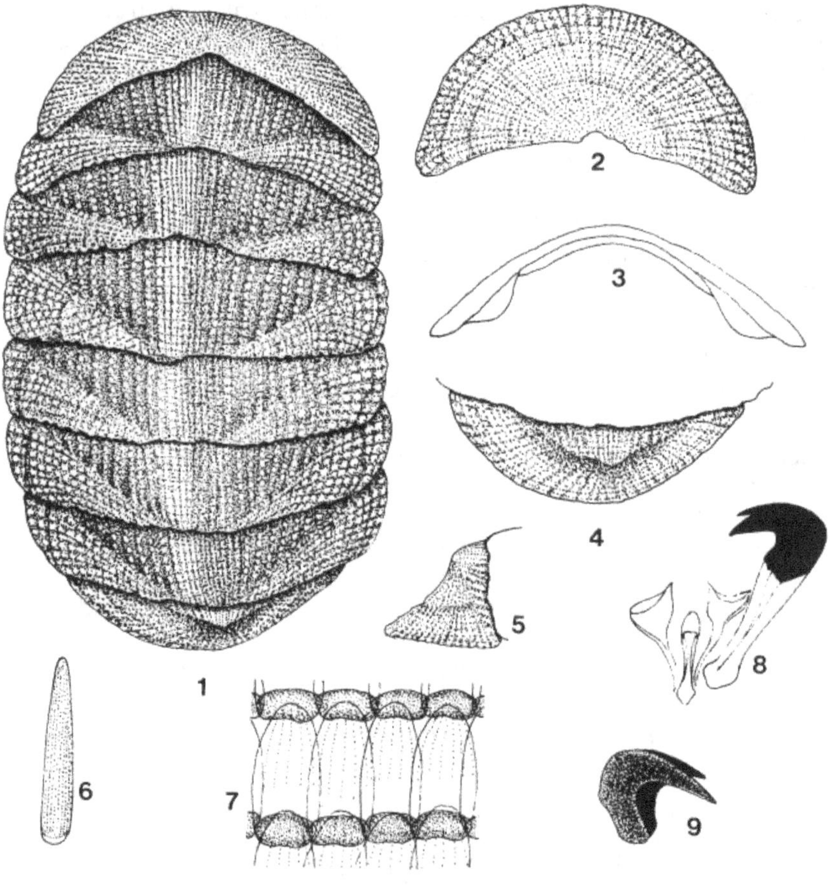

Fig. 65. *Leptochiton (L.) sperandus* (Iredale & Hull, 1925).

1, whole animal, dorsal view, × 19; 2, valve I, dorsal view, × 19; 3, valve IV, anterior view, × 19; 4, ventral VIII, in situ, dorsal view, × 19; 5, do, lateral view, × 19; 6, dorsal girdle spicule, × 190; 7, ventral girdle scales, × 375; 8, central, first lateral and major lateral radula teeth, × 190; 9, cusp of major lateral radula tooth, × 225.

1-9, holotype.

Colour of tegmentum yellowish to brownish, possibly by environmental deposits.

Perinotum dorsally clothed with small, striated scales and smooth spicules, up to 128 μm long. Ventrally the girdle is covered with 9 concentric rows of elongated, finely striated, imbricating scales, 54 × 24 μm, the base truncated, distally obtusely pointed.

Radula with a narrow central tooth, bearing a weakly developed, rounded cusp, first lateral tooth very narrow at the base, abruptly widening distally, major lateral with a bidentate cusp, the denticles sharply pointed, the outer one smallest.

Distribution. Only known from the type locality.

Leptochiton (L.) collusor (Iredale & Hull, 1925)

(Fig. 66. Map 16)

Type: SAMA D 11288.
Type locality: South Australia, Gulf St Vincent, dredged.

Lepidopleurus profundus Ashby, 1923a:221, pl. 16:fig. 2, 2a.
Non: *Lepidopleurus profundus* May, 1923.
Parachiton collusor Iredale & Hull, 1925:346, pl. 39:fig. 22; Cotton & Godfrey, 1940:481, fig. 463; Macpherson & Gabriel, 1962:3; Cotton, 1964:23, fig. 15; Kaas, 1977:85.

Description. Animal of rather large size for the genus, up to 20 mm long, elongate, width less than half the length, elevated (dorsal elevation c. 0.45), semicarinated, valves not beaked, colour pinkish or varying from creamish to deep brown, girdle narrow, spiculose.

Head valve raised, broad, closely rayed with about 100 very narrow, radial rows of minute, coalescing granules, and marked with 3-4 concentric growth lines near the margin.

Intermediate valves rectangular, front margin slightly concave in the central part, evenly convex on the sides, posterior margin straight, apices inconspicuous, side slopes a little convex, lateral areas raised, decorated with about 20 radiating ribs similar in character to those of the anterior valve, and crossed by up to 9 concentric growth lines. Central areas sculptured with longitudinal rows of semi-fused, polished, flattened granules, the rows very close together on the jugal area, posteriorly converging and getting further apart on the pleural areas, until the interspaces are twice the width of the ribs.

Tail valve large, semicircular, mucro distinct, submedian, post-mucronal slope slightly concave, antemucronal area sculptured like central part of intermediate valves, postmucronal area like head valve.

Articulamentum white, apophyses small, wide apart and triangular in the intermediate valves, trapezoid in the tail valve, sinus concave.

Girdle dorsally covered with almost smooth, elongate, needle-like spicules of c. 170 × 20 μm, ventrally with elongate, faintly longitudinally striated spicules of 96 × 16 μm.

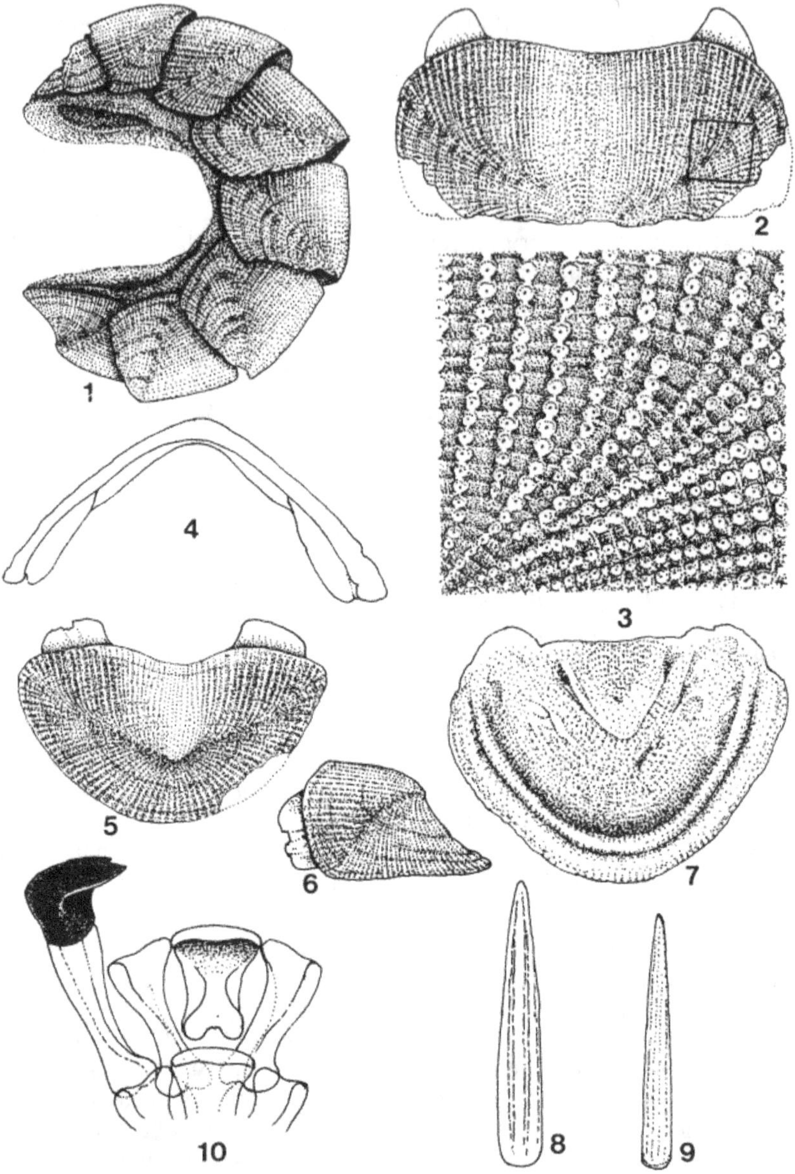

Fig. 66. *Leptochiton (L.) collusor* (Iredale & Hull, 1925).

1, whole animal, dried and curled up, lateral view, × 4.8; 2, valve IV, slightly damaged, dorsal view, × 9.6; 3, detail of sculpture on valve IV (the black square in 2), × 40; 4, camera lucida sketch of valve IV, anterior view, × 9.6; 5, valve VIII, dorsal view, × 9.6; 6, do, lateral view, × 9.6; 7, do (another specimen), ventral view, × 9.6; 8, ventral girdle spicule, × 400; 9, dorsal girdle spicule, × 200; 10, central, first lateral and major lateral radula teeth, × 100.

1-10, specimens from Gulf St Vincent, S Australia, SAMA D12159.

Central tooth of radula short, squarish, bearing a well developed flat cusp, first lateral tooth somewhat narrowed in the middle, not cusped, major lateral strongly cusped, with a curved, sharply pointed main denticle and a smaller interior one.

Distribution. *L. collusor* seems to be distributed along the coasts of South Australia and Victoria, in depths of 16 to 41 m.

New Zealand Species

Leptochiton (L.) inquinatus (Reeve, 1847)

(Fig. 67. Map 27)

Type: BMNH 42.11.7.99-102.
Type locality: Van Diemen's Land (in error, = New Zealand, Auckland Harbour).

Chiton inquinatus Reeve, 1847b:pl. 23 spec. and fig. 154.
Lepidopleurus iredalei Ashby, 1921a:157, pl. 8:figs 3a-b; Ashby, 1923a:217.
Terenochiton inquinatus; Iredale & Hull, 1929:76, pl. 9:figs 3-6 (bibliography and synonymy); Powell, 1979:22, pl. 5:fig. 1.
Non: *Lepidopleurus inquinatus;* Sykes, 1896, et auct = *Leptochiton liratus* (H. Adams & Angas, 1864).

Description. Animal of moderate size, up to 16 mm long, width about half the length, elongate oval, parallel-sided, little elevated (dorsal elevation 0.29), back evenly rounded, not carinated, valves not beaked, lateral areas decidedly raised, tegmentum uniformly granulose, colour yellowish brown to chestnut, girdle narrow.

Head valve semicircular, posterior margin very widely V-shaped, slightly notched in the middle, sculptured with numerous radiating rows of extremely fine granulose lirae and 6-8 strong, concentric growth marks.

Intermediate valves broadly rectangular, side margins evenly rounded, front and hind margins straight and parallel, apices inconspicuous, lateral areas sculptured like head valve, the granulose lirae 25-30 in number, central areas with 60 somewhat stronger, longitudinal granulose lirae, posteriorly converging towards the outer margins.

Tail valve semicircular, mucro slightly swollen but not prominent, a little behind the centre, postmucronal slope more or less straight, postmucronal area sculptured like head valve, antemucronal area like central areas of intermediate valves.

Articulamentum whitish, thin, the rows of granules on tegmentum shining through in jugal parts, apophyses small, triangular, trapezoid in valve VIII, sinus finely dentate by tegmental longitudinal lirulae.

Girdle dorsally clothed with white, closely imbricating, curved scales, with 8-12, generally 10, strong, narrow riblets, narrower than the interspaces; the mid-girdle scales are 112 μm broad, 100 μm high, the sutural ones are somewhat narrower and longer: width 80 μm, length 112 μm, while those near the outer margin are wider and shorter: width 120 µm, length 80 µm. There are intersegmental groups of 5-8 stout, gothic candle-shaped, finely longitudinally striate spicules, closely ringed at the base, 210 × 32 µm, surrounded by a few smaller, strongly distally ribbed spicules, 105 × 20 µm, embedded in short chitinous cups. Along the margins 2 types of conical spi-

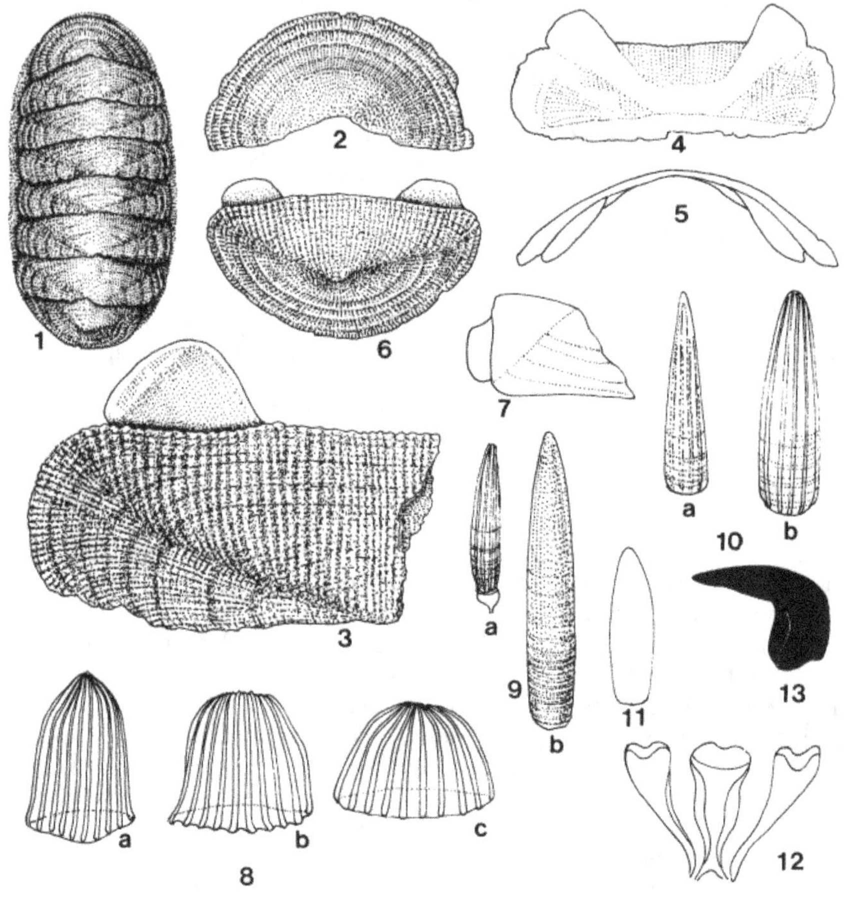

Fig. 67. *Leptochiton (L.) inquinatus* (Reeve, 1847).

1, whole specimen, dorsal view, × 4.5; 2, valve I, dorsal view, × 9; 3, left half of valve IV, dorsal view, × 19; 4, valve IV, ventral view, × 9; 5, do, anterior view, × 9; 6, valve VIII, dorsal view, × 9; 7, do, lateral view, × 9; 8, dorsal girdle scales, a sutural scales, b from mid-girdle, c supra-marginal, × 190; 9, intersegmental girdle spicules, a small, strongly ribbed one, in chitinous sheath, b large, finely striated one, × 190; 10, marginal spicules, a finely ribbed type, b strongly ribbed type, × 190; 11, ventral girdle scale, × 190; 13, cusp of major lateral radula tooth, × 190.

1-3, 6-7, 12-13, specimens from New Zealand, B. J. Weeding don., K 2263; 4-5, 8-11, specimen from New Zealand, Auckland region, J. Coles leg., VB 2672a.

cules can be noted: sharply pointed, finely ribbed ones, 144 × 32 μm, and bluntly pointed, more strongly ribbed ones: 152 × 44 μm. Ventrally the girdle is covered with smooth, elongated, bluntly pointed scales, 112 × 32 μm, shorter at the inner, longer at the outer margin.

Central and first lateral teeth of radula broadening to a more or less strong blade, conspicuously notched in the middle, major lateral with a strong cusp, bearing one sharply pointed, curved denticle.

Dimensions of lectotype: 9.5 × 4.5 mm. Largest specimen observed: 16 × 7.5 mm (VB 2672b).

Distribution. Intertidal and on continental shelf, all around New Zealand.

Observations. The type set comprises 3 specimens mounted on a tablet and 1 specimen in a glass tube, valves VII and VIII loose (illustrated by Reeve). The specimen on the right side of the tablet is not a *L. inquinatus,* but a juvenile *Ischnochiton* specimen, probably *I. circumvallatus* (Reeve, 1847). The specimen on the left, the largest one in the series, is here designated as the lectotype, the one in the middle and the one in the tube, as paralectotypes.

Leptochiton (L.) norfolcensis (Hedley & Hull, 1912)

(Fig. 68-69. Map 27)

Type: AM C-33111 (one complete specimen, 5.75 × 3.7 mm, lectotype + 1 disarticulated paralectotype).
Type locality: Norfolk Island.

Lepidopleurus norfolcensis Hedley & Hull, 1912:273, pl. 11:figs 1, 1a-c.
Terenochiton norfolcensis; Iredale & Hull, 1929:77, pl. 9:figs 7-10.
Lepidopleurus catenatus Hedley & Hull, 1912:273, pl. 11:figs 2, 2a-c (Type: AM).
Terenochiton catenatus; Iredale & Hull, 1929:78, pl. 9:figs 14-17.
Lepidopleurus (Terenochiton) subtropicalis Iredale, 1914:28, pl. 2:figs 10, 17 (Holotype: CM; paratype: RMNH 55486).
Terenochiton subtropicalis; Iredale & Hull, 1929:78, pl. 9:figs 11, 12.

Description. Animal small, up to 6.5 × 4 mm, elongate oval, almost parallel-sided, moderately elevated (dorsal elevation 0.32), subcarinated, valves not beaked, girdle rather wide, scaly with intersegmental and marginal spicules.

Head valve almost as broad as the intermediate valves, semicircular, anterior slope slightly concave, especially towards the apex, posterior margin widely V-shaped, notched in the middle. Intermediate valves broadly rectangular, side margins evenly rounded, hind margin straight, apices inconspicuous, side slopes almost straight, lateral areas little raised, hardly differentiated from the central areas. Tail valve rather small, oval, length width = 0.5, mucro submedian, postmucronal slope deeply concave.

Tegmentum uniformly granulose; the small, cylindrical, flat-topped granules arranged in concentric as well as in radiating rows on the end valves and

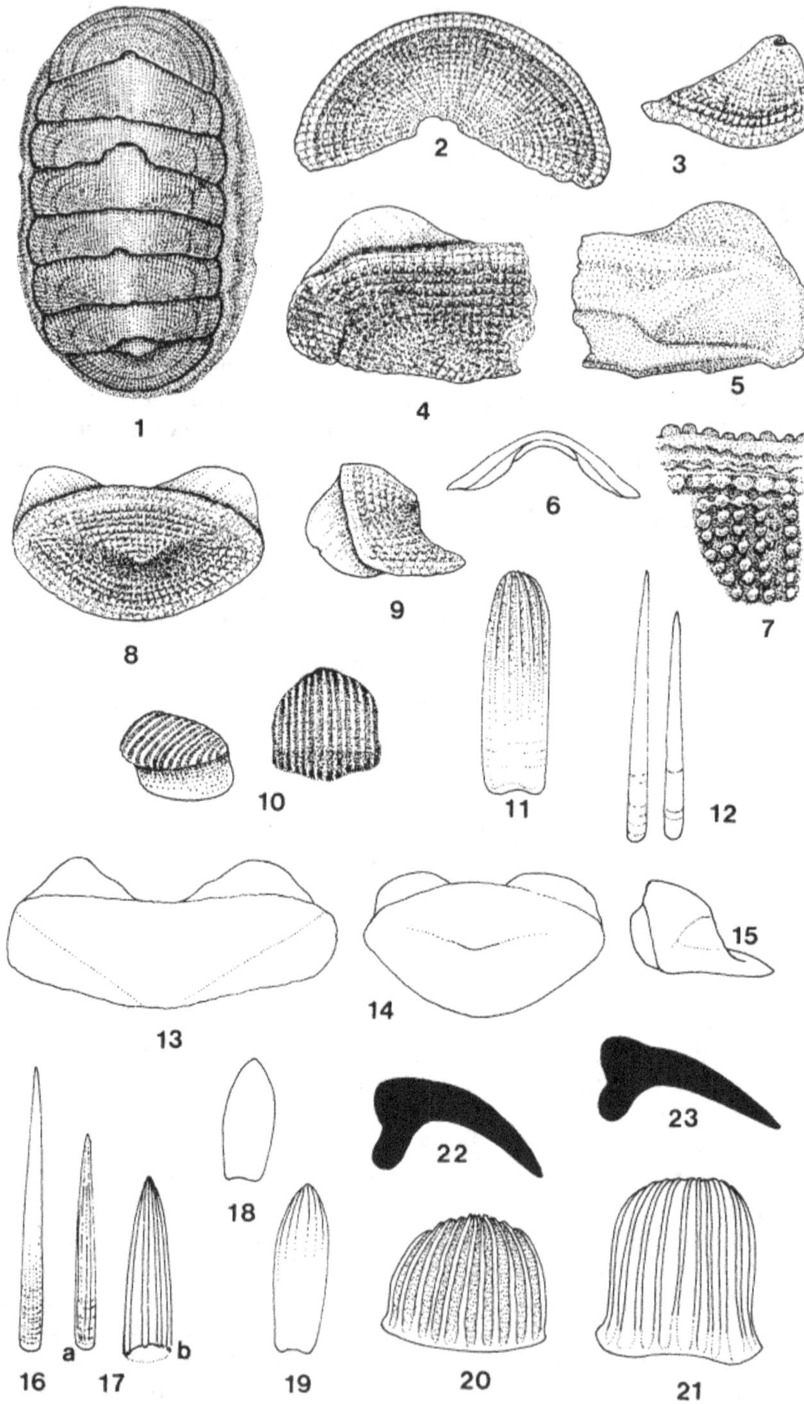
1
2
3
4
5
6
7
8
9
10
11
12
13
14
15
16
a
17
b
18
19
20
21
22
23

the lateral areas of the intermediate valves, in concentric and at the same time longitudinal series on the central areas, the latter posteriorly converging at the sides. There are c. 64 radiating rows of granules on the head valve, 8 on the lateral areas, while the longitudinal series on the central areas are about 25 in number, the interstices finely latticed, especially towards the front margin. On all valves the granules are growing larger towards the anterior and side margins. On the end valves and lateral areas a few irregular growth marks can be observed.

Articulamentum white, strongly developed, thickened along the side margins, markedly turned over at the posterior margin, apophyses short, evenly rounded, unusually wide for the genus, the sinus proportionally narrow, shallow.

Colour of tegmentum whitish, the exposed parts yellowish or brown by environmental deposits.

Girdle dorsally covered with strongly bent, short scales, 40 × 40 µm, with 10-12 fine ribs, narrower than the interstices, and intersegmental groups of 2-4 delicately striated needle-like spicules, up to 200 µm long. There is a marginal fringe of 150 µm long, slender spicules, the visible half marked with 4 fine but distinct ribs. Ventrally the girdle is paved with concentric rows of imbricating, elongate scales, the base truncated, distally bluntly pointed, the infra-marginal ones 80 × 24 µm, with 5-6 weak riblets on the distal half.

Cusp of major lateral radula teeth long, unidentate, sharply pointed.

Distribution. Only known from the type localities: Norfolk Island, "under loose stones at low tide" for *L. norfolcensis,* Lord Howe Island, "on the underside of smooth stones" for *catenatus,* and Sunday Island, Kermadec Group, "on the underside of embedded dirty stones below low water mark" for *subtropicalis.*

Observations. We have been able to compare the lectotype and one paralectotype of *norfolcensis* with three dried specimens (one disarticulated) of *catenatus* (J. R. Penprase leg. in VB, 2886a) from Dawson Point, W coast of

Fig. 68. *Leptochiton (L.) norfolcensis* (Hedley & Hull, 1912).

1, whole animal, dorsal view, × 9; 2, valve I, dorsal view, × 19; 3, do, lateral view, × 19; 4, left half of intermediate valve, dorsal view, × 19; 5, do, ventral view, × 19; 6, valve IV, anterior view, × 9; 7, sculpture on central area of intermediate valve, × 38; 8, valve VIII, dorsal view, × 19; 9, do, lateral view, × 19; 10, dorsal girdle scales, × 375; 11, ventral girdle scales, × 375; 12, intersegmental girdle spicules, × 190; 13, camera lucida sketch of valve III, dorsal view, × 19; 14, valve VIII, dorsal view, × 19; 15, do, lateral view, × 19; 16, intersegmental spicule, × 190; 17, a marginal spicule, × 190, b top of same, × 600; 18, ventral girdle scale on central part of the girdle, × 375; 19, do, infra-marginal, × 375; 20, dorsal girdle scale on central part of girdle, × 375; 21, do, sutural, × 375; 22-23, cusps of major lateral radula teeth, × 750.

1, 10-12, holotype; 2-9, paratype; 13-23, specimen from Dawson Point, Lord Howe Id, J. R. Penprase leg., labelled *"Lepidopleurus catenatus* Hedley & Hull, 1912", VB 2886a.

Lord Howe I., and one paratype of *subtropicalis* (H. B. Preston, 1912, RMNH 55486). The shape and sculpture of the valves of the three nominate species, obviously occupying the same ecological niches, do not differ substantially. Conchologically they are absolutely alike. There is only a difference in the dorsal girdle scales. We counted 12 narrow, sharp riblets (occa-

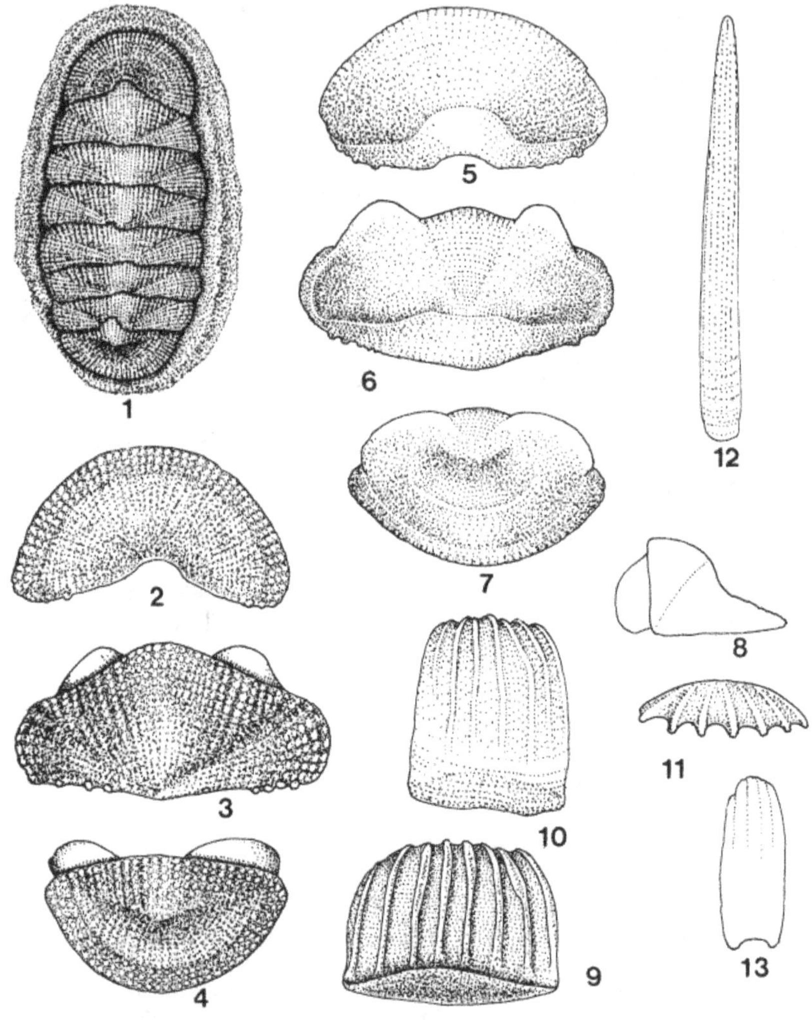

Fig. 69. *Leptochiton (L.) norfolcensis subtropicalis* (Iredale, 1914).

1, whole specimen, dorsal view, × 9.6; 2-4, valves I, II, VIII respectively, dorsal view, × 20; 5-7, do, ventral view, × 20; 8, valve VIII, lateral view, × 20; 9, dorsal scale, from mid-girdle, × 400; 10, do, sutural, × 400; 11, do, view from the top, × 400; 12, intersegmental spicule, × 400; 13, ventral girdle scale, near outer margin, × 400.

1-13, paratype.

sionally 11 or 10) on the scales of *norfolcensis* and *catenatus,* only 8 (occasionally 7 or 9) on those of *subtropicalis* (fig. 69).

The type of *norfolcensis* is an old, dried specimen and shows no marginal spicules. These may well be lost, as Hedley and Hull in their original diagnosis did mention "scales and spicules, the latter developing a marginal fringe". The authors stated for *catenatus:* "Girdle narrow, beset with minute scales," but did not mention the presence of intersegmental and marginal spicules, which are, however, clearly present in the specimens we could study.

Until we are able to study more material from the three localities, we prefer to regard *subtropicalis* as a subspecies of *norfolcensis,* only differing in the ribbing of the dorsal girdle scales.

Leptochiton (L.) otagoensis (Iredale & Hull, 1929)

(Fig. 70. Map 28)

Type: ?
Type locality: New Zealand, Otago, near Cape Saunders.

Terenochiton otagoensis Iredale & Hull, 1929:77, pl. 9:fig. 13; Beu, 1977:39, fig. I; Powell, 1979:22; Rainer, 1981:18.

Description. "Shell very small, elongate oval, elevated, semi-keeled, side slopes curved, girdle scaly. Colour reddish-buff, darker medially.

Sculpture consists of minute pustules, rayed on end valves, and lateral areas of median valves, and forming strings longitudinally on the pleural areas. On the anterior valve fifty closely packed radials may be counted and about forty on the postmucronal area of the posterior valve; twelve radials are visible on the lateral areas, while about fifteen longitudinal chains of pustules are easily seen on each side of the jugum, closer packed jugally and widely raying laterally. The mucro is at the anterior third, the post-mucronal slope a little concave.

Girdle scales minute, oval, striate.

Dimensions: length 7 mm, breadth 3.5 mm (type)" (Iredale & Hull, 1929).

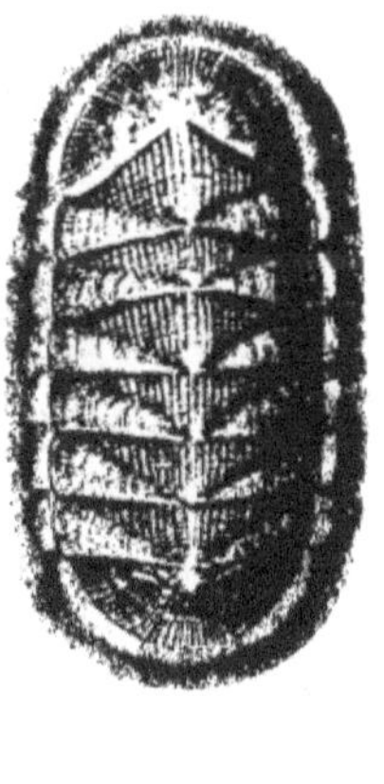

Fig. 70. *Leptochiton (L.) otagoensis* (Iredale & Hull, 1929).

Whole specimen, dorsal view, approximately × 6. After Iredale & Hull, 1929.

Distribution. Otago, New Zealand: near Cape Saunders (type) and Shag Point (Iredale & Hull); intertidal and subtidal on the shelf, Otago to Stewart Id. (Beu).

Observations. Despite our efforts we have not been able to locate the type, nor could we procure any specimens of this species. Powell (1979:22) states that it is very similar to *finlayi* in shape and sculpture, except that the dorsal area is not so acutely carinated and there are no projecting spine-like granules along the posterior edge of the intermediate valves.

Leptochiton (L.) fairchildi (Iredale & Hull, 1929)

(Fig. 71. Map 28)

Lectotype: AM C 46625.
Type locality: New Zealand, Macquarie Island.

Lepidopleurus kerguelenensis (emend); Hedley, 1916:34. Non *Leptochiton kerguelensis* Haddon, 1886.
Terenochiton fairchildi Iredale & Hull, 1929:79, pl. 9:figs 18-19; Cotton, 1937:12; Dell, 1964a:268; Powell, 1979:21.

Description. Animal rather small, up to 8 mm long, twice as long as wide, moderately elevated (dorsal elevation c. 0.40), valves solid, subcarinated, slightly beaked, side slopes convex, lateral areas little raised, tegmental sculpture obsolete, colour whitish, girdle rather wide.

Head valve semicircular, posterior margin widely V-shaped, tegmentum obsoletely sculptured with numerous very weak radiating ribs and several concentric growth marks. Intermediate valves broadly rectangular, anterior margin straight, side margins evenly rounded, lateral areas sculptured like head valve, central areas with a hardly discernible longitudinal ribbing, posteriorly converging near the outer sides. Tail valve semicircular, mucro submedian, clearly indicated, postmucronal slope decidedly concave directly behind the mucro, straightening near the posterior margin, antemucronal area sculptured like central area of intermediate valves, postmucronal area like head valve.

Articulamentum strongly developed, white, apophyses more or less triangular, widely separated by a straight sinus.

"Girdle scales small" (Iredale & Hull, 1929).

Dimensions: "... largest 5 mm long ..." (Hedley, 1916); 8 × 4 mm (type, Iredale & Hull, 1929).

Distribution. Only known from Macquarie Island, "on rocks" (Hedley, 1916); "depth 69 m" (Cotton, 1937).

Observations. The here figured specimen is obviously the same as the one pictured by Iredale & Hull, 1929. The small black dots, randomly disposed over the valves, are left out of our drawings, as they are caused either by some boring organism or by the affixion of small algae. They are no "shell-eyes", as Iredale & Hull suggested; these are never found in this group of Polyplacophora. Apparently the girdle and radula were not kept with the valves, which is to be regretted. Nevertheless the specimen, which is much alike *L. (L.) kerguelensis* (but for being almost twice as large and much more solid), is easily recognizable by its obsolete sculpture.

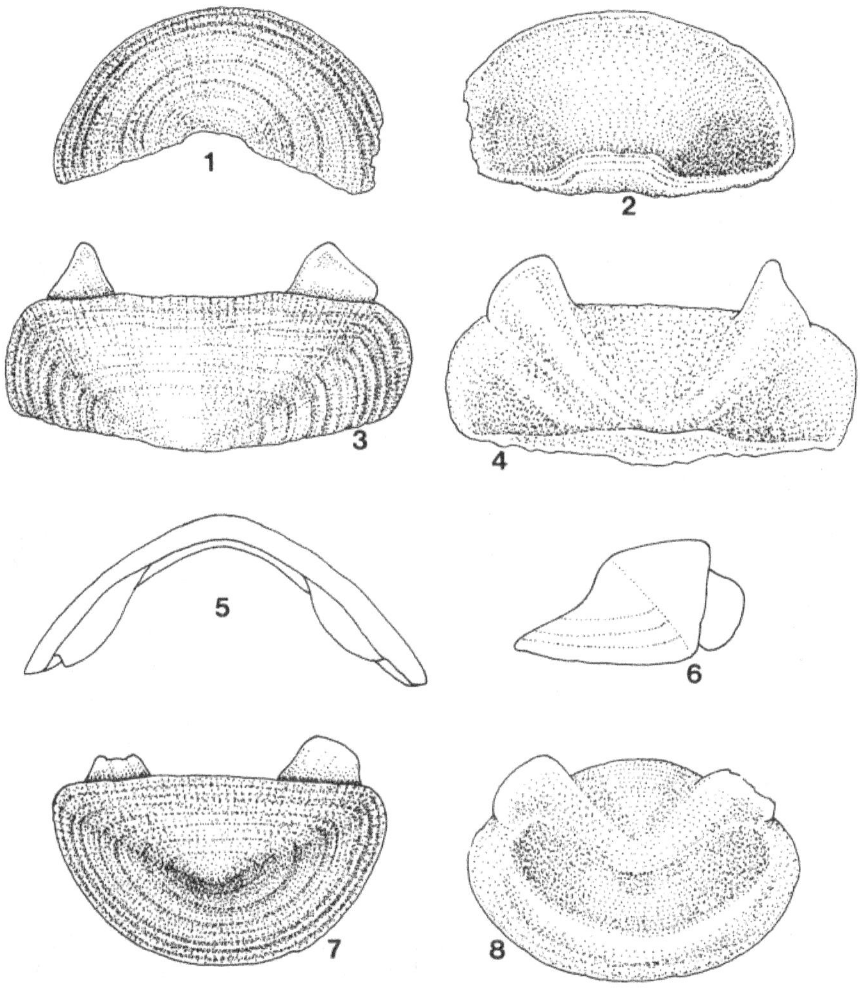

Fig. 71. *Leptochiton (L.) fairchildi* (Iredale & Hull, 1929).

1, valve I, dorsal view, × 20; 2, do, ventral view, × 20; 3, intermediate valve, dorsal view, × 20; 4, do, ventral view, × 20; 5, do, anterior view, × 20; 6, valve VIII, lateral view, × 20; 7, do, dorsal view, × 20; 8, do, ventral view, × 20.

1-8, lectotype.

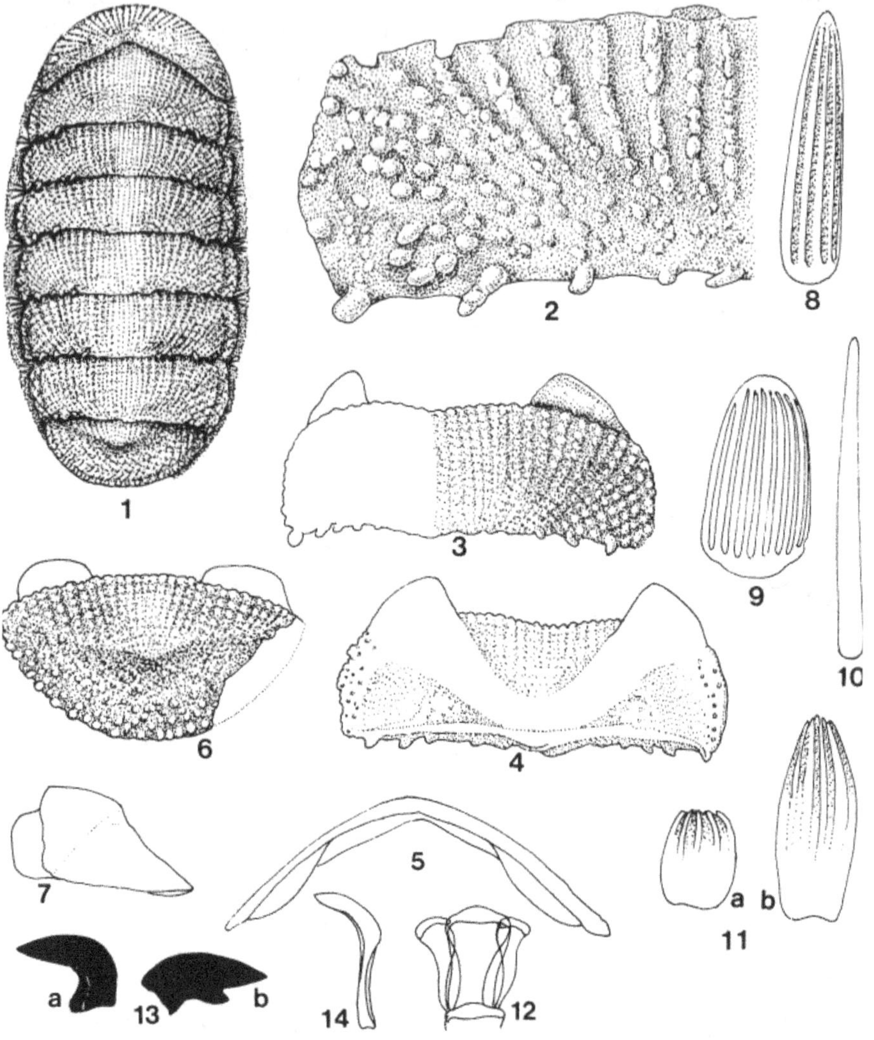

Fig. 72. *Leptochiton (L.) finlayi* (Ashby, 1929).

1, whole animal, × 9.6; 2, left half of valve V, dorsal view, in situ, × 40; 3, valve V, dorsal view, × 20; 4, do, ventral view, × 20; 5, do, anterior view, × 20; 6, valve VIII, dorsal view, × 20; 7, do, lateral view, × 20; 8, marginal girdle spicule, × 400; 9, dorsal scale, × 400; 10, intersegmental needle, × 200; 11, ventral girdle scales, a near inner margin, b near outer margin, × 400; 12, central and first lateral radula teeth, × 200; 13, cusp of major lateral tooth, a from the interior side, b from the exterior side, × 200; 14, spatulate uncinal tooth, × 200.

1-2, specimen from New Zealand, Stewart Id, Patterson Inlet, 13 m, P. Cresswell leg., VB 2677b; 3-14, idem, VB 2677a.

Leptochiton (L.) finlayi (Ashby, 1929)

(Fig. 72. Map 28)

Type: AM C 95165 (Holotype, 3 valves only).
Type locality: New Zealand, off Otago Heads, 108 m.

Lepidopleurus finlayi Ashby, 1929a:367, pl. 32:figs 5-7 (nom. nud.); 1929b:372; Beu, 1977:37, fig. G.
Terenochiton finlayi; Iredale & Hull, 1930:157, pl. 16:figs 2-5; Powell, 1979:21.

Description. Animal small, elongate oval, moderately elevated (dorsal elevation 0.29), subcarinated, side slopes convex, valves not beaked. Girdle narrow, clothed with elongate scales and fan-shaped bunches of sutural spicules.

Head valve semicircular, posterior margin widely V-shaped. Intermediate valves broadly rectangular, anterior and posterior margins about parallel, the former slightly curved inward, the latter little concave at both sides of the hardly perceptible apex, side margins more or less rounded, lateral areas hardly raised, sometimes with a conspicuous growth mark close to the side margin. Tail valve less than semicircular, anterior margin evenly convex, mucro pointed, a little anterior, postmucronal slope slightly concave.

Tegmentum finely granulose. On the head valve the granules are very small, closely packed, and arranged in numerous radial rows. On the lateral areas and the postmucronal area of the tail valve, the convex granules are considerably larger and more spaced in an irregular pattern, becoming radial towards the outer margin; at the posterior margin of the lateral areas, there are three or more equi-spaced protuberances standing up about four times the height of the larger grains, a most characteristic feature of this species. On the pleural areas and the antemucronal area of the tail valve, the granules are oriented in longitudinal rows, close together near the dorsal ridge, the spacing between the rows gradually increasing to almost three times the diameter of the granules near the side margins where the rows also become posteriorly converging.

Articulamentum white, weakly developed, apophyses small, sharply triangular, separated by a wide, serrated, slightly concave jugal sinus.

Colour of tegmentum pale brown, the jugal and latero-pleural areas in a somewhat lighter shade; in the illustrated complete specimen, a great many granules of the central areas are dark brown.

Girdle very narrow, dorsally clothed with elongate, round-topped scales, 68 × 37 µm, ornamented with 8-9 longitudinal riblets; at the junctions of the valves there are remarkable fan-shaped tufts of needle-like spicules, 216 × 16 µm; marginal spicules cone-shaped, bluntly pointed, longitudinally grooved, 83 × 21 µm. Ventrally the girdle is covered with distally ribbed scales, truncated at the base, varying in size from 34 × 26 µm near the inner margin, to 68 × 28 µm near the outer margin.

Central radula tooth widening to a well developed blade, first lateral tooth slender at the base, also widening to a small blade, major lateral with a curved, bidentate cusp, the inner denticle sharply pointed and much larger than the outer one.

Dimensions of the dry and curled holotype: 4 × 2.75 mm, of the illustrated whole specimen: 6.6 × 3.4 mm, of the disarticulated specimen: 5.2 × 2.6 mm.

Distribution. All around New Zealand and Stewart Island, from subtidal to quite deep water.

Leptochiton (L.) subantarcticus (Iredale & Hull, 1930)

(Fig. 73. Map 29)

Type: AM
Type locality: New Zealand, off Auckland Is, 171 m.

Lepidopleurus columnarius (non Hedley & May, 1908); Ashby, 1929b:372, pl. 32: fig. 9.
Parachiton subantarcticus Iredale & Hull, 1930:157, pl. 16:figs 6-7; Kaas, 1977: 85; Powell, 1979:22.
Lepidopleurus subantarcticus; Beu, 1977:39.

Description. Valve highly arched (dorsal elevation 0.71), the back evenly rounded, neither carinated nor beaked, the side slopes almost straight, central area with 25 chains of very fine pustules on each side of the jugum. Lateral areas narrow, hardly elevated, with 8-10 radial rows of weakly pronounced pustules, interrupted by a series of well marked, concentric growth lines. Apophyses small, broadly triangular, wide apart, the jugal sinus broad, shallow, slightly convex.

Distribution. Auckland Is, 171 m (Ashby); off Otago, deep water (Beu).

Observations. The species is based upon a single intermediate valve. Ashby, *l.c.*, described this valve as *L. columnarius* (Hedley & May, 1908),

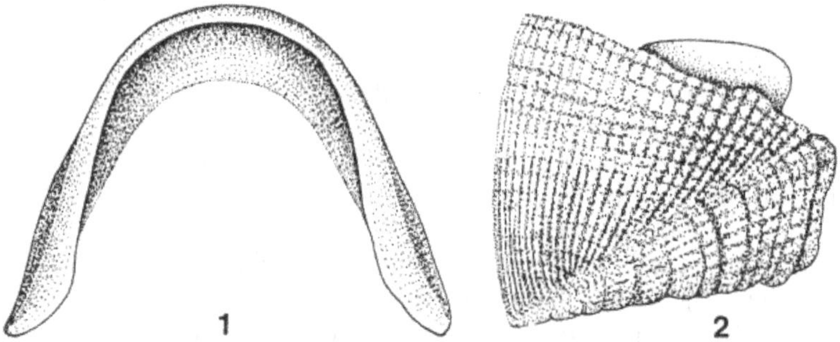

Fig. 73. *Leptochiton (L.) subantarcticus* (Iredale & Hull, 1930).
1, intermediate valve, anterior view, × 16; 2, do, dorsal view of right half, × 20.
1-2, holotype, after Iredale & Hull, 1930.

a Tasmanian species, which, however, differs in having the valves steeply Gothic arched, distinctly carinated. Moreover, the granulation of the tegmentum is much coarser in *columnarius*, the lateral areas distinctly raised, with 20 chains of pustules on either side of the jugum, without distinct growth marks (*fide* Iredale & Hull, 1930).

Beu, *l.c.*, adds to the description: "a small (to c. 12 mm) species, abundant in deep water dredgings off Otago, on Bryozoa, etc., south to the N. Z. subantarctic islands. Back very steeply angled, granules coarse, but evenly spaced and not clearly arranged in rows; placed in *Parachiton* by earlier workers, but the short tail valve is typical of *Lepidopleurus*" (spacing by authors). In the figures of Ashby, and of Iredale & Hull, the granulation of the central area is decidedly arranged in longitudinal rows, so we are uncertain whether Beu refers to the present species.

Subgenus *Pilsbryella* Nierstrasz, 1905

-Siboga Exp. 48: 13 [section of *Lepidopleurus* (Leach MS) Risso, 1826].
Type: *Lepidopleurus (Pilsbryella) setiger* Nierstrasz, 1905, OD.

Differs from *Leptochiton* s.s. by the perinotum which is clothed with spicules and scattered silky hairs, these productions being found even between the valves.

Distribution. Indonesian waters.

Leptochiton (Pilsbryella) setigera (Nierstrasz, 1905)

(Fig. 74. Map 22)

Lectotype (designated herein) and two paralectotypes, ITZ.
Type locality: Indonesia, Makassar Strait, "Siboga" sta. 88, 0°34.6'N 119°8.5. 'E, 1301 m.

Lepidopleurus (Pilsbryella) setiger Nierstrasz, 1905:11, pl. 1:fig. 7, pl. 3.:figs 60-65.

Description. Animal very small, up to 4 mm long, elongate oval, moderately elevated (dorsal elevation c. 0.31), subcarinated, side slopes about straight, valves thin, little or not beaked, girdle very narrow.

Head valve almost semicircular, posterior margin widely V-shaped, notched in the middle, sculptured with widely spaced, irregularly arranged granules of somewhat different sizes. Intermediate valves broadly rectangular, side margins evenly rounded, posterior margin straight to slightly beaked, apices not marked, lateral areas not raised, hardly perceptible, separated from the central area by a weak diagonal depression; sculpture of intermediate valves like head valve, sometimes forming a restricted quincunx-pattern, sometimes forming more or less longitudinal rows. Tail valve almost as wide as the preceding valves, posterior margin evenly rounded, mucro median, postmucronal slope little concave, complete valve irregularly sculptured like other valves.

Colour of tegmentum yellowish brown, jugal area of a somewhat lighter shade, granules darker brown.

Articulamentum weakly developed, apophyses small, broadly triangular, widely separated.

Girdle densely clothed with small, obtusely pointed, longitudinally striate, conical spicules-scales, 46 × 12.5 µm, and scattered, long silky hairs, 200 × 15 µm, that can be noticed even between the valves. There is a marginal fringe of smooth, sharply pointed spicules, up to 134 µm long.

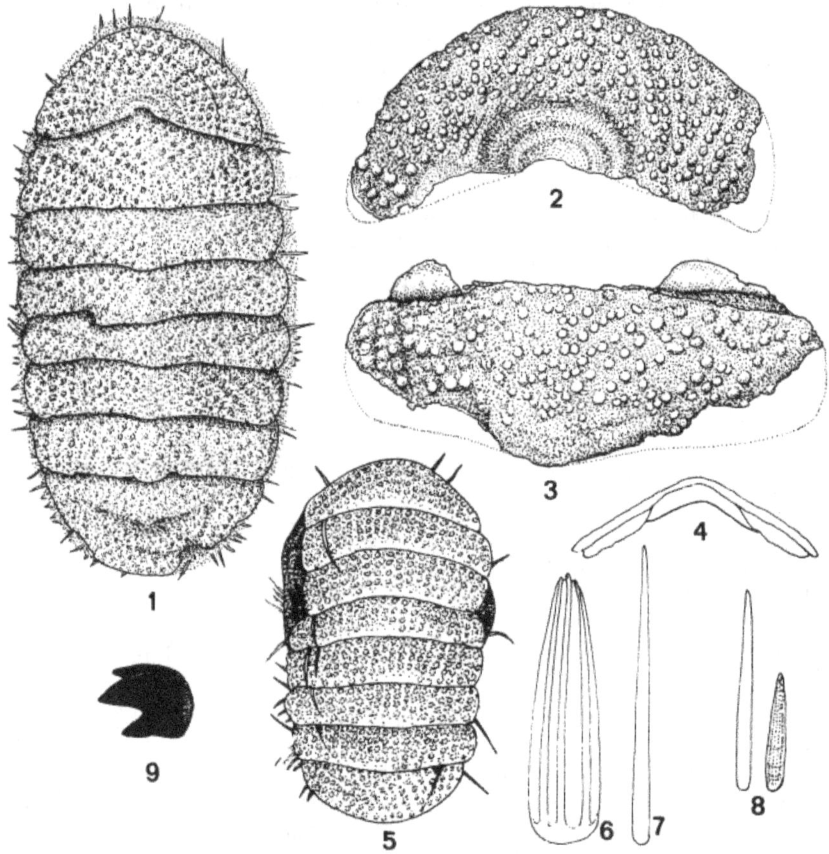

Fig. 74.*Leptochiton (Pilsbryella) setigera* (Nierstrasz, 1905).

1, whole animal, dorsal view, × 20; 2, valve I, dorsal view, × 40; 3, valve II, dorsal view, × 40; 4, do, anterior view, × 20; 5, whole specimen, dorsal view, × 14; 6, dorsal girdle spicule-scale, × 800; 7, dorsal needle, × 200; 8, marginal girdle spicules, × 200; 9, cusp of major lateral radula tooth, × 400.

1, paratype, 3.75 × 1.9 mm, "Siboga" sta. 12, Bali Sea, N of Bali, 7°15'S, 115°15.6'E, 289 m, on a piece of wood; 2-4, 6-9, slightly smaller paratype from same station; 5, holotype, after a pencil drawing in H. F. Nierstrasz, 1905.

Major lateral tooth of radula with a short tridentate cusp, the median denticle largest.

Gills merobranchial, 7-10 ctenidia on both sides.

The length of the three known specimens is about 3.5-4 mm, the width 2-2.5 mm.

Distribution. Only known from the type locality and from the Bali Sea, "Siboga" sta. 12, 7°15'S 115°15.6'E, 289 m (2 paralectotypes).

Observations. It must be noted that a study of the type material has revealed that the long silky hairs that characterize the subgenus can no longer be observed. This means that the taxonomic status of *Pilsbryella*, which we find rather questionable, can neither be confirmed nor rejected until fresh material can be studied.

Subgenus ***Parachiton*** Thiele, 1909

Zoologica Stuttg. 22:14 [subgenus of *Lepidopleurus* (Leach MS) Risso, 1826].
Type: *Lepidopleurus (Parachiton) acuminatus* Thiele, 1909, OD.

Differs from *Leptochiton* s.s. by the much larger posterior valve and the posterior position of the mucro. Girdle wholly spiculose.

Distribution. Mediterranean Sea; Indo-Pacific Ocean from the Bay of Bengal to the Kermadec Is, S to Tasmania.

Leptochiton (Parachiton) africanus (Nierstrasz, 1906)

(Fig. 75. Map 26)

Type: RMNH 2783.
Type locality: Algeria, Oran.

Lepidopleurus africanus Nierstrasz, 1906:155, pl. 3:figs 1-9.
Leptochiton (Parachiton) africanus; Kaas, 1977:81, figs 1-6.
Lepidopleurus (Parachiton) africanus; Laghi, Russo & Dell' Angelo, 1980:1, figs 4a-9b.

Description. Animal small, 7 × 3 mm, elongate oval, parallel-sided, more than twice as long as wide, rather flat (dorsal elevation 0.30), the back evenly rounded, not carinated, lateral areas not raised, tail valve extraordinarily long, about a quarter of the length of the shell, the mucro almost posterior. Girdle rather narrow, densely clothed with smooth spicules of various sizes.

Head valve semicircular, intermediate valves broadly rectangular, regularly arched, not carinated, the second valve decidedly longer than the others, posterior margins straight, apices not indicated, lateral areas only discernible by difference in sculpture, like the head valve marked with 6 or 7 faint concentric lines of growth. Tail valve a regular half oval in outline, the length 3/4 of the width, mucro somewhat swollen, rounded, posterior, at 7/8 of the length of the valve, the hind slope short, steep, straight, but for a little excavation directly behind the mucro.

Tegmentum finely granulose all over. On the head valve and postmucronal area of the tail valve the small, little elevated, cylindrical pustules are arranged in radiating rows, only interrupted by the lines of growth. On the lateral areas of the intermediate valves the arrangement is irregularly quincuncial; on the central and antemucronal areas the granules are more or less oval, arranged in 40 longitudinal series, forming fine lirae, on the jugum somewhat converging posteriorly, on the sides sometimes anastomosing, the granules more clearly pronounced towards the side margins.

Colour of the tegmentum yellowish white, sometimes with black-brown spots on the apex of the seventh and the mucro of the tail valve, the grooves between the lirulae on the central areas light yellowish brown.

Articulamentum weakly developed, white, the apophyses small, more or less triangular, wide apart, the jugal sinus broad, slightly convex or flat.

Girdle narrow, yellowish brown, dorsally rather closely armed with cylindrical, smooth calcareous spicules of different sizes, those near the sutures largest, 160 μm long, 19 μm thick. Towards the outer margin they become

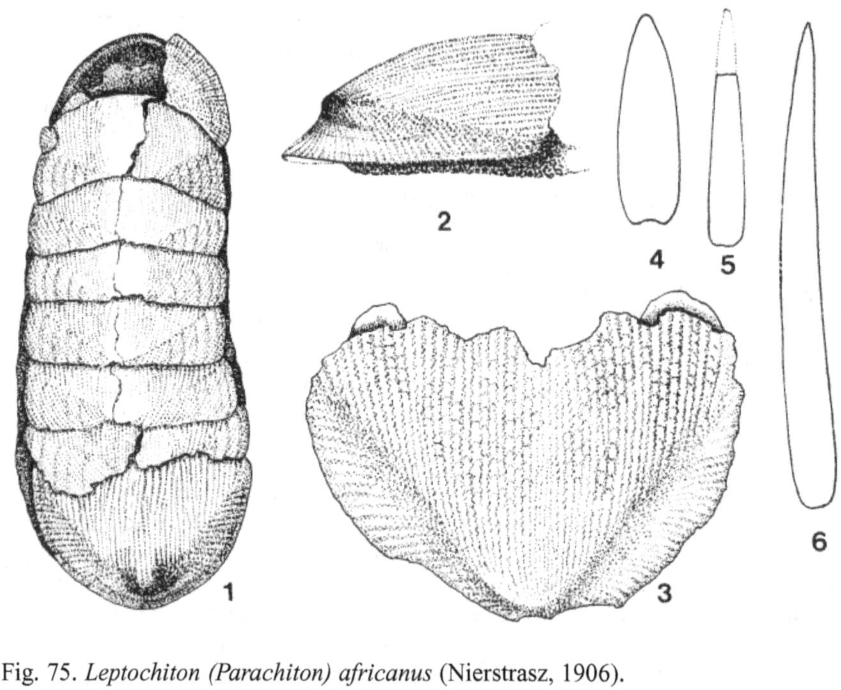

Fig. 75. *Leptochiton (Parachiton) africanus* (Nierstrasz, 1906).

1, whole specimen, dorsal view, × 9.6; 2, valve VIII, lateral view, × 16; 3, another valve VIII, dorsal view, × 20; 4, ventral girdle scale, × 400; 5-6, small and large dorsal girdle spicules, × 400.

1-6, after Kaas, 1977.

shorter and relatively stouter, 80 μm long, 15 μm thick. Ventral side of the girdle covered with radiating rows of flat, thin, elongate oval, imbricating scales, 68 × 20 μm, truncate or a little emarginate at the base, distally tapering to a blunt point.

Gills merobranchial, adanal with interspace; 11 ctenidia on both sides.

Distribution. Algeria: Oran (locus typicus); Gallipoli in the Golfo di Taranto, Italy; Calella on the Costa Brava, Spain; off Calvi, Corsica, France, 30 m (one tail valve only); Tuscan Archipelago, Capraia Id, 60 and 120 m (loose valves).

Observations. *L. (P.) africanus* is the only living representative of the subgenus outside the Indo-Pacific Ocean. It is very much alike (if not identical with) *L. (P.) thielei* Sulc (1936:7, pl. 1:figs 4-5) from Miocene deposits in the Vienna Basin, which only differs in being a trifle larger. Baluk (1971:454, pl. 1:fig. 8) reported *thielei* from Miocene deposits in Central Poland, Laghi, Russo & Dell'Angelo from Pliocene deposits in Valle Andona (Asti, Italy).

Leptochiton (Parachiton) acuminatus (Thiele, 1909)

(Fig. 76. Map 43)

Holotype: ZMHU; paratype: ZMHU 101921.
Type locality: Bismarck Archipelago, Duke of York Island (= Neu Lauenburg).

Lepidopleurus (Parachiton) acuminatus Thiele, 1909:13, pl. 1:figs 61-73.
Leptochiton (Parachiton) acuminatus; Kaas, 1977:82.

Description. Animal of medium size, up to 17 mm long, elongate oval, almost parallel-sided, moderately elevated (dorsal elevation 0.31), back evenly rounded, tail valve large; girdle narrow, densely spiculose.

Head valve semicircular, posterior margin slightly V-shaped. Intermediate valves broadly rectangular, anterior margin varying from slightly convex in valve II to about straight in the other valves, side margins evenly rounded, posterior margin almost straight, apices inconspicuous, lateral areas definitely raised, with 6 to 7 weak growth folds, all valves of about equal width. Tail valve roughly triangular in outline, large, almost as wide as the intermediate valves, length: width ratio c. 0.68, mucro small, pointed, practically terminal, postmucronal slope very short, concave, postmucronal area narrow, slightly denticulate along the posterior margin.

Tegmentum minutely granulose all over, granules oval to heart-shaped, joining each other in regular rows separated by narrow, shallow grooves. These rows are longitudinally oriented on the central areas of the intermediate valves, 120 in number, and on the antemucronal area of the tail valve, radially on the lateral areas and the postmucronal area.

Articulamentum whitish, somewhat translucent, apophyses small, triangular, wide apart.

Colour of tegmentum whitish with irregular grayish stains.

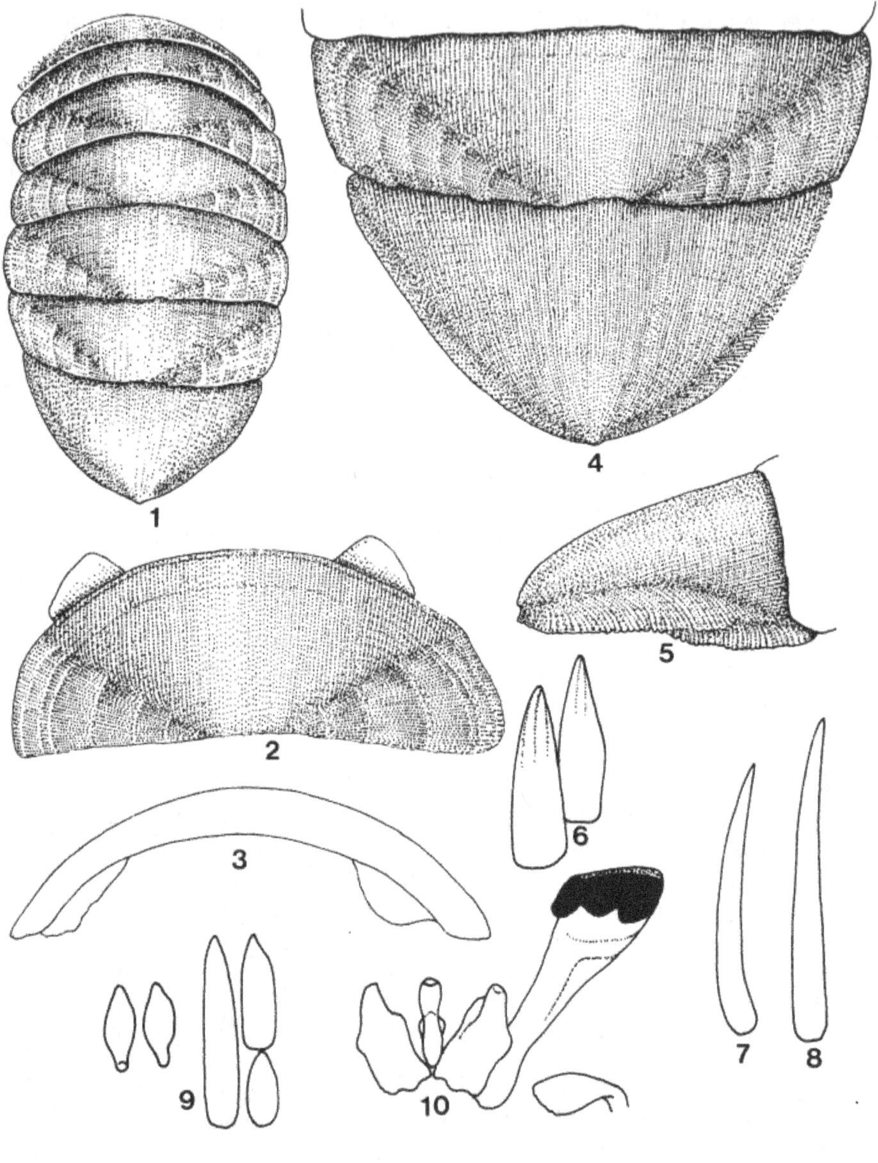

Fig. 76. *Leptochiton (Parachiton) acuminatus* (Thiele, 1909).

1, whole animal, somewhat curled, the girdle folded under, valve I missing, dorsal view, × 4.8; 2, valve II, dorsal view, × 9.6; 3, camera lucida sketch of II, anterior view, × 9.6; 4, valves VII and VIII, in situ, dorsal view, × 9.6; 5, valve VIII, in situ, lateral view, × 9.6; 6, dorsal girdle scales, × 115; 7, dorsal spicule, × 115; 8, marginal spicule, × 115; 9, ventral scales, × 115; 10, part of radula, × 115.

1-6, paratype, ZMHU 101921; 6-10, after Thiele, 1909.

Girdle narrow, covered with conical spicules, distally sculptured with weak longitudinal grooves, about 200 × 60 μm, interspersed with many long, more or less curved, smooth needles, also found along the margin, 300 to 400 μm long. Ventral scales smooth, leaf-shaped, about 100 × 40 μm, to elongate, up to over 200 μm long.

Radula with a narrow central tooth bearing a weakly developed cusp, first lateral tooth rather wide, truncated at the base, distally tapering to a small indistinct cusp, major lateral long and narrow, widening to a tridentate cusp, the denticles bluntly pointed and of about equal length.

Dimension: length 17 mm, breadth 8 mm (Thiele, 1909); estimated length and breadth of figured paratype: 16 × 8 mm.

Distribution. Only known from the type locality.

Leptochiton (Parachiton) mestayerae (Iredale, 1914)

(Fig. 77. Map 29)

Type: CM M5491.
Type locality: Kermadec Is, N coast of Sunday Island, 27 m.

Parachiton mestayerae Iredale, 1914a:27, pl. 1:fig. 1; Oliver, 1915:557; Iredale & Hull, 1929:80, pl. 9:fig. 20.
Leptochiton (Parachiton) mestayerae; Kaas, 1977:84.
Non: *Lepidopleurus (Parachiton) mestayerae;* Leloup, 1981d:25, fig. 12, pl. 1:fig. 2 (= *L. (P.) eugenei* sp. nov.).

Description. Animal of small to medium size, up to 15 mm long, elongate oval, parallel-sided, about twice as long as wide, moderately elevated (dorsal elevation 0.30), subcarinated, the sides slightly convex; end valve large; girdle rather wide, densely spiculose.

Head valve more than semicircular, rather long, the length of both end valves together more than one third the length of the shell. Posterior margin widely V-shaped, notched at the top. Intermediate valves broadly rectangular, valve II somewhat longer than the others, posterior margin almost straight, the apices inconspicuous, lateral areas hardly or not raised, differentiated from the central areas by the sculpture alone, side margins evenly rounded, all valves of equal width. Tail valve long, about one fifth of the total length of the shell, with a small, sharp, posterior mucro, at about three quarters of the length of the valve, the posterior slope steep, straight, but for a little excavation directly behind the mucro.

The tegmentum is finely granulose all over, the granules small, roundish to pear-shaped, little elevated, regularly arranged in quincunx on the head valve, the lateral areas of the intermediate valves, and the postmucronal area of the tail valve. On the central and antemucronal areas the granules are slightly more pronounced, especially on the pleurae, elongate oval, forming 70 closely set, longitudinal rows, often anastomosing on the sides, the interstices very narrow.

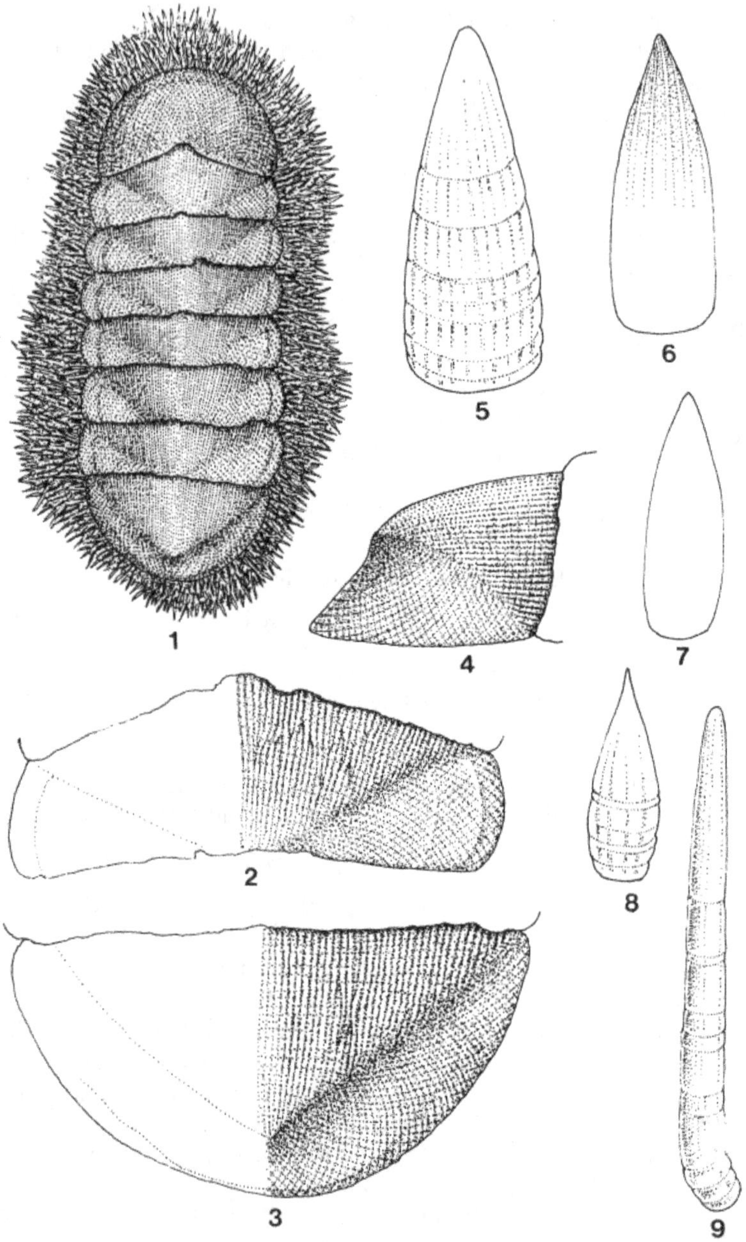

Fig. 77. *Leptochiton (Parachiton) mestayerae* (Iredale, 1914).

1, whole specimen, dorsal view, × 7.2; 2, valve II, dorsal view, × 19; 3, valve VIII, dorsal view, × 19; 4, do, lateral view, × 19; 5-8, dorsal girdle spicules, × 375; 9, dorsal needle, × 150.

1-9, holotype.

Articulamentum white to pinkish; sutural laminae small, somewhat quadrangular, wide apart.

Colour of tegmentum yellowish to roseate, with longitudinal white streaks, and small brownish spots at both sides of the apices.

Girdle rather wide, densely covered with short, conical, smooth to longitudinally coarsely grooved spicules, up to 150 × 50 μm, interspersed with many long, straight, almost smooth needles, up to 350 μm long, 35 μm thick. There is a beautiful fringe of still longer, glassy, transparent needles, up to 450 × 40 μm.

Dimensions: length 11 mm, breadth 6 mm (holotype). "A much larger specimen curled up before it could be preserved." (Iredale.)

Distribution. Only known from the type locality: Kermadec Is, Sunday I., N coast (locus typicus), Denham Bay and Coral Bay (Oliver, 1915), on and under stones, from low water mark to a depth of 46 m.

Leptochiton (Parachiton) puppis (Hull, 1923)

(Fig. 78. Map 25)

Type: AM.
Type locality: New South Wales, Port Jackson, Vaucluse, Bottle and Glass Reef, subtidal.

Parachiton puppis Hull, 1923:158, pl. 24:figs 1-5; Iredale & Hull, 1925:344, pl. 39: figs 15-19.
Lepidopleurus puppies; Ashby, 1924a:322.
Leptochiton (Parachiton) puppies; Kaas, 1977:85.

Description. Animal of moderate size, up to 16 × 8 mm, elongate, moderately elevated (dorsal elevation 0.35), back evenly rounded, uniformly, finely granulose. Colour chalky white when alive, becoming pale buff when dried. Girdle narrow, scaly-spiculose.

Head valve semicircular, posterior margin almost straight. Intermediate valves broadly rectangular, not carinated, side margins evenly rounded, posterior margin straight, apices inconspicuous, lateral areas more or less raised, with 4-5 obvious growth lines. Tail valve unusually large, almost as wide as the intermediate valves, the length about 3/4 of the breadth, mucro almost terminal, slightly projecting posteriorly, postmucronal slope steep, straight.

Tegmental sculpture consists of rows of minute, coalescing granules, arranged in radiating rows on the head valve, c. 100 in number, the lateral areas of the intermediate valves, up to 20 in number, the granules less distinct, and in forwardly curved series on the postmucronal area of the tail valve. On the central areas of the intermediate valves and the antemucronal area of the tail valve, the longitudinal, on the sides slightly posteriorly converging rows number 68 on valve II, 72 on valve III, 74 on valve VIII. The interspaces between the fine striae are marked by rows of punctures.

Articulamentum pearly white, apophyses small, triangular, wide apart.

Girdle dorsally clothed with more or less imbricating, round-topped scales, 36 × 32 μm, with 6-8 ribs, those near the outer margin smaller, 24 × 22 μm, with 3-4 ribs, those along the suture much longer, 68 × 38 μm, with also 6-8 ribs. Interspersed at random, there are stout, smooth, gothic candle-like spicules, 128 × 34 μm. There is no marginal fringe. The ventral covering of the girdle has not been examined so far.

Gills merobranchial, adanal without interspace, 10 ctenidia on each side.

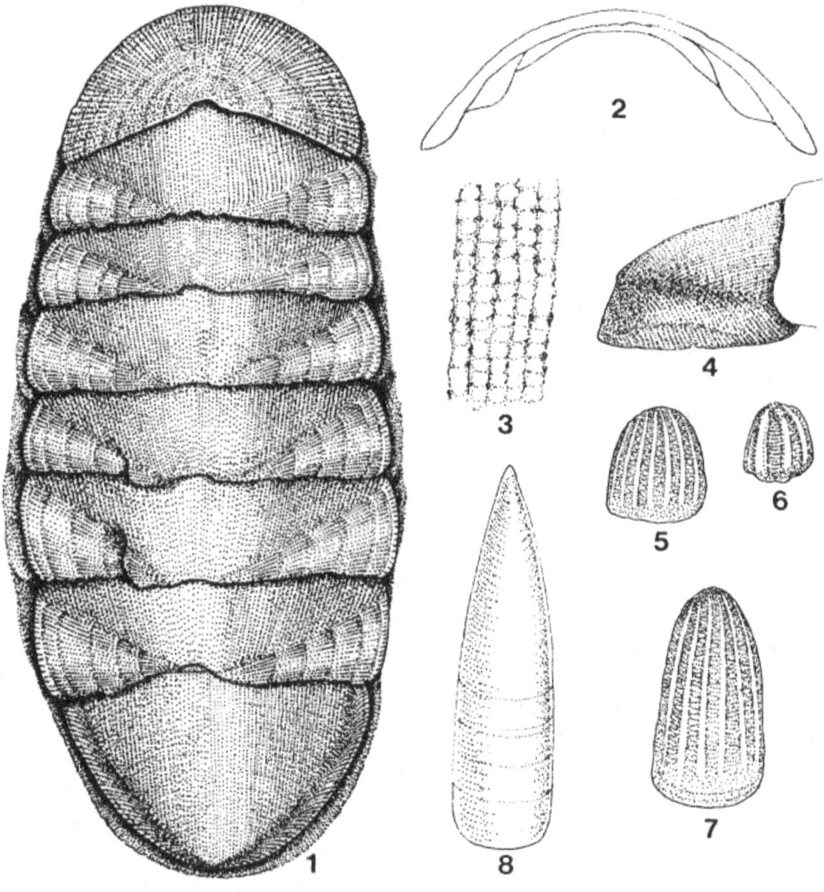

Fig. 78. *Leptochiton (Parachiton) puppis* (Hull, 1923).

1, whole animal, dorsal view, × 9.6; 2, valve IV, anterior view, × 9.6; 3, granulation of centro-pleural area of valve IV, × 40; 4, valve VIII, lateral view, × 9.6; 5, dorsal girdle scale, × 340; 6, do, near outer margin, × 340; 7, do, along suture, × 340; 8, dorsal spicule, × 340.

1-8, specimen from New South Wales, Port Jackson, Vaucluse, Bottle & Glass Reef, 5.I. 1923, W. Barnes leg., K 2824.

Dimensions: 12 × 5.5 mm, 16 × 8 mm (Hull, 1923); 13.5 × 6 mm (Ashby, 1924); 11.8 × 5.4 mm (figured specimen, K 2824).

Distribution. Only known from the type locality, "under stones in sand or mud below low water mark", and from Port Stephens, N.S.W. (Ashby, 1924).

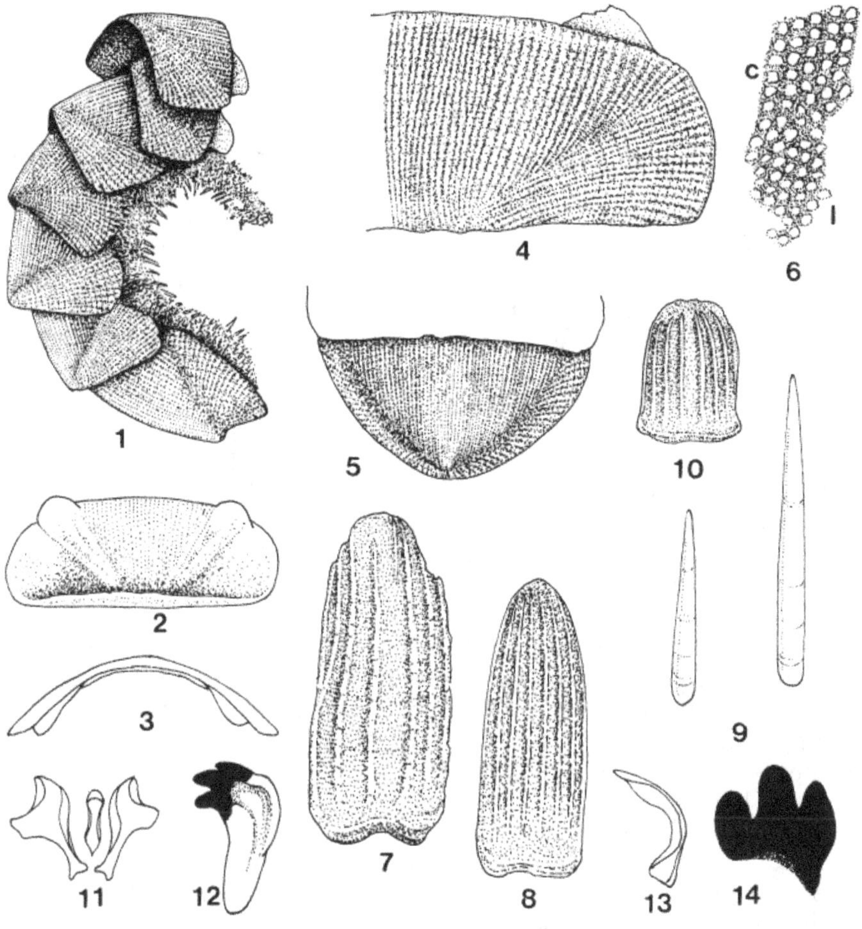

Fig. 79. *Leptochiton (Parachiton) litoreus* (Iredale & Hull, 1925).

1, whole animal, side view, valve I missing, valves II, III loose, × 9.6; 2, valve II, ventral view, × 9.6; 3, do, anterior view, × 9.6; 4, detail of valve II, dorsal view, × 20; 5, valve VIII, in situ, dorsal view, × 9.6; 6, sculpture of valve II: c = central area, l = lateral area, × 40; 7, dorsal girdle scale, × 400; 8, do, along suture, × 200; 9, marginal spicules, × 100; 10, ventral girdle scale, × 400; 11, central and first lateral radula × 100; 12, major lateral tooth, × 100; 13, spatulate uncinal tooth, × 100; 14, cusp of major lateral tooth, × 200.

1-14, holotype.

Leptochiton (Parachiton) litoreus (Iredale & Hull, 1925)

(Fig. 79. Map 25)

Type: AM C29622.
Type locality: Torres Strait, Murray Id, beach.

Parachiton litoreus Iredale & Hull, 1925:344, pl. 39:fig. 14 (not 13!).
Leptochiton (Parachiton) litoreus; Kaas, 1977 85.

Description. Animal small, 7 × 4 mm, curled, elongate oval, rather depressed (dorsal elevation 0.23), back evenly rounded, lateral areas not raised, only indicated by difference in sculpture. Central areas with longitudinal lirulae of small squarish granules, lateral areas with radiating granulose lirulae; mucro of tail valve almost terminal, little elevated; girdle densely clothed with ribbed scales and a fringe of long spicules. Valves thin and brittle.

Head valve (missing in the holotype) with c. 100 radial rows of granules (Iredale & Hull, 1925). Intermediate valves about rectangular, anterior margin almost straight, neither umbonated nor carinated, back evenly rounded, lateral areas not raised, central areas with 64 longitudinal rows of little elevated, squarish granules, the rows sometimes anastomosing, somewhat diverging anteriorly; lateral areas with 18-20 radial series of granules. Tail valve rather long, length: width ratio 0.55, mucro small, somewhat swollen, almost posterior, back slope steep, concave directly behind the mucro; antemucronal area with 60 longitudinal series of pustules, postmucronal area with radial series of granules.

Tegmentum cream-coloured, articulamentum white, the granulation of the tegmentum showing through. Apophyses small, widely separated, bluntly triangular.

Girdle narrow, whitish, dorsally densely clothed with elongate, round-topped, flat, slightly curved scales, 100 μm long, 45 μm wide, those along the sutures about twice as large, 200 μm long, 72 μm wide; all with 7-9 narrow, sharp, longitudinal riblets. Ventral side of girdle with similar, but smaller and relatively shorter, imbricating scales in radiating rows. There is a marginal fringe of long, slender, conical needles, 350-420 μm long, 30-36 μm thick at the base, very finely longitudinally scratched.

Radula with an elongate central tooth bearing a small curved blade, first laterals without a cusp, triangular, abruptly widening distally. Major laterals with a tridentate cusp, the denticles bluntly rounded, the central one longest.

Distribution. Only known from the type locality.

Leptochiton (Parachiton) capricornicus (Iredale & Hull, 1925)

(Fig. 80. Map 25)

Type: AM C21771. One complete specimen, dried, curled up, all valves broken along median axis (= lectotype); 5 loose valves, more or less eroded (1 head, 2 intermediate, 2 tail valves, paralectotypes)

Type locality: Queensland, Capricorn Group, Masthead Island, 31-36 m.

Parachiton capricornicus Iredale & Hull, 1925:345, pl. 39:fig. 13 (not 14!).
Leptochiton (Parachiton) capricornicus; Kaas, 1977:85.

Description. Animal small, elongate, pale pink, glossy (especially the lateral areas), moderately elevated (dorsal elevation 0.30), the back evenly rounded, not carinated, side slopes convex. Valves not beaked, tail valve extraordinarily long, the mucro far posterior. Girdle densely clothed with spicules and needles of different sizes.

Head valve semicircular, tegmentum covered with small, roundish, very

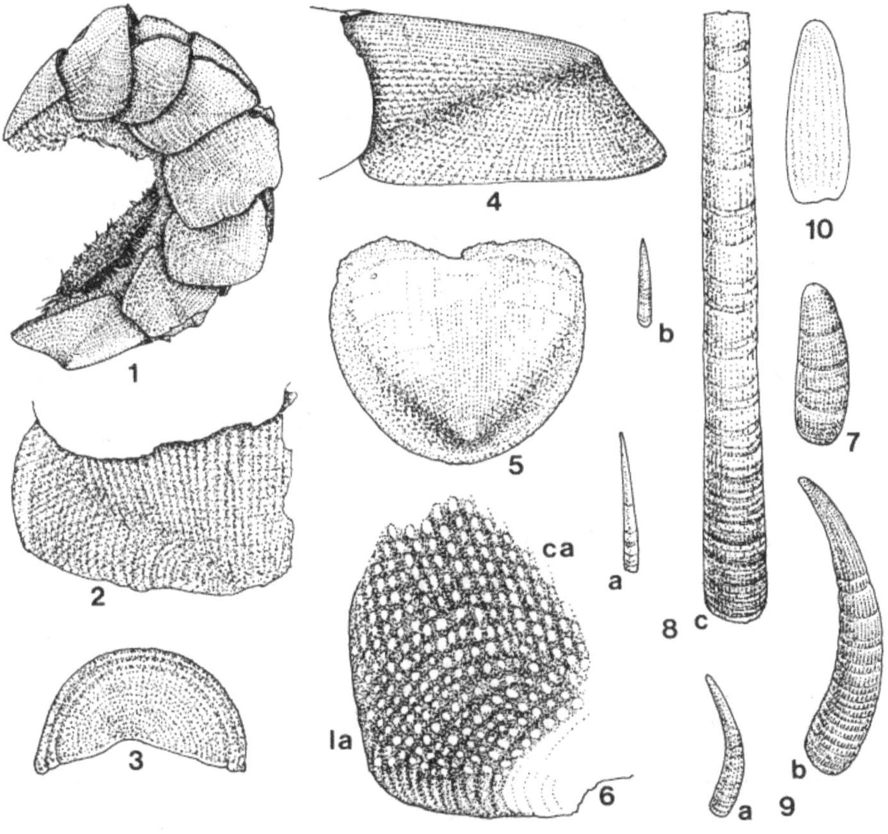

Fig. 80. *Leptochiton (Parachiton) capricornicus* (Iredale & Hull, 1925).

1, whole specimen, curled up, all valves broken along median axis, lateral view, × 9.6; 2, part of valve IV, in situ, dorsal view, × 20; 3, valve I, dorsal view, × 9.6; 4, valve VIII, in situ, lateral view, × 20; 5, valve VIII, strongly eroded, dorsal view, × 9.6; 6, part of valve V, dorsal view, la = lateral area, ca = central area, × 40; 7, dorsal girdle spicule, × 200; 8a-c, dorsal spines, × 200; 9a-b, sutural spicules, × 200; 10, ventral girdle spicule, × 200.

1-2, 4, 6-10, lectotype; 3, 5, paralectotypes.

little raised granules arranged in quincunx. Intermediate valves rectangular, posterior margin almost straight, not umbonated; lateral areas hardly or not raised, differentiated from the central areas by the sculpture alone. On the lateral areas the granules are arranged in curved series, forming a quincuncial pattern, on the central areas in 40-45 longitudinal lines, somewhat converging posteriorly on the pleurae, the granules oval, interstices narrow, shallow. Tail valve abnormally long (length: width ratio 0.70), with a small, pointed, almost terminal mucro, the posterior slope nearly straight. Postmucronal area sculptured like head valve, antemucronal area like the central areas.

Articulamentum weakly developed, white, apophyses small, wide apart.

Girdle narrow, densely covered with small, bluntly pointed, longitudinally grooved and concentrically ringed, cylindrical spicules, up to 140 × 40 µm; those along the sutures relatively longer, rather strongly curved, varying in size from 90 × 16 µm to 200 × 40 µm. Among the spicules, scattered at random, many long, striated glassy needles are found, differing in size, small ones up to 90 × 8 µm, large ones (all broken in the lectotype) 700 × 40 µm by estimate. The marginal fringe is formed by equally large needles.

Length 12 mm, width 4 mm (according to Iredale & Hull). The largest loose tail valve (paralectotype) is, however, 1.5 times larger than that of the lectotype.

Distribution. Only known from the type locality.

Observations. *L. (P.) capricornicus* differs from *litoreus* in its finer granulation and glossy appearance, and in a much longer tail valve. In our opinion figs 13 and 14 on plate 39 of Iredale & Hull (1925) were accidently interchanged.

Leptochiton (Parachiton) lifuensis (Hull & Risbec, 1931)

(Fig. 81. Map 32)

Type: ?
Type locality: Loyalty Is, Lifu and Mouac.

Parachiton lifuensis Hull & Risbec, 1931:373, figs 1-9, pl. 31:figs 1-7.
Leptochiton (Parachiton) lifuensis; Kaas, 1977:85.

Description. "Shell small, narrow, round-backed, moderately elevated, side slopes rounded, not keeled. Colour, creamy buff, sometimes maculated with reddish. Sculpture uniformly grain-striate.

Anterior valve very finely closely rayed, and having a distinct marginal growth ridge.

Median valves: Lateral areas raised, clearly differentiated, radially grain-striate, the rays crossed by eight to ten low growth lines or folds, terminating in a marginal ridge; central areas with about forty straight longitudinal rows of granules, sometimes coalescing, the granules larger than those on the lateral areas.

Posterior valve large; mucro sharp and situated almost above the posterior margin; postmucronal slope steep, convex; postmucronal area very small; antemucronal area very large; the sculpture of the whole valve resembling that of the anterior valve, but granules larger, increasing in, size, and radiating outwards towards the margin.

Girdle densely clothed with fine elongate glassy spicules, projecting at margins.

Interior pearly white; sutural laminae large, distant.

Dimensions: 15 × 6 mm (type, dried and somewhat curled)." (Hull.)

Gills merobranchial, adanal; ca 17 ctenidia on both sides.

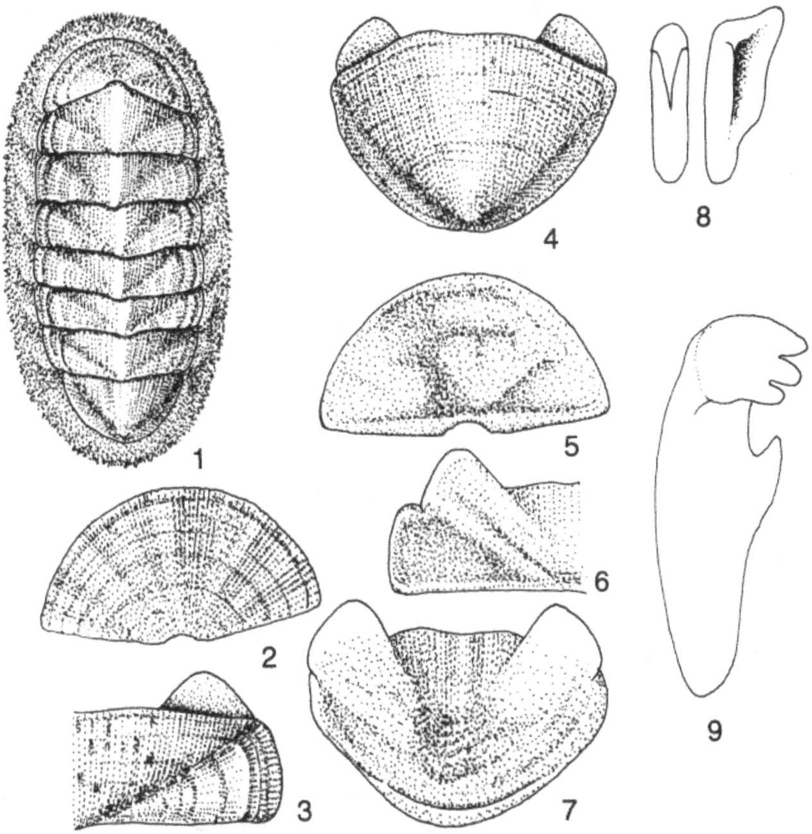

Fig. 81. *Leptochiton (Parachiton) lifuensis* (Hull & Risbec, 1931).

1, whole animal, dorsal view, × 6; 2, valve I, dorsal view, × 15; 3, right half of an intermediate valve, dorsal view, × 15; 4, valve VIII, dorsal view, × 15; 5, valve I, ventral view, × 15; 6, right half of an intermediate valve, ventral view, × 15; 7, valve VIII, ventral view, × 15; 8, central and first lateral radula teeth, × 250; 9, major lateral radula tooth, × 250.

1-9, syntypes, Loyalty Islands, Lifu, under dead coral or stones, below low water mark, all after Hull & Risbec, 1931.

Radula with a narrow central tooth, bearing a very slightly recurved cusp. Major lateral with a tridentate cusp, the denticles bluntly rounded, the base showing a supplementary apophysis (After Risbec).

Distribution. Only four specimens were procured in the Loyalty Islands: Lifu (3 specimens, Hull), Ile Mouac (1 specimen, Risbec). They were found under dead coral or stones, below low water mark.

Observations. As we have been unable to locate the type of the present species the original description is given above. Hull remarks: "This shell is characteristic of the Australasian species of the genus (i.c. *Parachiton*), and is closely allied to *P. puppis* Hull, from Port Jackson, although quite easily separated, especially in life, by the distinctive coloration and coarser sculpture."

Leptochiton (Parachiton) textilis (Powell, 1936)

(Fig. 82. Map 29)

Type: BMNH 19621073.
Type locality: New Zealand, off Three Kings Is, "Discovery" sta. 934, 92 m.

Parachiton textilis Powell, 1936:219, pl. 48:figs 4-6; Beu, 1977:39, fig. F; Powell, 1979:22, pl. 13:fig. 4.
Leptochiton (Parachiton) textiles; Kaas, 1977:85.

Description. As only several loose valves were ever found, Powell's original description is cited here.

"Median valves narrow, not beaked, round-backed and high arched, lateral areas very slightly raised; the sculpture of fine, regular quincuncial punctation, central areas with crowded delicately granulate fine closely spaced longitudinal lirae which curve inward slightly at the dorsal ridge, a few short lirae being intercalated above to fill the space occasioned by this slight crowding. There are about 100 of the longitudinal lirae across the whole of a valve, those towards the sides becoming oblique. Posterior valve large and long, almost as long as wide; mucro very close to the lower margin; ante-mucronal area extremely large, triangular, rounded, low-arched and sculptured similarly to the central areas of the median valve, longitudinal lirae totalling about 70 across the valve. Post-mucronal area radially quincuncially lirate. Insertion plates absent, sutural laminae small. Colour uniformly pale buff.

Length 2.5 mm; breadth 7.5 mm; height 3.5 mm; median valve (paratype).
Length 2.5 mm; breadth 3.8 mm; height 1.3 mm; posterior valve (hol type)."

Distribution. New Zealand, Three Kings Is, "Discovery" sta. 934, 92 m; N. Island, Mercury Bay, scallop beds, 36.5 m.

Observations. Powell wrote: "This is the second New Zealand *Parachi-*

ton to be described, the other (*P. subantarcticus*, Iredale & Hull, 1930, Aust. Zool. 6, pt. 2, p. 157) being from 95 fathoms off the Auckland Islands. The Auckland Islands species is based on a single median valve which differs from that of the Three Kings Islands shell in having the longitudinal sculpture coarser and in the form of chains of pustules."

We have placed *Parachiton subantarcticus* Iredale & Hull, 1930, in *Leptochiton* s.s., as long as no complete specimen is available. The only tail valve of *textilis* we could study (fig. 82:4) agrees fairly well with Powell's description, although it is somewhat eroded, the apophyses worn away and the longitudinal lirae only indicated by rows of punctures. It is more triangular in outline than Powell's holotype, and it is faintly mottled with pale brownish on the pleurae of the antemucronal area. From Powell's figures may be concluded that the dorsal elevation amounts to 0.42, and that the anterior margin of the intermediate valves is decidedly curved inward, forming a wide, concave jugal sinus.

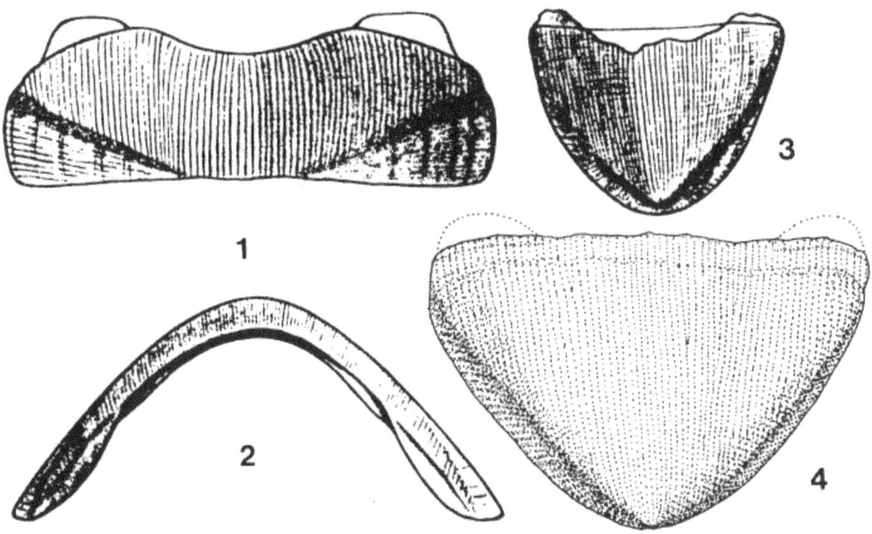

Fig. 82. *Leptochiton (Parachiton) textilis* (Powell, 1936).

1, intermediate valve, dorsal view, × 9; 2, do, anterior view, × 9; 3, valve VIII, dorsal view, × 9; 4, another valve VIII, × 10.8.

1-2, paratype; 3, holotype; 1-3, after Powell, 1936; 4, N New Zealand, Mercury Bay, scallop beds, 36.5 m, XII. 1976, A. Penniket leg., VB 2871a.

Leptochiton (Parachiton) verconis (Cotton & Weeding, 1939)

(Fig. 83. Map 25)

Type: SAMA D 11689.
Type locality: South Australia, St. Francis Island.

Parachiton verconis Cotton & Weeding, 1939:183, pl. 7:fig. 2; Cotton & Godfrey, 1940:482, fig. 464; 1940a:22.
Leptochiton (Parachiton) verconis; Kaas, 1977:85.

Description. The original description reads: "The posterior valve 5 × 5 mm, colour a light buff. Anti-mucronal area exceptionally long and covered with about eighty rows of small coalescing granules, giving, where touching, a net-work appearance. Post-mucronal area very short and concave under the mucro, with short rows of coalescing granules and suggestions of growth ridges."

Distribution. Only known from the type locality.

Observations. The present "species" is based upon a single tail valve. The measurements given by the authors are dubious, the length of the valve being 0.8 of the breadth. Cotton & Weeding (1939:183) remark: "The nearest related species appears to be *P. puppis* Hull, a Peronion species found in a restricted area at Vaucluse, Watson Bay, N. S. W. The number of minutely grained striae in the Flindersian species is less than in the Peronian, giving it a slightly coarser appearance."

We compared the above description and the given picture with a fine specimen of *L. (P.) puppis* (see fig. 78) collected by W. Barnes in 1923 at Bottle and Glass Reef, Vaucluse, Port Jackson, N. S. W. (K 2824). The tail valve has ca 80 granulose lirae on the antemucronal area, which is the same number as stated by Cotton & Weeding for their *P. verconis*. In our opinion the latter may best be regarded as a Flindersian subspecies of the Peronian *puppis* (if not identical with it!), a question which cannot be decided until a complete specimen of *verconis* turns up.

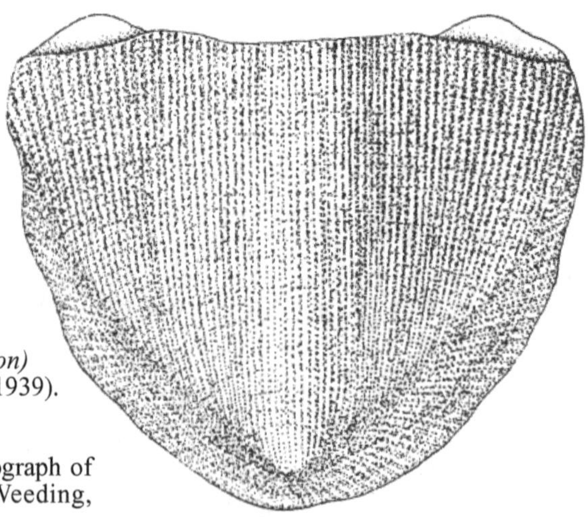

Fig. 83. *Leptochiton (Parachiton) verconis* (Cotton & Weeding, 1939).

Valve VIII, dorsal view, × 15.

Drawn after the original photograph of the holotype in Cotton & Weeding, 1939.

Leptochiton (Parachiton) ronaldi sp. nov.

(Fig. 84. Map 38)

Type: BMNH 1952.11.17.320 (Winckworth collection).
Type locality: Bay of Bengal, Andaman Islands, Dakshin-Andaman, Port Blair.

Diagnosis. Animal of small to medium size, up to 13 × 7 × 3.5 mm, elongate oval, parallel-sided, about twice as long as wide, moderately elevated (dorsal elevation 0.32), the back rounded, not carinated; intermediate valves not beaked, lateral areas not raised; girdle narrow, covered with ribbed scales and small spicules.

Description. Head valve semicircular, median valves broadly rectangular, posterior margin almost straight, apices not indicated, valve II longer than the others, back rounded, side slopes regularly convex, lateral areas not raised, differentiated from the central areas by the sculpture alone. Tail valve disproportionally long, about 1/4 of the total length of the shell, semi-elliptical, the length 3/4 of the breadth, the mucro almost terminal, pointed, slightly swollen, the hind slope steep, short, a little concave.

Tegmentum evenly granulose, the granules roundish, only weakly pronounced, arranged in radiating rows on the head valve and on the postmucronal area of the tail valve, in quincunx on the lateral areas. Central areas and antemucronal area with juxtaposed longitudinal granulose lirae, about 60 on the central areas, 45-50 on the antemucronal area. There are five to six very faint concentric lines of growth on end valves and lateral areas.

Colour of tegmentum dirty white with two small reddish brown spots at the posterior margins of the valves on both sides of the jugum.

Articulamentum white, the apophyses small, somewhat rounded, wide apart, the jugal sinus shallow, wide, almost flat.

Girdle narrow, densely covered with not imbricating, elongate, conical to ovoid scales, always sculptured with 6-8 fine, sharp riblets, 84 × 52 μm, the marginal scales largest, up to 124 μm long. Among the scales, distributed at random, small, striated spicules occur, 36 μm long. Ventral side of the girdle covered with radiating rows of imbricating flat, smooth, elongate scales, 64 × 25 μm, the base truncated, distally bluntly pointed.

Distribution. Only known from the Andaman Islands: Dakshin-Andaman, off Port Blair.

Observations. The holotype specimen, labelled "*Lepidopleurus andamanicus* Smith" by the late Ronald Winckworth, M.A. (1884-1950), was sent to us from the British Museum (Natural History). From Smith's decription of *andamanicus,* and the figure in Annandale & Stewart (1910:pl. 23:fig. 4), it became clear that Winckworth had misindentified the specimen. As it appears to be a true representative of the subgenus *Parachiton,* different from all other known species of this taxon, we dedicate this new species to its distinguished discoverer.

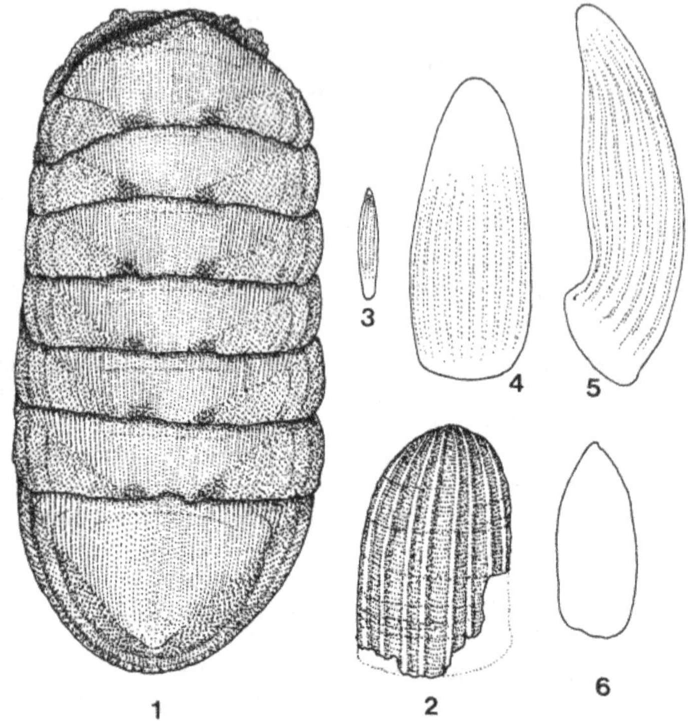

Fig. 84. *Leptochiton (Parachiton) ronaldi* sp. nov.

1, whole animal, dorsal view, valve I missing, girdle folded under, × 9.6; 2, dorsal girdle scale, × 400; 3, dorsal spicule, × 400; 4-5, marginal scales, × 400; 6, ventral girdle scale, × 400.

1-6, holotype.

Leptochiton (Parachiton) eugenei sp. nov.

(Fig. 85. Map 45)

Type: MNHN.
Type locality: Malagasia (Madagascar), Tuléar, lagoon, S Passage, 13 m.

Lepidopleurus (Parachiton) mestayerae; Leloup, 1981d:25, fig. 12, pl. 1:fig. 2 (non *Parachiton mestayerae* Iredale, 1914).

Diagnosis. Animal of small size, up to 10 × 3.5 mm (holotype c. 6 × 2 mm), elongate, more than twice as long as wide, back evenly rounded, little elevated (dorsal elevation 0.28), valves brittle, not beaked, lateral areas not raised, mucro of tail valve at two thirds of the length. Tegmentum granulated in longitudinal chains on central areas, in quincunx on end valves and lateral areas. Girdle narrow, covered with elongate, finely longitudinally striated scales. Colour chalky white.

Description. Head valve semicircular, posterior margin widely V-shaped, tegmentum evenly sculptured with minute granulae arranged in quincunx, and, especially near the outer margin, some very fine, close set, concentrical growth lines. Intermediate valves broadly rectangular, about 3 times as wide as long, side margins rounded, front and hind margins straight and parallel-sided, apices inconspicuous, lateral areas not raised, neatly marked by the difference in sculpture which is equal to that of the head valve, while the central areas are sculptured with c. 45 fine, longitudinal rows of more or less squarish granules. Tail valve large, as wide as the intermediate valves, the length about two thirds of the width, anterior margin straight to slightly convex, mucro not prominent, situated at the posterior third of the valve, posterior slope regularly declining, about straight, postmucronal area sculptured like head valve, antemucronal area like central area of intermediate valves.

Articulamentum weakly developed, more or less translucent, apophyses very small, triangular, situated close to the side margins, leaving a very wide, straight to slightly convex jugal sinus; apophyses of tail valve somewhat wider, almost trapezoid, the sinus much narrower.

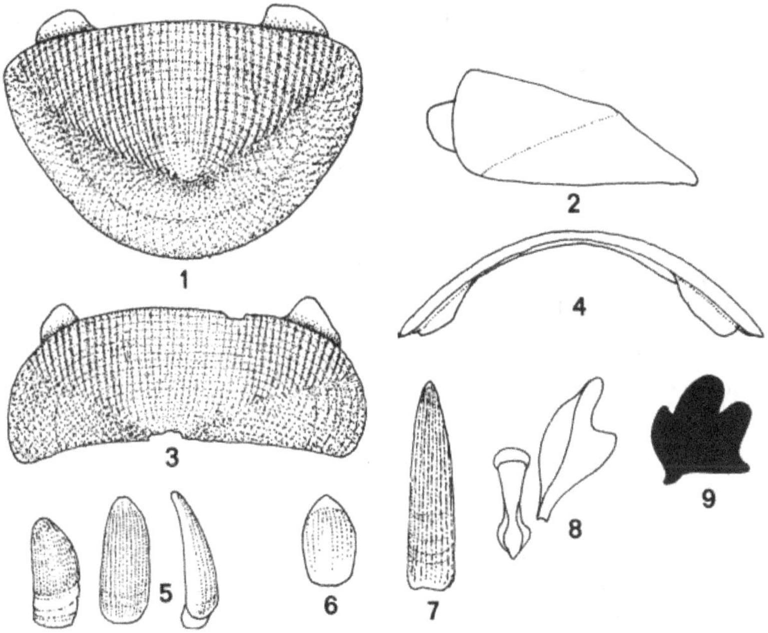

Fig. 85. *Leptochiton (Parachiton) eugenei* sp. nov.

1, valve VIII, dorsal view, × 20; 2, do, lateral view, × 20; 3, valve V, dorsal view, × 20; 4, do, anterior view, × 20; 5, dorsal girdle scales, × 200; 6, ventral girdle scale, × 200; 7, marginal spicule, × 200; 8, central and first lateral radula teeth, × 280; 9, cusp of major lateral radula tooth, × 230.

1-9, paratype from Malagasia (Madagascar), Tuléar, lagoon, S. Passage, 13 m. B. Thomassin leg., 10.IV.1972, K 4995.

Girdle narrow, yellowish to ochraceous, covered with elongate oval scales, 70-100 μm long, 30-35 μm wide, dorsally convex and sculptured with 12 fine, longitudinal grooves, ventrally flat. There is a fine marginal fringe of elongate, obtusely pointed spicules, 140-160 μm long, 28-30 μm wide, truncated at the base, sculptured with 8-10 sharp, longitudinal riblets on the visible half. Ventral scales convex-sided, distally pointed, 60 × 36 μm in mid-girdle, shorter near the inner margin, longer towards the outer margin.
Central tooth of radula more than twice as long as wide, narrowed in the middle, with a thin blade, first lateral tooth wing-shaped, major lateral with a strong tridentate cusp, denticles round-topped, the central one longer than the others.

Distribution. Only known from the reefs off Tuléar, SW Malagasia; Great Reef, inner and outer slope; lagoon, S. Passage; Ifaty Reef, S. Great Passage, outer slope, 0-24 m. All specimens collected by B. Thomassin, 1965-1972.

Observations. This species was misidentified by Leloup (1981d:25) as *Lepidopleurus (Parachiton) mestayerae,* described by Iredale (1914a:27) from the Kermadec Is., which is characterized by a wide girdle, densely beset with long needles. We name it *eugenei* in honour of the late Dr Eugène Leloup of Bruxelles.

Leptochiton (Parachiton) fornix sp. nov.

(Fig. 86. Map 43)

Material. Papua-New Guinea, Madang province, Mililat (11 km N of Madang): handdredging on sandy slope from -20 m till -15 m, 5.III. 1979 (sta. 35A), J. Pierret leg., IRSN IG 26080/1 (holotype); barrier reef, handdredging from -25 m till –12 m, sand and coral rubble, 17.II.1979 (sta. 31), J. Pierret leg., IRSN IG 26080/1 (para-type); Wongat, small reef, handdredging on sandy slope with Halimeda, from -20 m till –12 m, 28.II.1979 (sta. 34), J. Pierret leg., IRSN IG 26080/1 (paratype); II-III. 1979 (sta. 50), J. Pierret leg., IRSN IG 26080/1 (paratype), now disarticulated; dredging in Bay, 1978, J. Pierret & D. Zimmerman leg., IRSN IG 25930/1 (paratype).

Diagnosis. Animal of medium size, maximum length 18 mm, holotype 16 × 7.5 mm, elongate oval, length about twice the width, back evenly rounded, dorsal elevation 0.36, valves thin, not beaked, girdle relatively wide. Colour of tegmentum translucent white, the holotype with two small brown spots on the apical area of valves I-VII.

Description. Head valve semicircular, ornamented with extremely fine, radially directed, close set, granular series, interrupted by numerous fine, concentric growth lines. Intermediate valves broadly rectangular, side margins obtusely rounded, anterior and posterior margins about straight, parallel-sided, apices inconspicuous, lateral areas hardly raised, sculptured like head valve, with some additional fine, concentric ribs, central area with 80-90 parallel, longitudinal series of somewhat larger granules, dentating the

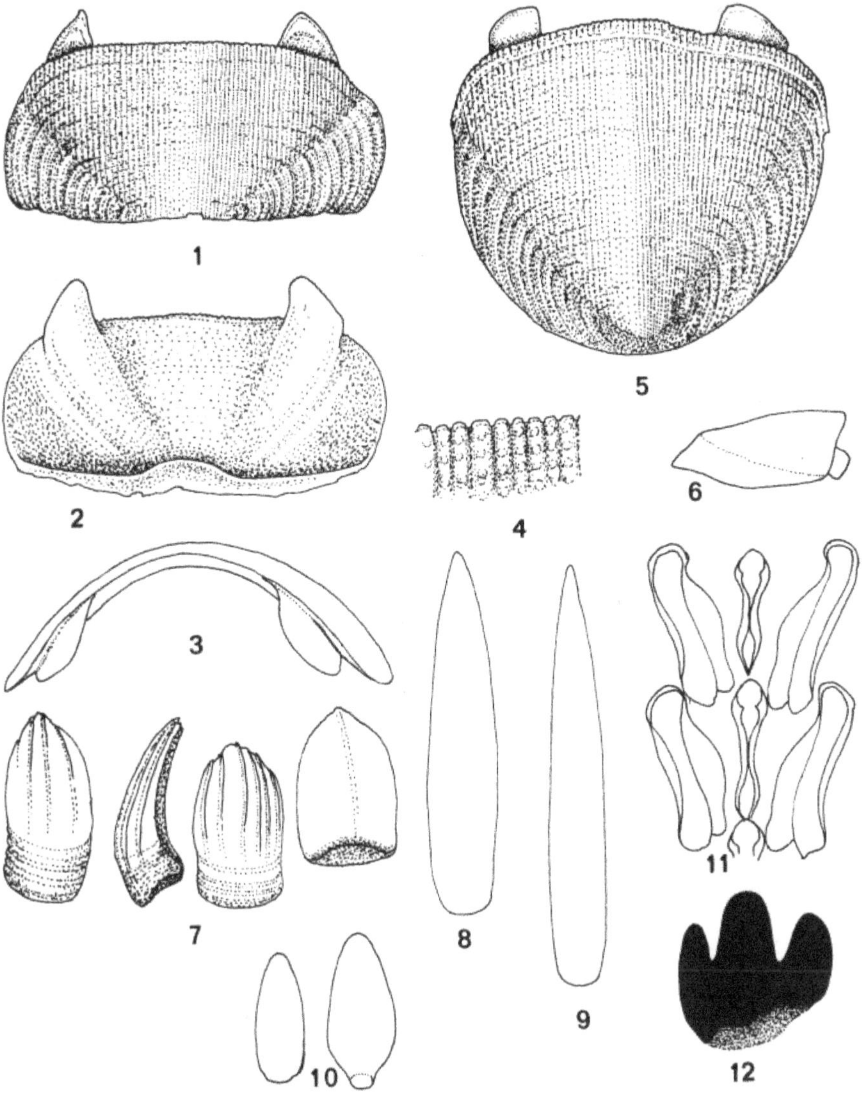

Fig. 86. *Leptochiton (Parachiton) fornix* sp nov.

1, valve IV, dorsal view, × 9.6; 2, do, ventral view, × 9.6; 3, do, anterior view, × 9.6; 4, do, tegmental sculpture of anterior margin, × 40; 5, valve VIII, dorsal view, × 9 6; 6, do, lateral view, × 4.8; 7, dorsal girdle scales, x 200; 8, dorsal spicule, × 200; 9, marginal spicule, × 200; 10, ventral scales, × 200; 11, central and first lateral radula teeth, × 200; 12, head of major lateral tooth, × 200.

1-12, paratype from Mililat, 11 km N of Madang, Papua New Guinea, II/III 1979, J. Pierret leg., IRSN, I.G. 26080.

anterior margin of the valve, separated by rather deep, narrow grooves. Tail valve semi-elliptical, almost as long as wide, anterior margin slightly convex, mucro subterminal, not prominent, antemucronal area triangular, somewhat swollen, sculptured like central areas, postmucronal area like lateral areas, posterior slope straight, not steep.

Articulamentum weakly developed, glossy white, apophyses very small, sharply triangular, rectangular in the tail valve, sinus wide, straight, finely denticulate.

Girdle light yellowish to orange yellow, dorsally covered with rather large, imbricating, elongate, obtusely pointed, calcareous scales, up to 125 × 60 µm, dorsally convex, with 4-6 sharp, longitudinal riblets, ventrally excavated, strongly curved near the truncated base; between the scales, and more particularly on the perinotum tongues, there are white and yellowish brown, smooth, straight, bluntly pointed spicules, up to 250 µm long, varying in diameter from 32 to 48 µm. There is a marginal fringe of somewhat longer, more slender, regularly tapering spicules, 300 × 35 µm. Ventrally the girdle is paved with smooth, oval, rather flat scales, up to 100 × 48 µm, mostly with a narrow, bud-like base.

Central tooth of radula narrowly elongate, somewhat pinched in the middle, without a distinct blade, first lateral tooth slightly aliform on the inner side, with a thin blade, major lateral with a tridentate cusp, denticles round-topped, the central one longer than the others.

Gills merobranchial, abanal, about 15-16 ctenidia on each side.

Distribution. Only known from the type-locality, 12-25 m.

Observations. This new species is most closely allied *L. (P.) acuminatus* (Thiele, 1909), from Duke of York, also in the Bismarck Sea. *L. (P.) fornix* differs principally in the position of the mucro (terminal in *acuminatus,* subterminal *in fornix),* in the clothing of the girdle (scales and spicules in *fornix,* scales only in *acuminatus),* and in the granulation of the valves (much finer in *acuminatus).*

Genus ***Oldroydia*** Dall, 1894

Nautilus 8:90 [section of *Lepidopleurus* (Leach MS) Risso, 1826].
Type: *Lepidopleurus (Oldroydia) percrassus* Dall, 1894, OD.

Shell elongate, valves heavy, lateral areas not elevated, girdle encroaching at the sutures, clothed with spines and slender spicules.

Distribution. West coast of N. America; Japan.

Oldroydia percrassa (Dall, 1894)

(Fig. 87. Map 39)

Type: USNM 107274.
Type locality: California, Santa Barbara Channel, off San Pedro, 135 m.

Lepidopleurus (Oldroydia) percrassus Dall, 1894:90.
Hanleya spicata Berry, 1919b:8, figs 5-6, pl. l:fig. 3.
Oldroydia percrassa; Ferreira, 1979b:160, fig. 20 (bibliography and synonymy).

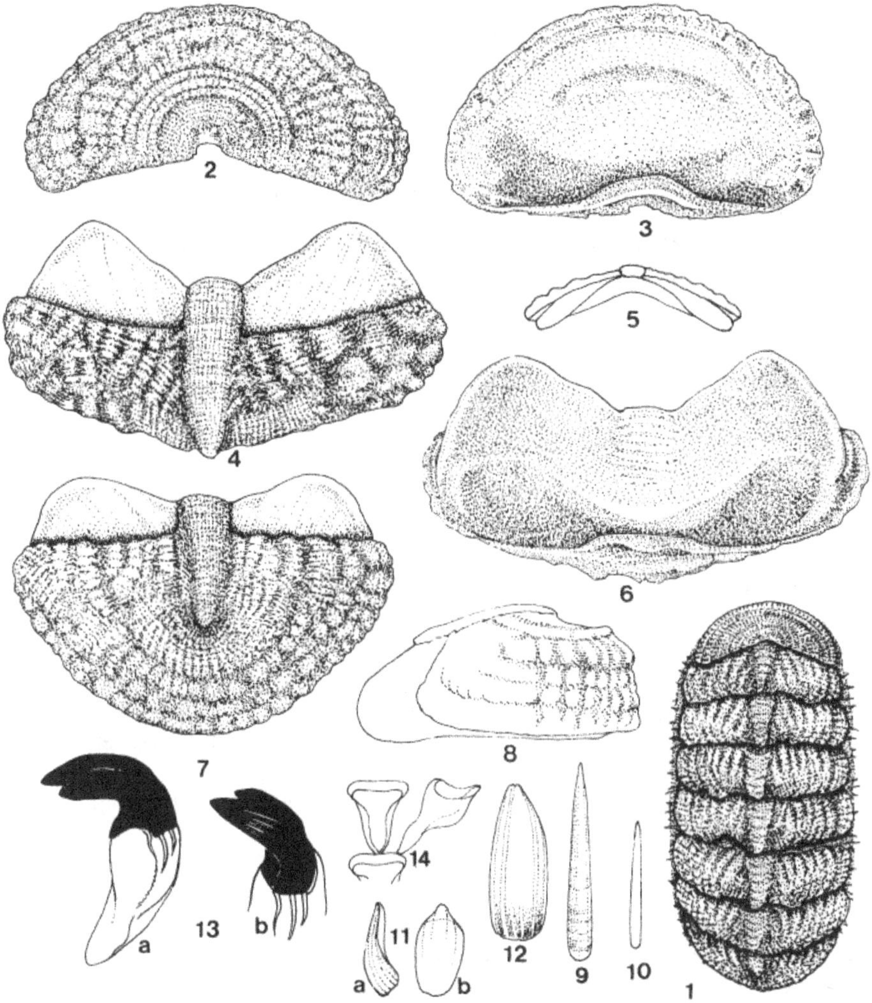

Fig 87. *Oldroydia percrassa* (Dall, 1894).

l, whole animal, dorsal view, × 4.2; 2, valve I, dorsal view, × 8.4; 3, do, ventral view, × 8.4; 4, intermediate valve, dorsal view, × 8.4; 5, do, anterior view, × 4.2; 6 do, ventral view, × 8.4; 7, valve VIII, dorsal view, × 8.4; 8, do, lateral view, × 8.4; 9, dorsal girdle spine, × 70; 10, dorsal spicule, × 175; 11, ventral girdle scales at inner margin, a lateral view, b ventral view, × 350; 12, do, from mid-girdle, × 350; 13, major lateral radula tooth, a from the reverse side, b in natural position, × 87; 14, central and first lateral radula teeth, × 87.

1-14, specimens from San Diego, California, 22 m, II.1973, B. W. Myers leg., VB 2818.

Description. Animal of moderate size, up to 28 mm long, elongate oval, about twice as long as wide, rather flat (dorsal elevation c. 0.26), subcarinated, valves solid, decidedly beaked, jugal area much projecting anteriorly, end valves almost as wide as intermediate valves. Girdle rather narrow, spiculose.

Head valve somewhat less than semicircular, sculptured with more or less radiating rows of irregular granulae, crossed by several concentric growth ridges. Intermediate valves with the short tegmentum widely V-shaped, side margins evenly rounded, jugal area neatly indicated, characteristically projecting forward for almost half the length of the tegmentum, obsoletely sculptured with fine, longitudinal grooves, lateral areas not elevated, hardly perceptible, irregularly sculptured with flat, coarse granules and concentric growth ridges, pleural areas with 6-8 flat, posteriorly diverging ribs, much wider than the interstices which are transversely crossed by many fine, sharp lamellae. Tail valve semicircular, anterior margin about straight, mucro subcentral, highly elevated, not very prominent, postmucronal area concave, sculptured like head valve, antemucronal area sculptured like pleural areas of intermediate valves.

Colour of tegmentum mostly a uniform tan or light brown, some much lighter, creamy, almost white.

Articulamentum white, thick, heavily calloused, apophyses very large, broadly triangular, tops rounded, especially in valve VIII, separated by the protruding, anteriorly rounded jugum.

Girdle of a dark brown colour, encroaching at the sutures so as to form cuticular bridges between the valves, closely beset with slender, smooth, straight spicules, 100 × 8 µm, and randomly dispersed, much stronger, straight spines, 380 × 50 µm. Ventrally the girdle is clothed with pointed scales, 60 × 24 µm, somewhat smaller at inner margin, 36 × 20 µm.

Central tooth of radula very narrow at the base, distally broadening to a narrow, straight blade, first lateral tooth wing-like, major lateral with a bidentate cusp, both denticles sharply pointed, the outer one somewhat smaller than the inner one.

Distribution. West coast of North America, from Monterey Bay, California, to San Benito Island, Baja California, Mexico, and in the Gulf of California at Salsipuedes Channel. The bathymetric range extends from intertidal to 730 m, with a median depth of 40 m.

Oldroydia bidentata Is. Taki, 1938

(Fig 88 Map 34)

Type: Lost during World War II.
Type locality: Japan, Sagami Bay, 1-2 mi. off the coast of Zyogasima, 100-140 m.

Oldroydia bidentata Is. Taki, 1938b-81, pl 1.

Description. Animal elongate oval, about twice as long as wide, rather flat (dorsal elevation 0.32), subcarinated; the valves long, decidedly beaked, jugal area much projecting anteriorly. Girdle rather wide, encroaching at the sutures.

Head valve thick, solid; tegmentum densely sculptured all over with minute, flat granules, strongly marked with concentric lines of growth. Intermediate valves roundly arched in the jugal part, strongly protruded anteriorly, sharply beaked, marked with longitudinal threads and strong, transverse

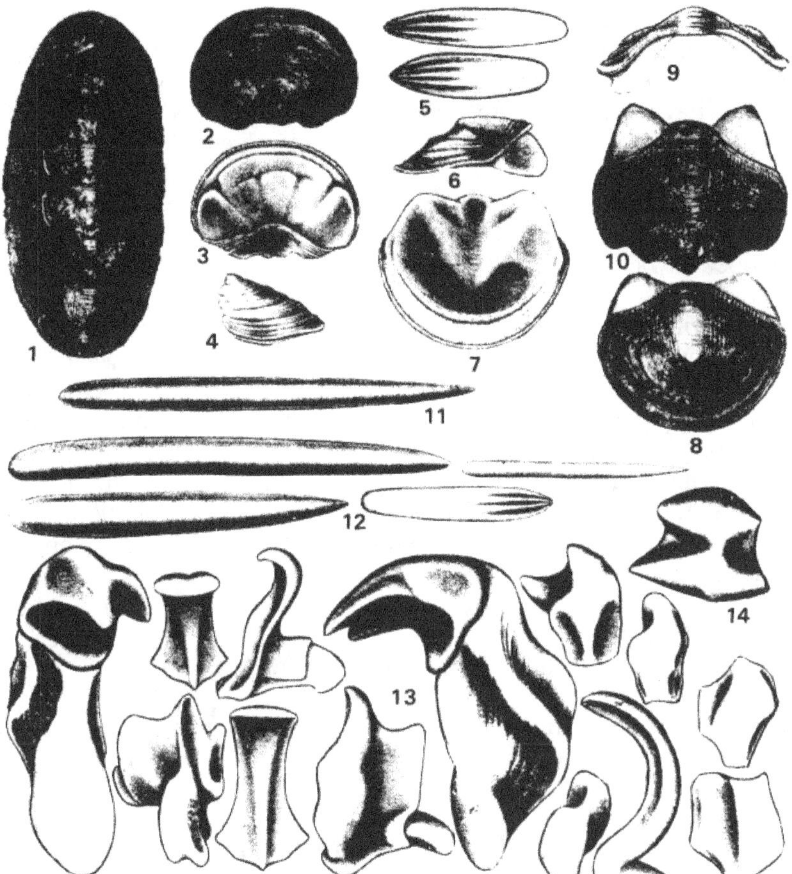

Fig. 88. *Oldroydia bidentata* Is. Taki, 1938.

1, whole animal, dorsal view, × 1; 2-4, valve I, dorsal, ventral, and lateral view resp., × 2.2; 5, ventral girdle scales, × 200; 6-8, valve VIII, lateral, ventral, and dorsal view resp., × 2.2; 9-10, valve V, anterior and dorsal view resp., × 2.2; 11, marginal spine, × 110; 12, dorsal girdle spicules, × 110; 13, isolated radula elements, × 60; 14, cusp of major lateral radula tooth, × 60.

1-14, after Is. Taki, 1938.

lines of growth; pleural areas nearly straight, sculptured with longitudinal series of coarse granules, sometimes coalescent, forming a continuous rib; lateral areas little elevated, moderately arcuated, sculptured with minute granules and strong concentric lines of growth. Tail valve larger than head valve, mucro central, distinct but not elevated; jugum protruded anteriorly, sculptured like in the intermediate valves, though the fine longitudinal threads are composed of closely set, very minute, flat granules; pleural areas sculptured like in the intermediate valves; postmucronal area sculptured like the head valve, posterior slope almost straight.

Colour of tegmentum pale olive-buff; inside white.

Articulamentum of head valve forming a narrow insertion plate on both sides, not on the anterior margin; four distinct shallow grooves radiating from the apex. Eaves wide, smooth. Tegmentum widely folded under posteriorly. Intermediate valves with a weakly developed valve callus at both sides of the median axis, no radial grooves. Apophyses triangular, acute in front, thickened at outer edge like the insertion plate. Jugal sinus wide, convex. Jugal region of tail valve striated with transverse short lines; callus well developed at both sides, articulamentum slightly thickened at the posterior edge.

Girdle vinaceous buff, rather wide at the sides, encroaching at sutures, dorsally covered with small, hyaline, strongly ribbed, somewhat flattened spines, 12 μm long, intermingled with small and large needle-shaped, smooth, sharply pointed needles. Marginal spines like the large dorsal ones, 50 μm long. Ventral side of girdle with imbricating, flat, distally pointed scales, strongly marked with 2 or 3 ribs near the tips, 13-53 μm long.

Central tooth of radula oblong, shallowly sinuated at both sides, well cusped with arcuate edge. Major lateral with a bidentate cusp, the outer denticle larger and more acute than the inner one.

Gills merobranchial, abanal, ranging from 7th valve to near the anus, 13 ctenidia at both sides.

Length 45 mm, breadth 23 mm. (Altered after Taki, 1938.)

Distribution. Only known from the type locality: Japan, Sagami Bay, 100-140m.

Genus *Hanleyella* Sirenko, 1973

-Zool. Zh. 52(10): 1569.
Type: *Hanleyella asiatica* Sirenko, 1973, M.

Shell ovate, valves thin, lateral areas not elevated, tegmentum granulose, the granules getting larger towards the outer margins, apophyses rather large, sloping gently towards the jugal sinus, girdle clothed with minute spicules and large hyaline spines.

Distribution. North Pacific Ocean.

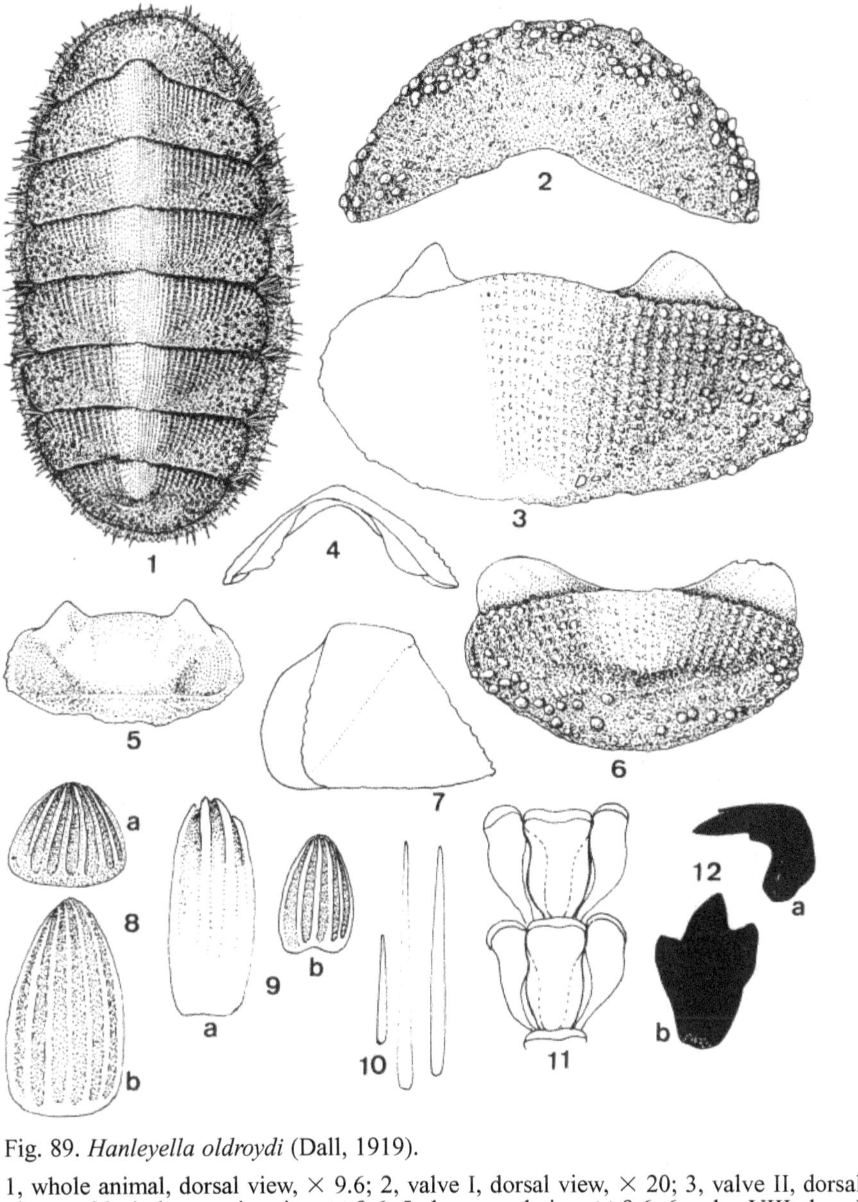

Fig. 89. *Hanleyella oldroydi* (Dall, 1919).

1, whole animal, dorsal view, × 9.6; 2, valve I, dorsal view, × 20; 3, valve II, dorsal view, × 20; 4, do, anterior view, × 9.6; 5, do, ventral view, × 9.6; 6, valve VIII, dorsal view, × 20; 7, do, lateral view, × 20; 8, dorsal girdle scales, a near the outer margin, b near the suture, × 400; 9, ventral scales, a near outer margin, b near inner margin, × 400; 10, dorsal girdle spicules, × 400; 11, central and first lateral radula teeth, × 200; 12, cusp of major lateral radula tooth, a in natural position, b viewed from the top, × 200.

1-12, specimens from British Columbia, Hope Island, off Cape James, 137 m, 21.V.1962,I. Mc T. Cowan leg., VB 2728.

Hanleyella oldroydi (Bartsch MS, Dall, 1919)

(Fig. 89. Map 42)

Type USNM 218767.
Type locality: California, Monterey Harbour.

Lepidopleurus (Leptochiton) oldroydi Bartsch MS, Dall, 1919:500.
Hanleyella oldroydi; Ferreira, 1979b:158, figs 17-19, 39-40 (bibliography).

Description. Animal rather small, elongate oval, up to 8.2 mm long, 4.3 mm broad, rather elevated (dorsal elevation c. 0.41), valves carinated, not beaked, lateral areas slightly raised, colour white, girdle narrow, clothed with scales and spicules.

Head valve somewhat less than semicircular, posterior margin widely V-shaped, tegmental sculpture of rather bold pustules, randomly arranged, mostly close to the outer margin. Intermediate valves broadly rectangular, side margins evenly rounded, apices inconspicuous, side slopes almost straight, lateral areas clearly indicated by the difference in sculpture; jugal area with 12 close, longitudinal series of very faint granules, pleurae each with 15 rows of neatly separated, round, flat-topped granules, longitudinally arranged, slightly converging posteriorly, the interstices about as wide as the series, lateral areas sculptured like head valve. Tail valve more or less oval, width of tegmentum more than twice the length, mucro not prominent, a little anterior, submedian when seen from above, postmucronal slope very steep, about straight, antemucronal area sculptured like central areas of intermediate valves, postmucronal area like head valve.

Articulamentum whitish, apophyses small, subtriangular, rounded in valve VIII, sinus wide, straight-edged, evenly convex in valve II.

Girdle densely clothed with hardly imbricating, flat, oval scales, 6-8 ribbed, 40 μm broad at the base, 50 μm long, sutural ones longer, up to 72 μm, supramarginal ones shorter, up to 36 μm high. Many long, smooth needles are scattered over the whole dorsal surface of the girdle, on the perinotum tongues between the valves, there are bunches of 5-6, while along the margin there is a fringe, of the same needles, all up to 400 μm long, 30 μm thick at the base. Ventrally the girdle is clothed with ribbed (4-6) scales, short near the inner margin, 30 × 24 μm, much longer near the outer margin, up to 72 × 28 μm, the latter only distally strongly ribbed.

Central radula tooth almost rectangular, with a narrow blade, major lateral tooth with a tridentate cusp, central denticle largest.

Distribution. Recorded from Kosciusko Island, Alaska, down to Cabo San Quintin, Baja California, Mexico. Bathymetric range from 18 to 455 m, with a median depth of about 100 m.

Observations. For the purpose of this study two perinotum-slides were prepared. After two days of drying all needle-like spicules had disappeared, probably by the continued action of the KOH. The same effect has been noted by Ferreira (1979b:159). This means that these spicules are not calcareous.

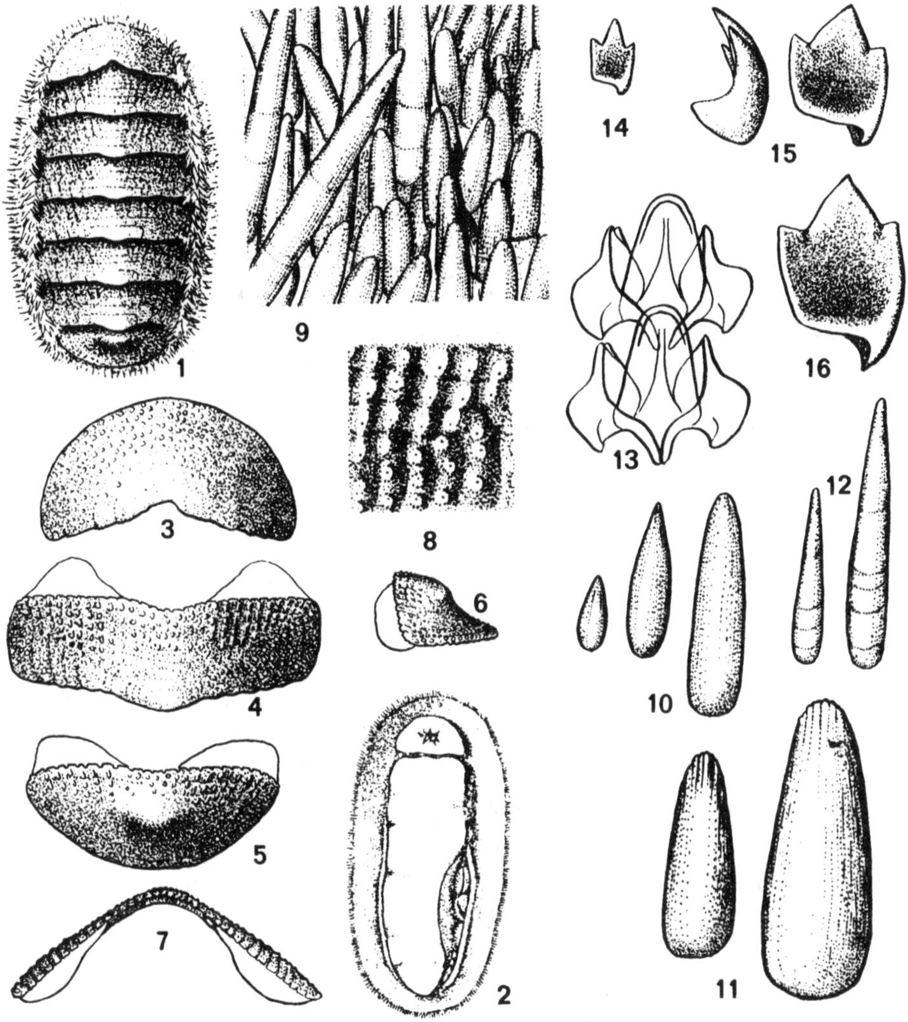

Fig. 90. *Hanleyella asiatica* Sirenko, 1973.

1, whole animal, dorsal view, × 6.7; 2, do, ventral view, showing young animals in the mantle cavity, × 6; 3, valve I, dorsal view, × 12; 4, valve V, dorsal view, × 12; 5-6, valve VIII, dorsal and lateral view, × 12;7, valve V, anterior view, × 12;8, sculpture of the pleural side of an intermediate valve, × 55; 9, part of the girdle, dorsal side, × 120; 10, ventral girdle spicules, × 250; 11, dorsal girdle scales, × 250; 12, dorsal girdle spicules, × 120; 13, central and first lateral radula teeth, × 190; 14-16, cusps of major lateral radula teeth, specimens of different sizes, × 190.

1-13, 15, holotype, 14, 16, paratypes; all after Sirenko, 1973.

Hanleyella asiatica Sirenko, 1973

(Fig. 90. Map 30)

Type: ZIAS.
Type locality Okhotsk Sea, Kurile Is, off Simoshir Id, 10 m
Hanleyella asiatica Sirenko, 1973c.1569, figs 1-16.

Description. Animal rather small, maximum 10.5 × 6 mm (holotype 6.5 × 4 mm), oval, moderately elevated (dorsal elevation c. 0.33), subcarinated, valves hardly beaked, white with dark sand-coloured deposits, girdle narrow, of same colour as tegmentum.

Head valve semicircular, posterior margin widely V-shaped, notched in the middle. Intermediate valves broadly rectangular, the width more than twice the length, anterior margin straight, with a slight jugal excavation in all valves, except in valve II, posterior margin somewhat projecting, side slopes almost straight, lateral areas little raised. Tail valve less than semicircular, anterior margin straight, mucro central, slightly projecting, postmucronal slope somewhat concave.

Tegmentum uniformly sculptured with fine, flat granules, arranged in more or less concentric rows on the head valve, the postmucronal area of the tail valve, and the lateral areas of the intermediate valves, in longitudinal rows on the central areas and the antemucronal area of the tail valve. On all valves the granules become obsolete in the apical part.

Articulamentum white, smooth, apophyses broadly triangular, with rounded top in valves II-VI, trapezoid in valves VII-VIII.

Girdle dorsally covered with elongate, longitudinally grooved, conical scale-spicules, up to 145 × 55 µm, interspersed with many slender, needle-like spicules, up to 260 µm long. Ventrally the girdle is clothed with elongate, more or less sharply pointed spicules, up to 105 × 25 µm.

Gills merobranchial, adanal, 7 ctenidia on each side.

Radula with a distally narrowing, rounded central tooth, first lateral tooth sharply pointed, with a broad downwardly directed side-lobe, major lateral with a tricuspid blade, median denticle largest.

Distribution. Found in varying depths, 10-130 m, from Bering Sea (Providence Bay, Anadyr Bay, NE Kamchatka) to the Kurile Is (Paramushir, Onekotan, Simoshir, Urup).

Family **HANLEYIDAE** Bergenhayn, 1955

-Kungl. Fysiogr. Sallsk. Handl. Lund 66(8):33.

= Leptoidea Dall, 1879 (partim); Holochiton Fischer, 1885 (partim); Lophyrochitonidae de Rochebrune, 1889 (partim).

Animals ovate to elongate, of different size, tegmental sculpture varies from almost smooth to granular, valve I with an insertion plate, valves II-VII with more or less developed insertion plates, or only both end valves with insertion plates; sutural laminae well developed; perinotum with spicules and longer spines.

Key to genera

1a Insertion plate on valve I, more or less developed on valves II-VII . . . . *Hanleya*
1b Insertion plates on I and VIII only . . . . *Hemiarthrum*

Genus *Hanleya* Gray, 1857

-Guide Moll. Br. Mus. 1:186.

= *Laminoplax* Ferreira, 1981.

Type: *H. debilis* Gray, 1857 = *Chiton hanleyi* Bean *in* Thorpe, 1844, M.

Animal small to large, tegmentum granulated, insertion plate on valve I well developed, more or less obsolete on II-VII, unslit. Girdle clothed with spicules and longer, randomly dispersed, spines.

Distribution. Atlantic Ocean from the Barents Sea to the Canary Islands and in the Mediterranean Sea; E coast of N America, from Maine S to Florida.

Hanleya hanleyi (Bean in Thorpe, 1844)

(Fig. 91. Map 18)

Type: WEM.
Type locality: England, Yorkshire, Scarborough.

?*Chiton mendicarius* Mighels & Adams, 1842:42, pl. 4:fig. 8 (type lost); Clench & Turner, 1950:308, pl. 42:fig. 6.
Chiton hanleyi Bean *in* Thorpe, 1844:263, fig. 57; Sparre Schneider, 1886:96.
Hanleya debilis Gray, 1857:186.
Hanleya hanleyi; Dautzenberg & Fischer, 1912:21 (bibliography and synonymy); Nierstrasz & Hoffmann, 1929:29, figs 26a-g, 52; Leloup & Voltz, 1938:11, fig. 11; Knudsen, 1949:4; Jakovleva, 1952:59, fig. 20, pl. 2:fig. 2; Muus, 1959:40, fig. 23; Malatesta, 1962:153, figs 9-10; Knudsen, 1970:3; Sabelli, 1972:97, figs 1-6; 1974b, figs 1-13; Laghi, 1977:99, figs 5-9; Kaas, 1979a:25; McKay & Smith, 1979:2.
Lepidopleurus carinatus Dall, 1927:11 (non Leach, 1852).
Hanleya dalli Kaas, 1957:83 (nom. nov. pro L. *carinatus* Dall).
Laminoplax dalli; Ferreira, 1981b:190, figs 1-6.

Description. Animal of small to medium size, up to about 22 × 13 mm, elongate oval, little elevated (dorsal elevation c. 0.30), subcarinated, side slopes straight, valves rather thick, little or not beaked, girdle of moderate width, spiculose.

Head valve semicircular, posterior margin widely V-shaped. Intermediate valves more or less rectangular, side margins evenly rounded, posterior margin practically straight, except for the protruding, well developed apex in the middle, lateral areas not raised, only marked by a difference in sculpture. Tail valve little less than semicircular, anterior margin slightly convex, mucro median, prominent, postmucronal slope concave.

Tegmentum uniformly sculptured with numerous, roundish granules, arranged without special order, and increasing in size towards the outer margins, on the head valve, the lateral areas of the intermediate valves and the postmucronal area of the tail valve, in longitudinal rows, fine and close set on the jugum, getting larger and posteriorly converging towards the side margins, on the pleural areas and the antemucronal area of the tail valve.

Colour of tegmentum whitish to cream, often covered with dull brown or cinereous environmental deposits.

Articulamentum white, apophyses rather large, broadly triangular with rounded top, separated by a slightly concave jugal sinus, about 1/3 of valve's width, anterior valve with an unslit, short, obsoletely striate insertion plate. II-VII with more or less obsolete insertion plates, VIII with only a thickened articulamental ridge close to the outer margin.

Girdle dorsally densely clothed with straight, smooth, needle-like spicules, 92 × 8 µm, and randomly dispersed, stronger, straight, weakly longitudinally striate spicules, stalked in short chitinous cups, varying in size from 122 × 18 µm to 256 × 20 µm. Ventrally the girdle is covered with elongate, sharply pointed, weakly longitudinally ribbed spicules, more or less truncated at the base, 92 × 24 µm in mid-girdle, 96 × 18 µm near the outer margin.

Central radula tooth only slightly longer than wide, with a thin blade, first lateral tooth narrow near the base, distally broadening and conspicuously notched in the middle of the anterior border, major lateral with a tridentate cusp, denticles sharply pointed, median one much longer than the others.

Gills merobranchial, adanal without interspace, 12 to 19 ctenidia on both sides.

Distribution. Atlantic Ocean from the Barents Sea to the Canary Islands and in the Mediterranean Sea, near Iceland, Greenland, and the E coast of N America. The bathymetric range extends from 15 to 555 m.

Observations. Of *Lepidopleurus carinatus* Dall, 1927 (non Leach, 1852), renamed *Hanleya dalli* by Kaas (1957:83), we have seen the type (USNM, 225149) from Eastport, Maine (coll. A. E. Verrill, 1870). It is a dried and strongly curled-up specimen on which all valves are broken along the median axis. The length is ca 18 mm when stretched and the specimen differs in no way from *H. hanleyi*. This makes it highly probable that *Chiton mendicarius* Mighels & Adams, 1842, is the same species. This name should have preference over *hanleyi*. Unfortunately the type of *mendicarius* was lost during the Great Fire of Boston in November 1872. As the original

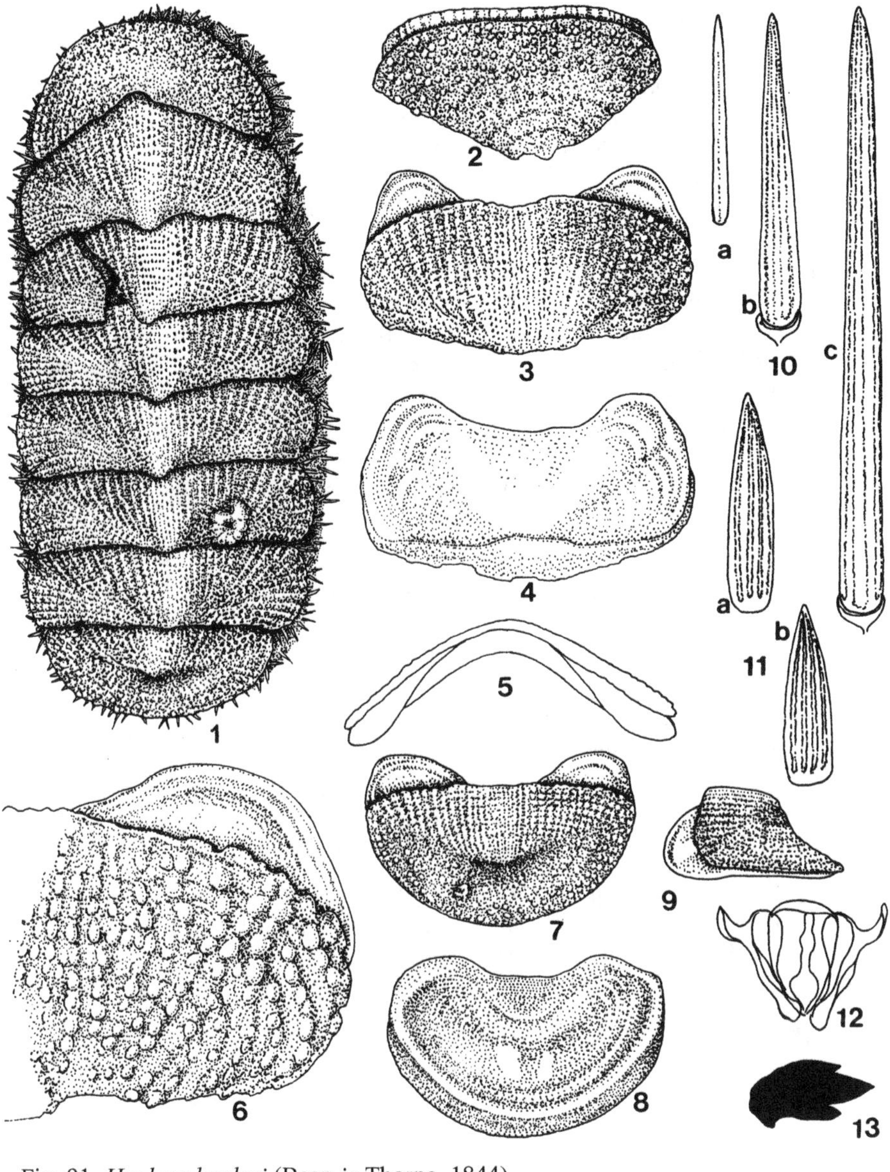

Fig. 91. *Hanleya hanleyi* (Bean *in* Thorpe, 1844).

1, whole animal, dorsal view, × 9.6; 2, valve I, antero-dorsal view, × 9.6; 3, valve V, dorsal view, × 9.6; 4, do, ventral view, × 9.6; 5, do, anterior view, × 9.6; 6, right half of do, × 20; 7, valve VIII, dorsal view, × 9.6; 8, do, ventral view, × 9.6; 9, do, lateral view, × 9.6; 10, dorsal girdle spicules, a small one, b-c large ones, × 320; 11, ventral spicules, a near outer margin, b from mid-girdle, × 320; 12, central and first lateral radual teeth, × 100; 13, cusp of major lateral tooth, × 100.

1,10-11, specimen from Dröbak, Oslofjord, Norway, K 2732; 2-9,12-13, specimen from Stjernsund, Norway, K 4956.

description and the figure of *mendicarius* are totally inadequate, we think it better to neglect this name.

Ferreira (1981b:95) created a new genus *Laminoplax* (Type: *Hanleya dalli* Kaas, 1957, by monotypy), placing it in the family Afossochitonidae.

Obviously Ferreira, unacquainted with such species as *H. hanleyi* and *H. nagelfar,* was misled by Thiele's diagnosis of the genus *Hanleya* (1929:7), "Schale ähnlich wie bei *Lepidopleurus,* doch mit einem unzerteilten vorderen Insertionsrande", which gives the impression that only the head valve is provided with an unslit insertion plate. In fact the sutural plates in *Hanleya* spread out with growth along the sides of the valves, until they reach the posterior margin, thus becoming true insertion plates. Only valve VIII bears a thickened articulamental ridge, at some distance of the outer margin, which cannot be called an insertion plate.

As *Hanleya dalli* falls into the synonymy of *H. hanleyi* (or, perhaps, *H. nagelfar,* if this is a good species), the taxon *Laminoplax* Ferreira, 1981, must be regarded as a junior objective synonym of *Hanleya* Gray, 1857, with the same type species.

Hanleya nagelfar (Lovén, 1846)

(Fig. 92. Map 19)

Lectotype: NHRM, type coll. 1329, animal without valves in alcohol, loose valves in separate box (designated herein).
Type locality: Norway, Finnmark.

Chiton nagelfar Lovén, 1846:158; Jeffreys, 1865:216; G. O. Sars, 1878:110.
Chiton abyssorum M. Sars MS, Jeffreys, 1865:216 (nom. nud.); M. Sars MS, G. O. Sars, 1878:109, pi. 7:fig. 4, pl. 34:fig. 3; Sparre Schneider, 1886:96; Oskarsson, 1944:9, fig. 1.
Hanleya abyssorum; Dall, 1879:319; Thiele, 1893:388, pl. 31:fig. 39; Burne, 1896:4, figs 1-3, pl. 2:figs 1-3; Thiele, 1909:14, pl. 2:fig. 4;Dons, 1933:151, fig. 3; Knudsen, 1949:4; Muus, 1959:41.
Hanleya hanleyi var. *abyssorum;* Pilsbry, 1892:18, pl. 4:figs 74-77; Nierstrasz & Hoffmann, 1929:30.
Hanleya nagelfar; Dons, 1933:151, figs 1-2.

Description. Animal large, up to 60 mm long, 35 mm broad, moderately

Fig. 92. *Hanleya nagelfar* (Loven, 1846).

1, whole specimen, dorsal view, × 1.6; 2, valve I, antero-dorsal view, × 4.8; 3, valve V, dorsal view, × 4.8; 4, do, anterior view, × 2.4; 5, valve VIII, lateral view, × 4.8; 6, do, dorsal view, × 4.8; 7, dorsal girdle needles, × 50; 8, ventral spicules, × 200; 9, do, in situ, × 50; 10, dorsal spicules and needle, in situ, × 50; 11, half radula row.

1, specimen from Brettingen, Norway, 150 m, 21.IV.1933, C. Dons leg., KNVSM; 2-6, specimen from Senja, Norway, 210-220 m, 26.VII.1951, T. Soot-Ryen leg., K 4767; 7-10, from a ca 60 mm long specimen, Norway, probably Trondheimsfjorden, K 4955; 11, after Thiele, 1893.

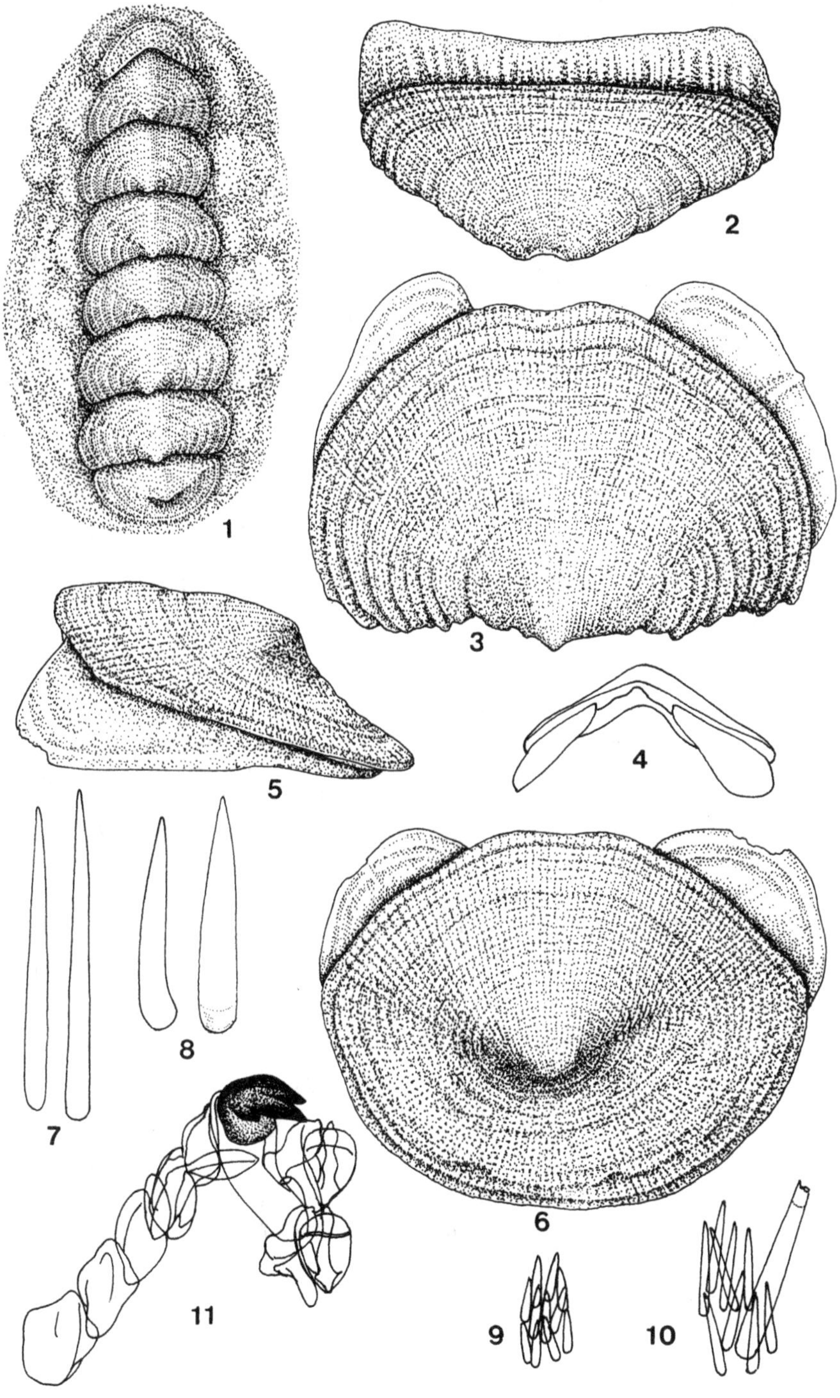
1
2
3
4
5
6
7
8
9
10
11

elevated (dorsal elevation c. 0.35), carinated, side slopes almost straight, intermediate valves of equal width, rather long (l:b= 0.70), with a slightly projecting, sharp apex. Tegmentum evenly longitudinally granulose, lateral areas not indicated. Girdle very wide and fleshy, encroaching at the sutures, closely beset with sharply pointed, glassy needles. Colour whitish to light roseate or brownish, the girdle a shade darker.

Head valve semicircular, sculptured with numerous radial rows of roundish to oval granules. There are a number of well marked, concentrical growth marks. Posterior margin of intermediate valves about straight, the apex slightly projecting to a small, sharp beak, the lateral areas, which are neither indicated nor raised, with 8-10 growth ridges (very faint on the central areas). The side margins and anterior margin half elliptical, with a small inward curve in the jugal area. Tegmentum sculptured with longitudinal series of small, roundish to oval granules, the interstices very narrow, strongly converging posteriorly towards the sides. Posterior valve transversely elliptical, with a small, somewhat posterior mucro, posterior slope slightly concave directly behind the mucro, almost flat towards the hind margin. Antemucronal area sculptured like the intermediate valves, postmucronal area like the head valve.

Articulamentum white, rather solid, forming a strongly developed, non fissured, exteriorly striate insertion plate on the head valve only, with a slight upward wave anteriorly. Apophyses rather wide, short, sinuate, placed sidewards; sinus convex near the apophyses, concave in the jugal area.

Girdle very wide and thick, encroaching at the sutures, densely covered with glassy, needle-like, smooth, spicules, 150-200 μm long, interspersed with much larger spines, up to 500 μm long. On the ventral side the spicules are smaller, uniformly up to 160 μm long, relatively thicker.

Radula with a short, oval, central tooth bearing a thin blade, first lateral tooth interiorly embracing the central tooth, with a rather wide, weakly developed cutting edge; major lateral with a tridentate cusp, the denticles sharply pointed, the central one largest.

Gills merobranchial, adanal without interspace; in a 56 mm long specimen 18 ctenidia on both sides.

Distribution. Only found off the Norwegian coast, from Bergen in the South to Finnmark in the North, and in the Eyjafjord, Iceland, always in a depth of at least 90 m; it is regularly dredged in Trondheimsfjorden.

Observations. *H. nagelfar* is a deep water species. Though fullgrown specimens are quite different from *H. hanleyi,* many authors are of the opinion that *nagelfar* is only a *hanleyi* of extraordinary size (Jeffreys, 1865: 216). Dons (1933:151) regarded *nagelfar* and *abyssorum* as specifically different, but from the types they appear to be absolutely identical. Strange enough *young nagelfar* are very rare. We have seen two specimens, dredged in 100-200 m off Röberg, Trondheimsfjorden, respectively 23 mm and 18 mm long, which are in sculpture exactly like *hanleyi,* but different from it in the very wide, light yellowish, encroaching girdle.

Thiele (1909:14) was the first who examined the girdle armature of both forms and observed that the spicules of *hanleyi* are rather strongly lengthwise ribbed, while those of *abyssorum* (= *nagelfar)* are invariably smooth. Nevertheless, Thiele concluded that Sparre Schneider (1886:97), who on basis of the sculpture could not separate *hanleyi* from *abyssorum,* might be right.

In our opinion the differences in shape of the valves and in the girdle armature are convincing enough to justify specific separation of both forms.

Besides, there is also a difference in the biology of the two, though much research has still to be done. *"H. nagelfar* lives on certain species of sponges. I have never collected it in the absence of these, and usually attached to them. *H. hanleya,* typical specimens, usually lives on rocks, shells and stones. It feeds mainly on detritus, occasionally on small sponges. There are no problems to separate them from *H. nagelfar.* It is a quite common species, and occurs from depths of 15 m and deeper, while *H. nagelfar* occurs deeper than 80 m in S Scandinavia." (A. Warén, Zoological Institute, University of Göteborg, in litt, 17.06.1978.) Warén also found an intermediate form on dead branches of *Lophohelia.* So the last word on Scandinavian *Hanleya* species has not yet been spoken.

Hanleya tropicalis Dall. 1881

(Fig. 93. Map 17)

Type: USNM 95172 (paratype).
Type locality: Florida, Sands Key, Pourtales, 230 m.

Hanleyia tropicalis Dall, 1879:319 (nom. nud.); 1881:53 (diagn.).
Hanleya tropicalis; Kaas, 1972:20, figs 13-18 (bibliography).

Description. Animal small, about 10 mm in length, 5 mm in width, rather highly elevated, carinated, girdle narrow, colour entirely white.

Head valve lacking in the studied paratype (?) specimen mentioned above. According to Dall (1881:53): "Anterior valve with a well marked mucro from which proceed concentric striae of growth, and in front of which the valve is a little concave". Intermediate valves rounded at the side margins, almost straight behind, apices indicated, side slopes nearly straight, angle of divergence about 100°. Tail valve triangular, posterior margin decidedly waved upward in the centre, mucro submedian, prominent, somewhat swollen.

Tegmentum sculptured with small, flattened nodules, more or less quincuncially arranged on the head valve, the postmucronal area of the tail valve, and the central areas of the intermediate valves. The lateral areas are marked with fine radiating striae. There are also irregularly distributed, opaque white pustular bubbles looking like attached grains of coral sand, in some places the pustules take on a decidedly cylindrical, elongated shape, looking like the stalks of decapitated mushrooms.

Articulamentum white, apophyses rather well developed, wide, more or less evenly rounded, jugal sinus about 1/3 of valve's width, anterior valve with an unslit insertion plate, other valves without insertion plates.

Girdle thin, narrow, dorsal side densely clothed with three kinds of vitreous spiculae: large slender ones, 620 μm long (not 1-1.5 mm as Dall states), small slender ones, 50-80 μm long, and thicker small ones of about the same size. Ventrally the girdle is covered by close set, sharply pointed spicules of about 125 μm long.

Distribution. Only known from the type locality.

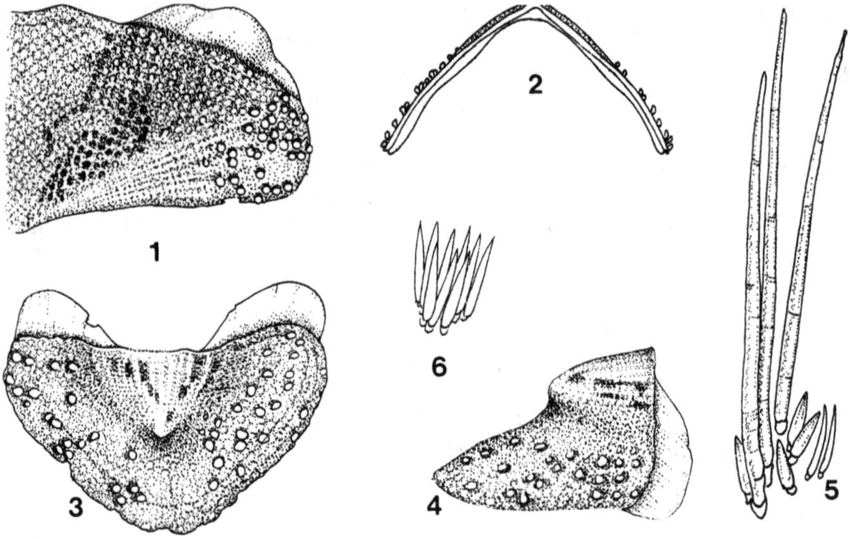

Fig. 93. *Hanleya tropicalis* Dall, 1881.

1, right half of valve VI, dorsal view, × 18; 2, valve VI, posterior view, × 13; 3, valve VIII, dorsal view, × 18; 4, do, lateral view, × 18; 5, dorsal girdle spines, × 100; 6, ventral girdle spicules, × 100.

1-6, paratype, after Kaas, 1972.

Genus ***Hemiarthrum*** Carpenter in Dall, 1876

-Bull.U.S.nat.Mus. 3:44.

Type: *Hemiarthrum setulosum* Carpenter *in* Dall, 1876, M.

Animal small, tegmentum practically smooth, insertion plates present on both end valves, unslit, girdle with scattered small spicules and longer spines, forming tufts at the sutures and around the end valves.

Distribution. Antarctic seas.

Hemiarthrum setulosum Carpenter *in* Dall, 1876

(Fig. 94. Map 40)

Type: USNM 11903.
Type locality: Kerguelen Island.

Hemiarthrum setulosum Carpenter in Dall, 1876:44; Castellanos, 1951:29, fig. 10; Carcelles, 1953:158; Leloup, 1956b:16, figs 3-4 (bibliography and synonymy); Dell, 1964a:268; 1971:216; Powell, 1979:22.
Chiton castaneus Couthouy MS, Gould, 1852:326, fig. 411c (Non *C. castaneus* Wood, 1815, nee Quoy & Gaimard, 1835); Thiele, 1909:40.
Acanthochiton couthouyi de Rochebrune, 1889:133.
Hemiarthrum hamiltonorum Iredale & Hull, 1932:137, pl. 8:figs 9-15; Cotton, 1937:14; Powell, 1955:134.

Description. Shell of small to moderate size, elongate oval, generally 9-12 mm long, exceptionally up to 15 mm, width about half the length, dorsal elevation varying from rather flat to moderately arched, not carinated, jugum evenly rounded, side slopes almost straight, girdle leathery with scattered small spicules and tufts of spines.

Head valve semicircular, posterior margin slightly curved inward. Intermediate valves more than twice as wide as long, anterior margin sinuate, concave in the jugal area, convex at the pleurae, posterior margin evenly convex to decidedly beaked, beaks rounded, apices not indicated, lateral areas not raised, not distinguishable from the central areas. Tail valve less than semicircular, width of tegmentum about three times the length, anterior margin slightly concave, mucro far behind the centre, not prominent, antemucronal slope convex.

Tegmental sculpture obsolete, if not eroded, central area of intermediate valves and antemucronal area of tail valve with a micro-sculpture of extremely fine, longitudinal grooves, some specimens with a weak, rough granulation near the outside margins; all valves with several concentric growth marks near the anterior margin, eaves very porous.

Articulamentum strongly developed, brownish, insertion plate of head valve rather wide, radially striate on the upper side, insertion plate of tail valve smooth, the tegmentum exceeding it; on valves I-VII the underfolded part of the tegmentum forms a wide and thick ridge on the inside of the posterior margin; apophyses large, each one about 1/3 the width of the valve, more or less triangular in outline, tops rounded, trapezoid in valve VIII, separated by a more or less concave sinus.

Colour of tegmentum chestnut brown, dark sepia, or almost black, girdle generally a shade lighter.

Girdle leathery with scattered, widely separated spicules of two kinds: drop-shaped ones, embedded in very short cups, length 30-56 µm, width 22 µm, and larger, conical ones, not cupped, up to 70 µm long; all spicules finely, longitudinally striate near the top. Intersegmentally there are tufts of 6-8 white, slightly curved, weakly longitudinally striate spicules, c. 180 × 24 µm; around the head valve there are 6, and behind the tail valve 2 such tufts, the total number thus adding up to 22. The marginal fringe is formed by strong, straight, longitudinally ribbed spicules of 124 × 30 μm. Ventrally the girdle is covered with elongate, smooth, sharply pointed spicules, 60 × 16 μm.

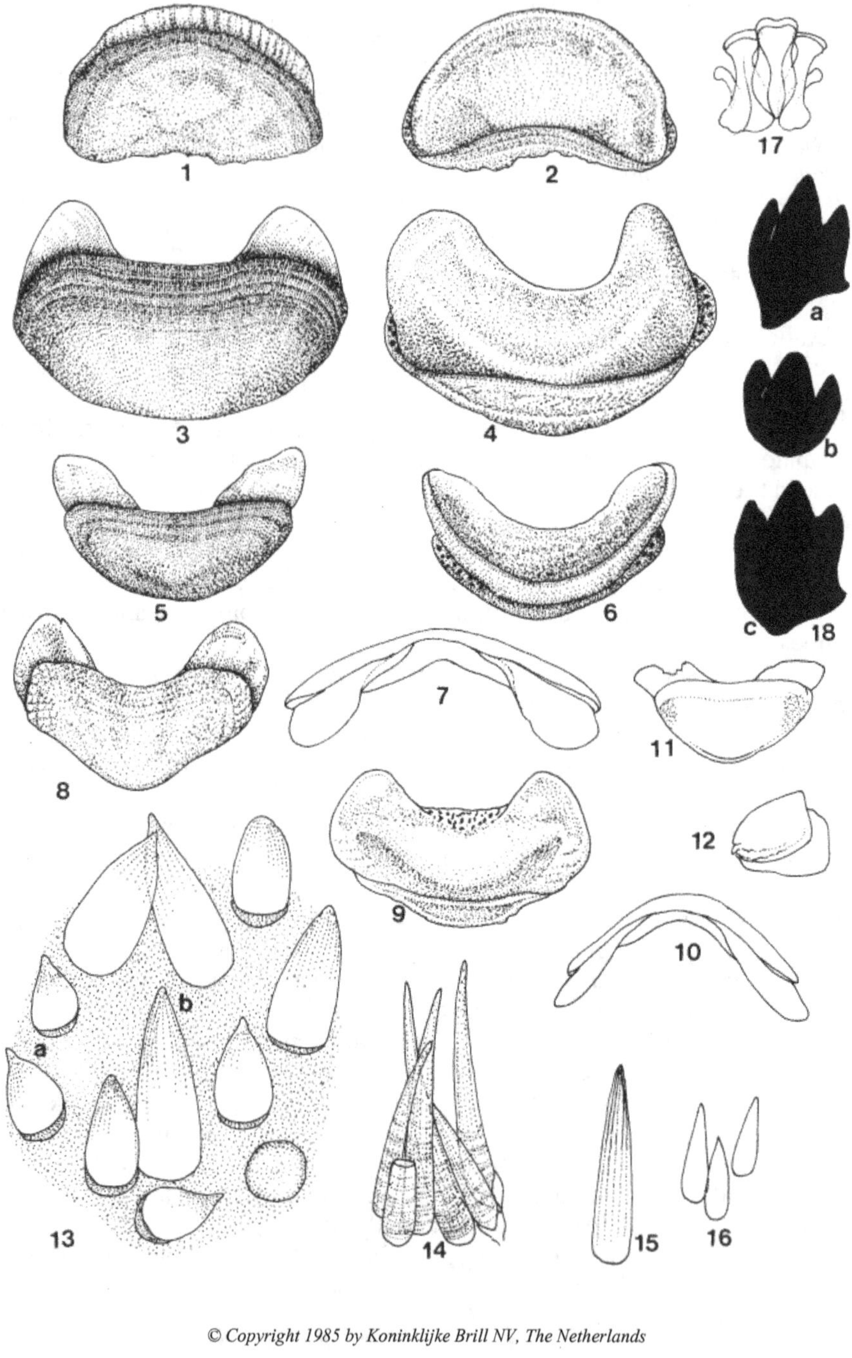

Central radula tooth broadening to a thin, weakly notched blade, first lateral somewhat pinched in the middle, with a narrow appendage on the outer side, major lateral with a tridentate cusp, the central denticle largest.
Gills merobranchial, abanal, about 6 on each side (Pilsbry, 1892:20).

Distribution. Between latitudes 45°S and 70°S all around the mainland of Antarctica: Patagonia, Tierra del Fuego, S. Georgia, S. Shetlands, S. Orkneys, Crozet, Kerguelen and Macquarie Islands. The species has been recorded from low water to a depth of 380 m.

Family AFOSSOCHITONIDAE Ashby, 1925

Proc. R. Soc. Vict. 37 (n.s.) (2): 179 (nom. transl. Bergenhayn, 1945, ex Afossochitoninae Ashby, 1925).

=*Leptoidea* Dall, 1879 (partim); *Holochiton* Fischer, 1885 (partim); *Eochiton* Fischer, 1885 (partim); *Microplaxinae* Ashby, 1921; *Choriplacinae* Ashby, 1928; *Glyptochitonidae* Starobogatov & Sirenko, 1975.

Key to genera

1a Intermediate valve twice as long as wide .... *Glyptochiton* (†)
1b Intermediate valves less than twice as long as wide 2
2a Tegmentum much reduced .... *Choriplax*
2b Tegmentum of normal proportions 3
3a Sculpture granulose .... *Afossochiton* (†)
3b Sculpture of widely spaced ribs .... *Lirachiton* (†)

Fig. 94. *Hemiarthrum setulosum* Carpenter *in* Dall, 1876.

1, valve I, dorsal view, × 19; 2, do, ventral view, × 19; 3, intermediate valve, dorsal view, × 19; 4, do, ventral view, × 19; 5, valve VIII, dorsal view, × 19; 6, do, ventral view, × 19; 7, intermediate valve, anterior view, × 19; 8, valve IV, dorsal view, × 19; 9, do, ventral view, × 19; 10, do, anterior view, × 19; 11, valve VIII, dorsal view, × 19; 12, do, lateral view, × 19; 13, part of girdle, dorsal view, a small spicules, b large spicules, × 375; 14, sutural tuft, × 190; 15, marginal spicule, × 190; 16, ventral spicules, × 190; 17, central and first lateral radula teeth, × 375; 18a-c, cusps of major lateral radula teeth, seen from different sides, × 375.

1-7, disarticulated valves from Macquarie Island, C. Hedley leg., AM C46625; 8-18, specimen from S Georgia, Hystadhullet, 10.II.1928, O. Olstad leg., "Norvegia" Exp., K 3700.

Genus *Choriplax* Pilbry, 1894

Nautilus 7:139.

=*Microplax* H. Adams & Angas, 1864, non Fieber, 1861 (Hemiptera).

Type: *Microplax grayi* H. Adams & Angas, 1864, OD.

Animal small, elongate, tegmentum much reduced, heart-shaped, strongly granulated. Girdle corneous, nude to the naked eye.

Distribution. New South Wales and South Australia.

Choriplax grayi (H. Adams & Angas, 1864)

(Fig. 95. Map 16)

Type: BMNH 1877.11.7.2.
Type locality: New South Wales, Sydney Harbour.

Microplax grayi H. Adams & Angas, 1864:194; Angas, 1865:58, pl. 2:fig. 16; 1867b:224; Pilsbry, 1892:21, pl. 6:figs 9-11.
Choriplax grayi; Ashby, 1921c:136; 1924c:331; Iredale & Hull, 1925:99, pl. 11:figs 24-27.
Choriplax grayi pattisoni Ashby, 1921c:137, pl. 9:figs la-c; Iredale & Hull, 1925: 100, pl. 11:fig. 28.
Choriplax pattisoni Ashby, 1924c:331; Cotton & Weeding, 1939:188; Cotton & Godfrey, 1940:540, fig. 542; Cotton, 1964:85, fig. 96.

Description. Animal rather small, ca 12 × 3.5 mm, back much elevated, exposed parts of the valves minute (in valve II 0.88 mm long, 1.34 mm broad), separated, the intervals almost as long as the visible part of the valves. Tegmentum strongly granulated, brownish. Girdle very large, membraneous.

Head valve (now missing in the holotype) evenly rounded, more than semicircular, posterior margin very widely V-shaped. Intermediate valves roughly rectangular, side margins almost straight, posterior margins concave, tegmentum much reduced, broadly heart-shaped, the point directed forward, lateral areas neatly indicated, somewhat depressed instead of raised, apex conspicuous, not reaching the posterior edge of the valve, directed backwards. Tail valve as long as wide, of half oval form, tegmentum almost circular, anteriorly pointed in the middle of the anterior margin, mucro antemedian, directed forwardly, postmucronal slope slightly convex.

Tegmentum basically sculptured with coarse granules. On the head valve the granules are roundish, regularly arranged in radiating rows, getting larger towards the outer margin. On the intermediate valves the more or less coalescing granules are also radially directed, with 4-5 almost longitudinal, broad ribs on the jugal area, and 2-3 narrow ribs of much smaller granules forming a neat separation between the central and lateral areas. On the tail valve the granules are also much smaller and regularly arranged in radiating fan-like

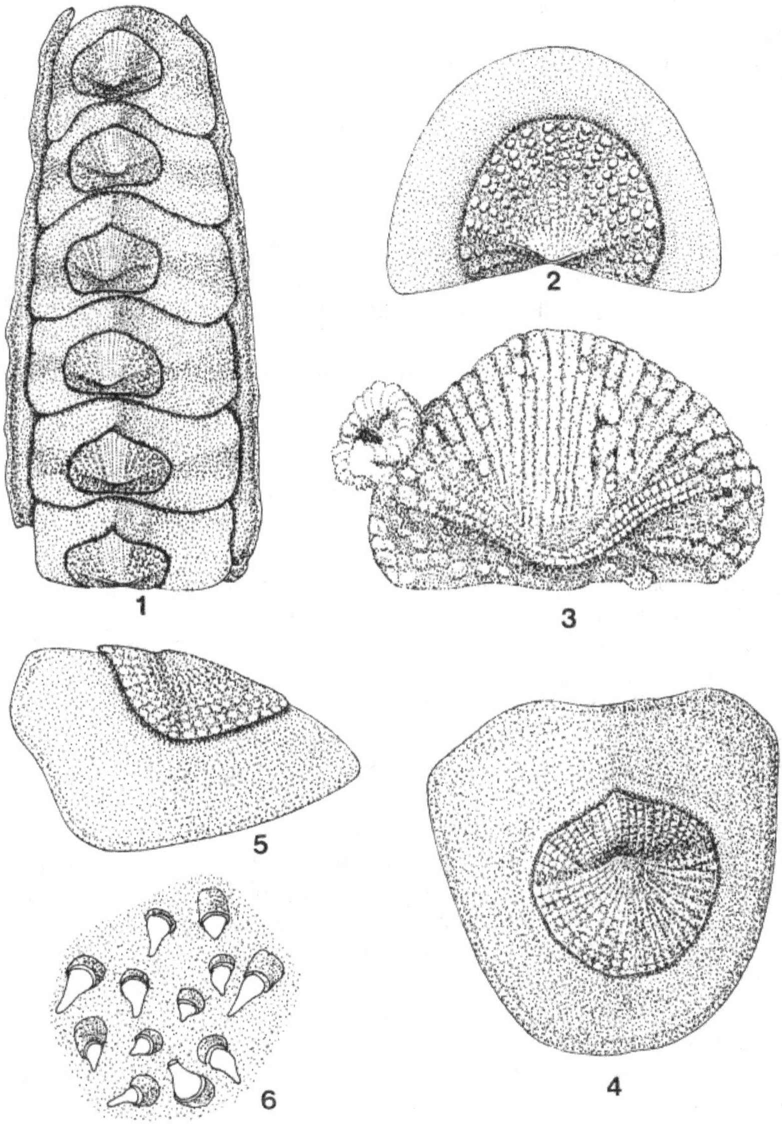

Fig. 95. *Choriplax grayi* (Adams & Angas, 1864).

1, whole animal, valves I and VIII missing, dorsal view, × 9.6; 2, valve I, dorsal view, c. × 20; 3, tegmentum of valve II, × 40; 4, valve VIII, dorsal view, × 20; 5, do, lateral view, × 20; 6, dorsal girdle spicules, in situ, × 320.

1-6, holotype; 2, enlarged after E. A. Smith *in* Pilsbry, 1892.

series on the postmucronal area, in posteriorly strongly converging longitudinal series on the antemucronal area.

Articulamentum very large, apophyses peculiarly anteriorly produced, jugal sinus extremely slight and far in front of the top of the heart-shaped tegmentum. Insertion plates present in all valves, very wide and unslit.

Girdle light yellowish, very thin, transparent, rather sparingly beset all over with 20-30 μm long, smooth, conical, calcareous spicules, abruptly widening towards the base which is enclosed in a yellowish chitinous cup embedded in the cuticula.

Distribution. Only two specimens are known. The holotype is from Port Jackson, New South Wales, where it was procured, according to Angas (1867), "adhering to a stone dredged in Watson's Bay". A second specimen was found by G. Pattison near Cape Banks Lighthouse, South Australia, washed ashore after a heavy storm amongst large kelp, described by Ashby as *Choriplax grayi pattisoni.*

Observations. Through the kind assistance of Mrs Solene Morris of the BMNH we had the opportunity to study the unique type of *C. grayi.* It is a dried and cleaned shell of which the end valves are disarticulated, the head valve missing, the girdle strongly curled-up. Pilsbry, quoting from Carpenter's MS, gives a measurement of 13 × 5 mm. We found the greatest width (girdle included) to be no more than 3.5 mm, the length of the 6 intermediate valves in situ 8.17 mm, so the total length of the animal may be estimated to c. 12 mm (length: width ratio 3.43). The tegmentum, according to Carpenter in Pilsbry, measures 1.5 × 1.75 mm. We found exactly 0.88 × 1.34 mm for valve II (1:w= 0.66). The specimen described by Ashby as *"pattisoni"* is much larger, 18 × 8.5 mm and proportionally wider (1:w= 2.28), elongate oval (not "broadly oval" as Ashby states). That the tegmentum of *pattisoni* should be "proportionally smaller" than that of *grayi,* as Ashby mentions, is only based on Carpenter's erroneous measurements. Ashby gives 1 × 1.5 mm (l:w= 0.67), which is almost 14% larger.

Carefully reading Ashby's elaborate description of *"pattisoni"* we found no other differences than the dimensions of the two specimens. There is no difference whatsoever in the shape of the valves, nor in the shape and sculpture of the tegmentum. As long as only two specimens of this remarkable animal, which may well be qualified as a "living fossil", are known, we thinck it premature to distinguish them as different species on such objectionable grounds.

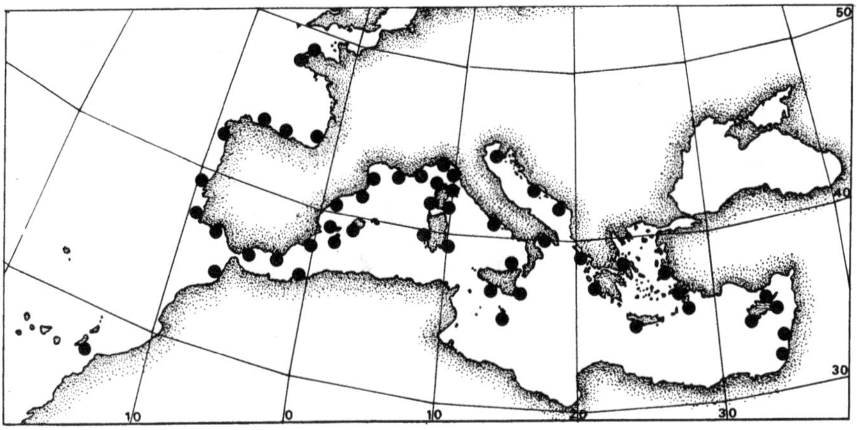

MAP 1. *Lepidopleurus cajetanus* (Poli).

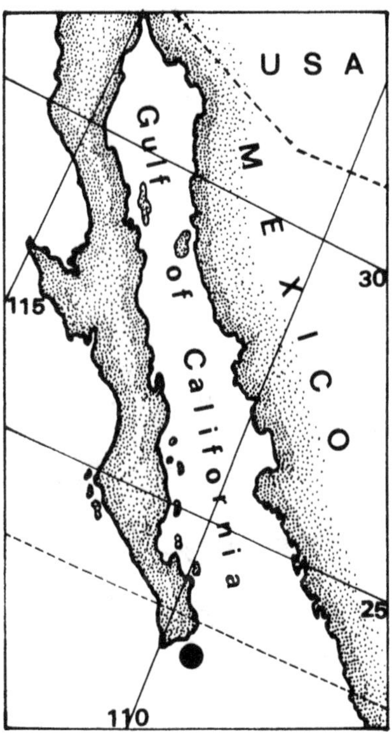

MAP 2. *Lepidopleurus scrippsianus* Ferreira.

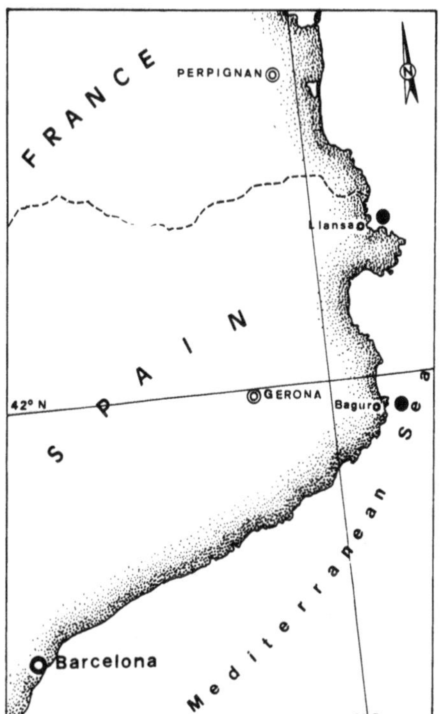

MAP 3. *Leptochiton (L.) geronensis* sp. nov.

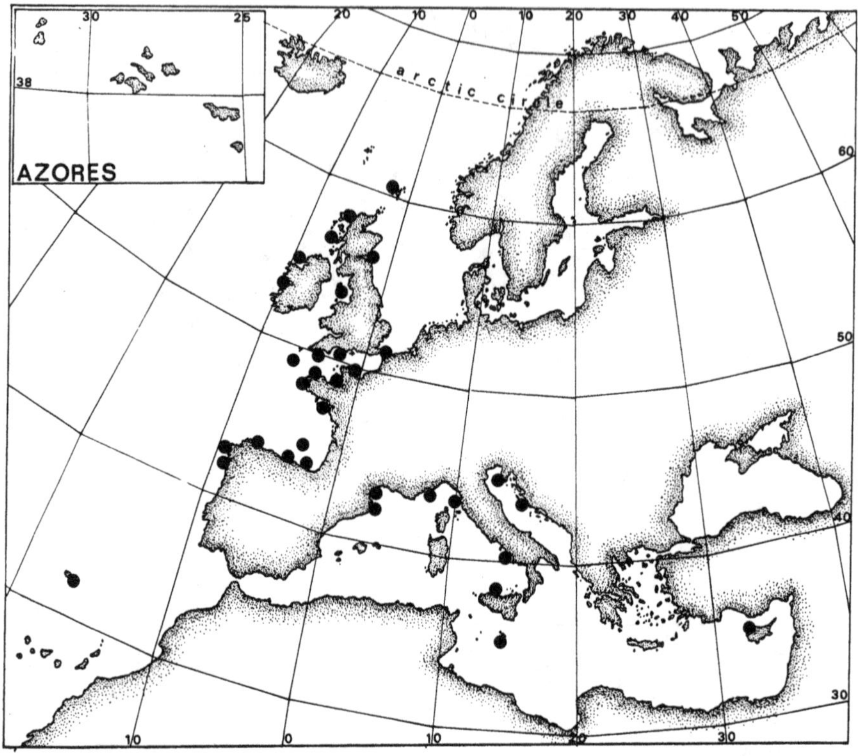

MAP 4. *Leptochiton (L.) cancellatus* (Sowerby II).

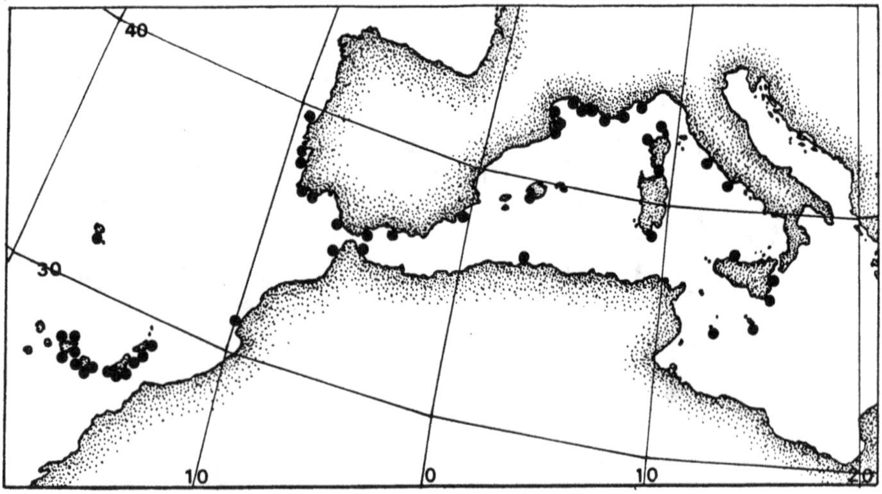

MAP 5. *Leptochiton (L.) algesirensis* (Capellini).

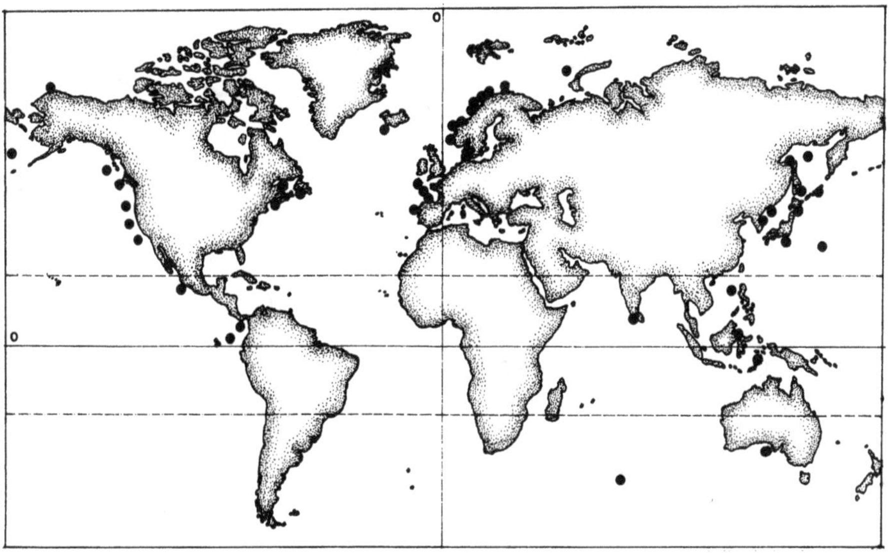

MAP 6. *Leptochiton (L.) alveolus* (Lovén).

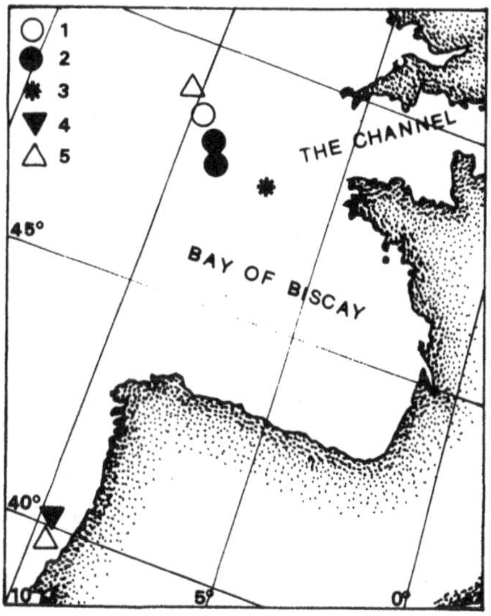

MAP 7. 1, *Leptochiton (L.) tenuis* Kaas; 2, *Leptochiton (L.) leloupi* Kaas; 3, *Leptochiton (L.) gascognensis* sp. nov.; 4, *Leptochiton (L.) thalattius* sp. nov.; 5, *Leptochiton (L.) rarinota* (Jeffreys).

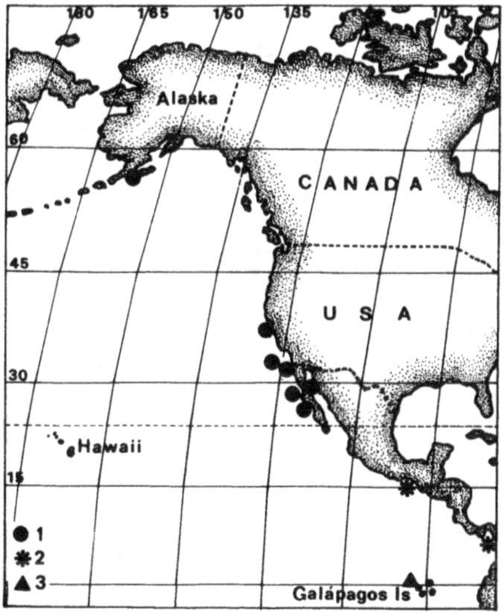

MAP 8. 1, *Leptochiton (L.) nexus* (Carpenter); 2, *Leptochiton (L.) incongruus* (Dall); 3, *Leptochiton (L.) albemarlensis* (Smith & Ferreira).

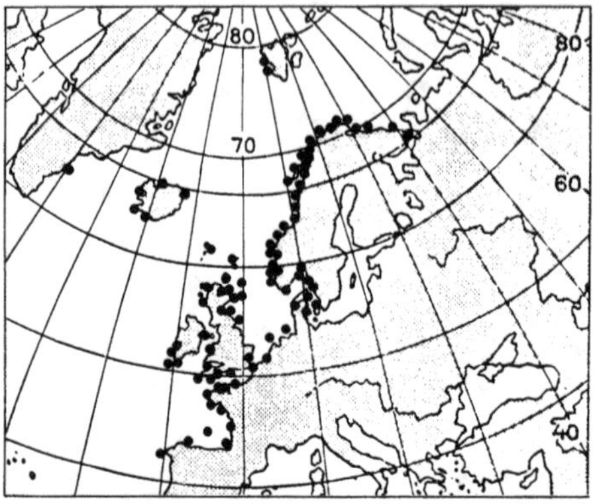

MAP 9. *Leptochiton (L.) asellus* (Gmelin).

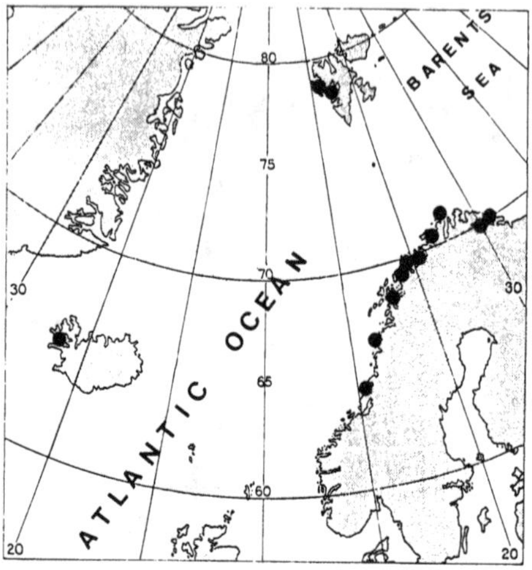

MAP 10. *Leptochiton (L.) arcticus* (Sars).

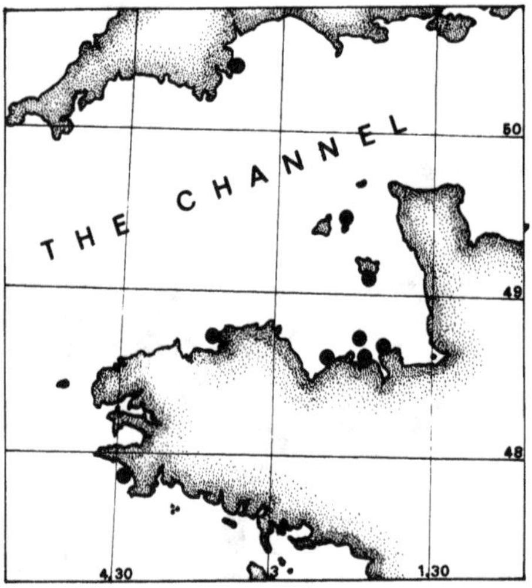

MAP 11. *Leptochiton (L.) scabridus* (Jeffreys).

MAP 12. *Leptochiton (L.) sarsi* Kaas.

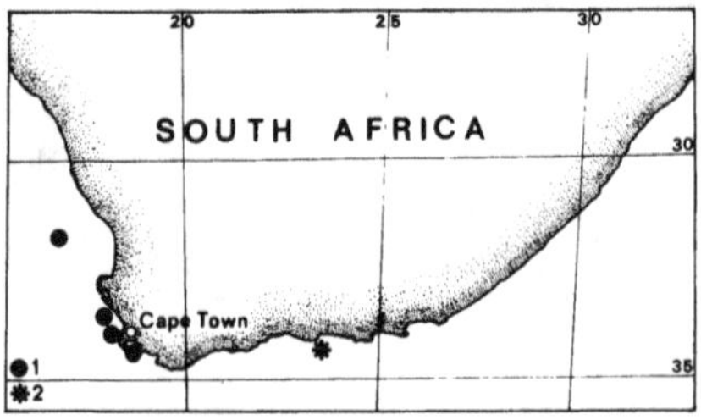

MAP 13. 1, *Leptochiton (L.) sykesi* (Sowerby III); 2, *Leptochiton (L.) chariessa* (Barnard).

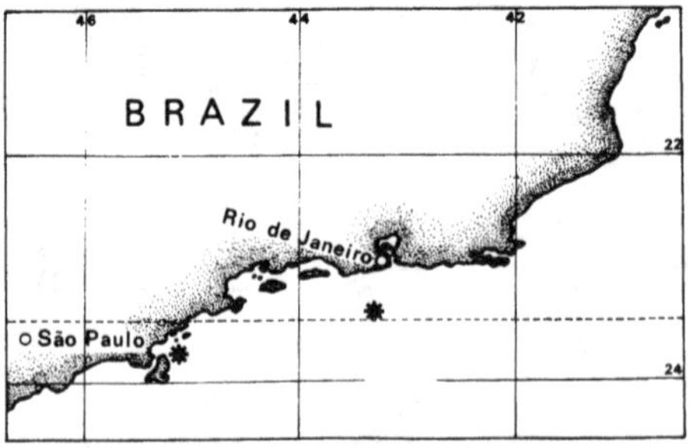

MAP 14. *Leptochiton (L.) darioi* (Righi).

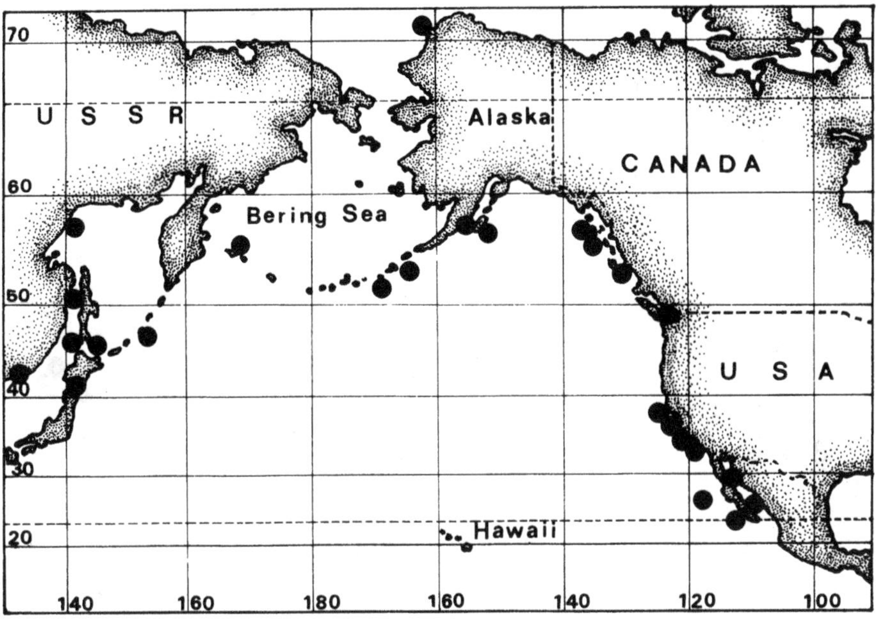

MAP 15. *Leptochiton (L.) rugatus* (Carpenter in Pilsbry).

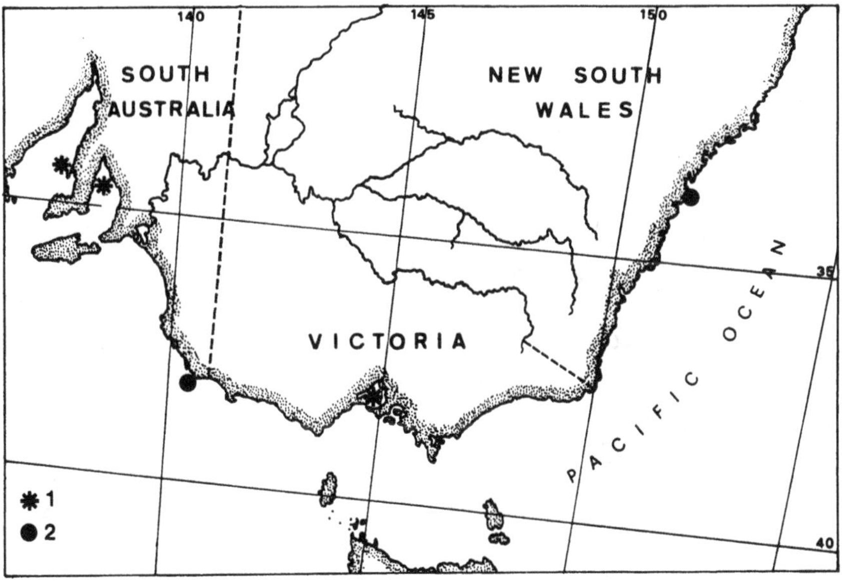

MAP 16. 1, *Leptochiton (L.) collusor* (Iredale & Hull); 2, *Choriplax grayi* (Adams & Angas).

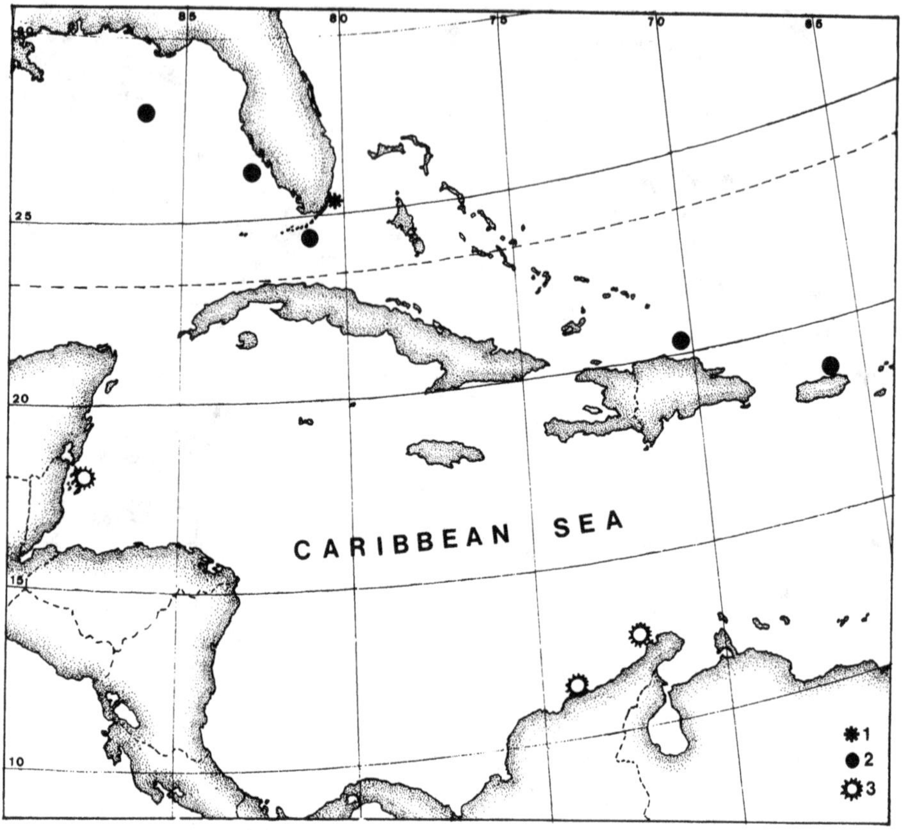

MAP 17. 1, *Hanleya tropicalis* Dall; 2, *Leptochiton (L.) pergranatus* (Dall & Simpson); 3, *Leptochiton (L.) binghami* (Boone).

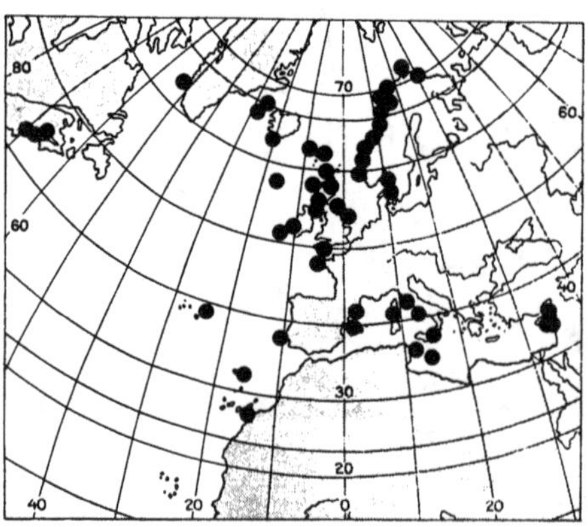

MAP 18. *Hanleya hanleyi* (Bean in Thorpe).

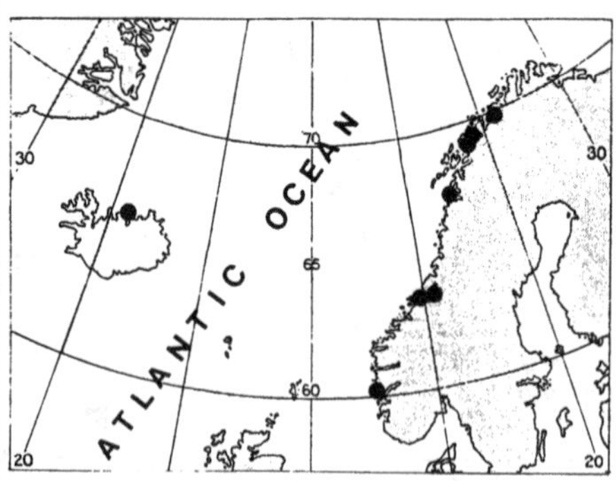

MAP 19. *Hanleya nagelfar* (Lovén).

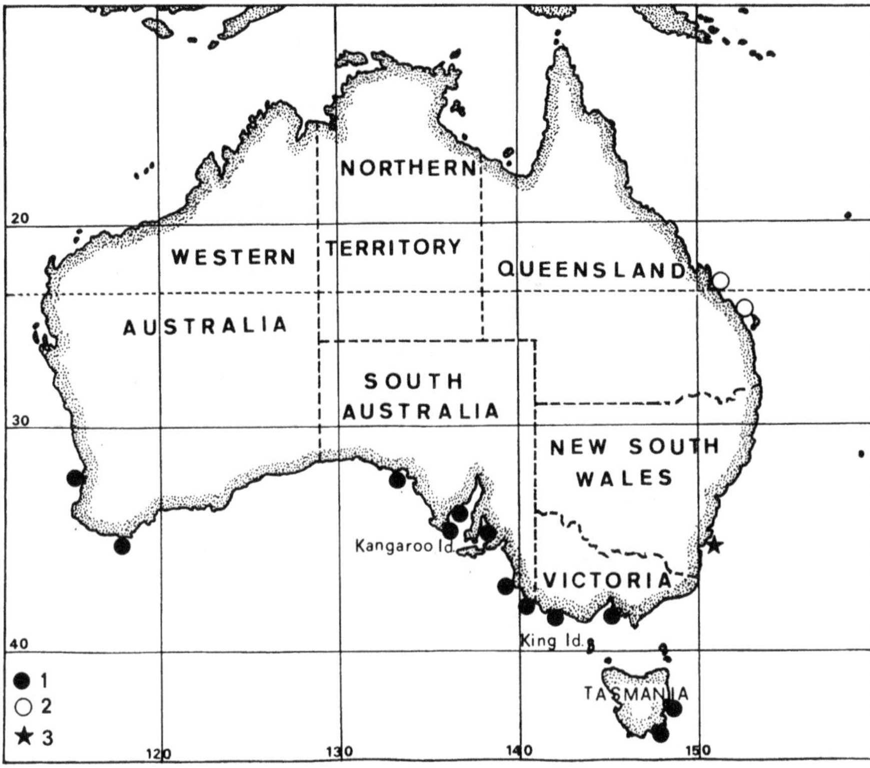

MAP 20. 1, *Leptochiton (L.) liratus* (Adams & Angas); 2, *Leptochiton (L.) liratellus* (Iredale & Hull); 3, *Leptochiton (L.) sperandus* (Iredale & Hull).

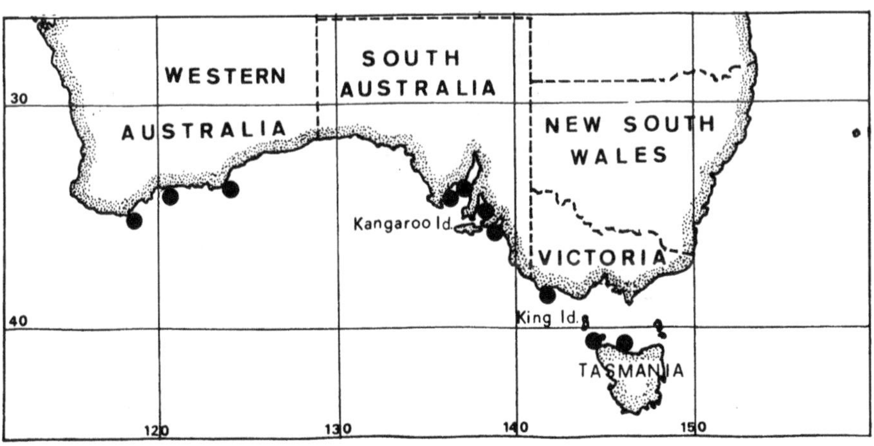

MAP 21. *Leptochiton (L.) matthewsianus* (Bednall).

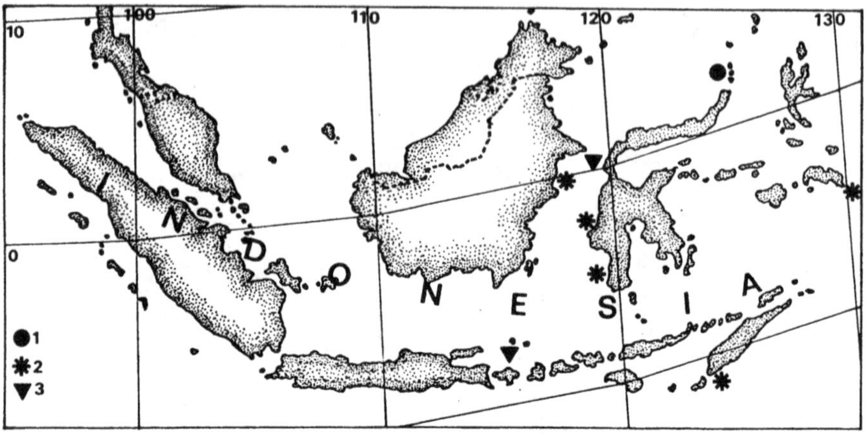

MAP 22. 1, *Leptochiton (L.) planus* (Nierstrasz); 2, *Leptochiton (L.) lineatus* (Nierstrasz); 3, *Leptochiton (Pilsbryella) setigera* (Nierstrasz).

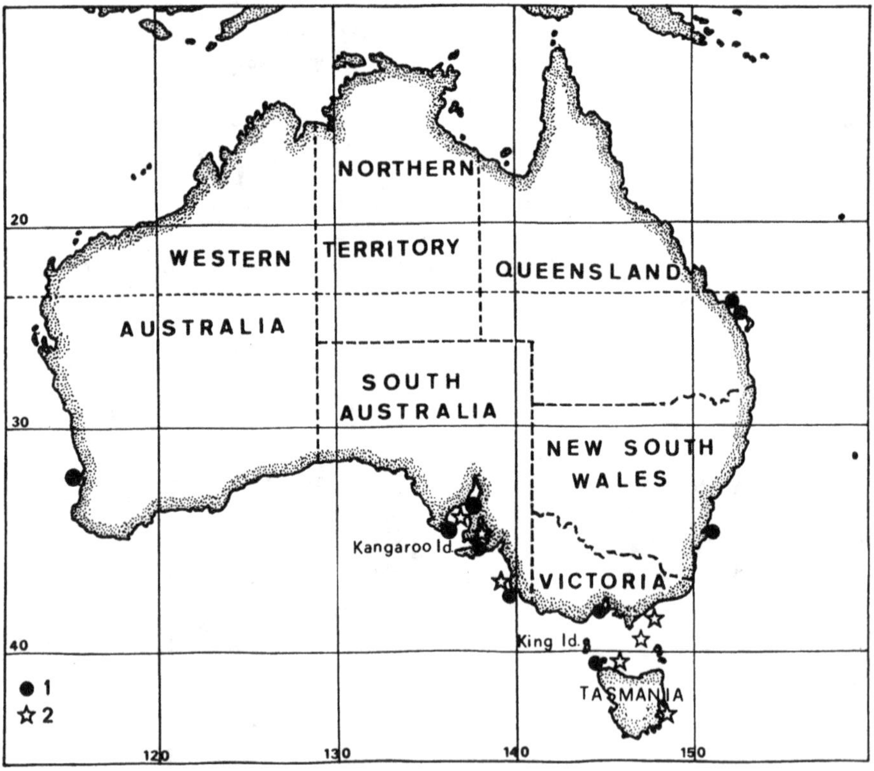

MAP 23. 1, *Leptochiton (L.) badius* (Hedley & Hull); 2, *Leptochiton (L.) columnarius* (Hedley & May).

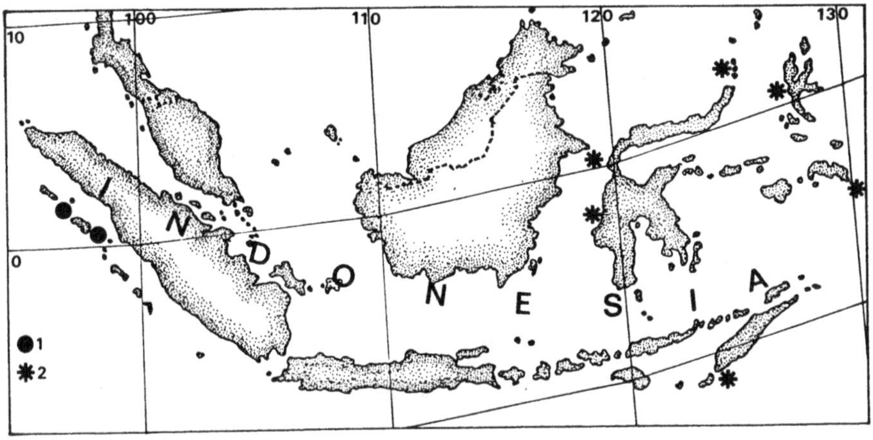

MAP 24. 1, *Leptochiton (L.) niasicus* (Thiele); 2, *Leptochiton (L.) rissoi* (Nierstrasz).

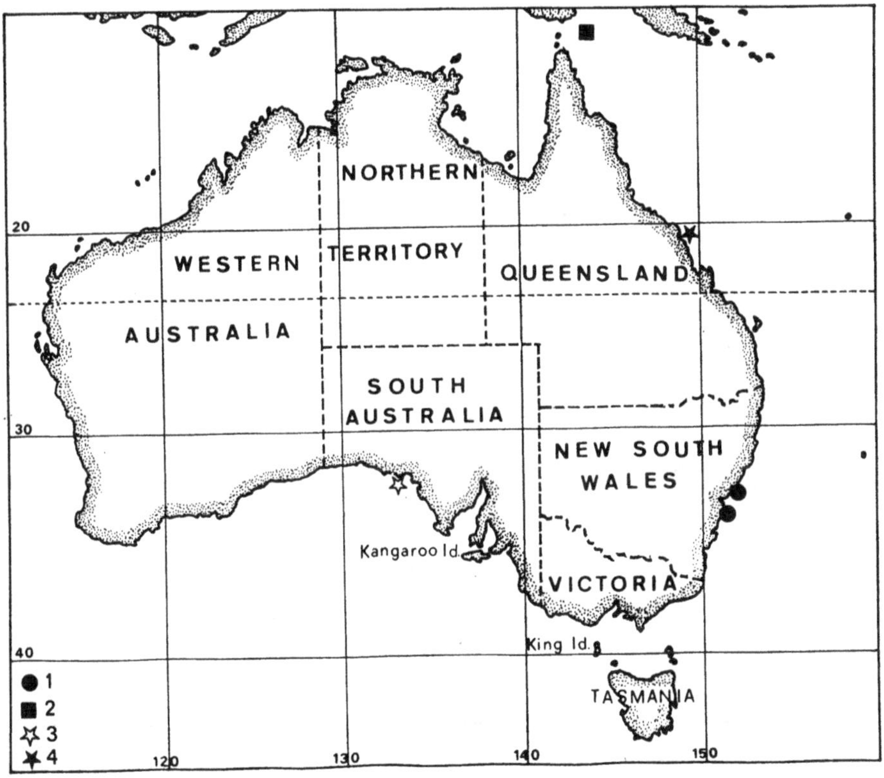

MAP 25. 1, *Leptochiton (Parachiton) puppis* (Hull); 2, *Leptochiton (Parachiton) litoreus* (Iredale & Hull); 3, *Leptochiton (Parachiton) verconis* (Cotton & Weeding); 4, *Leptochiton (Parachiton) capricornicus* (Iredale & Hull).

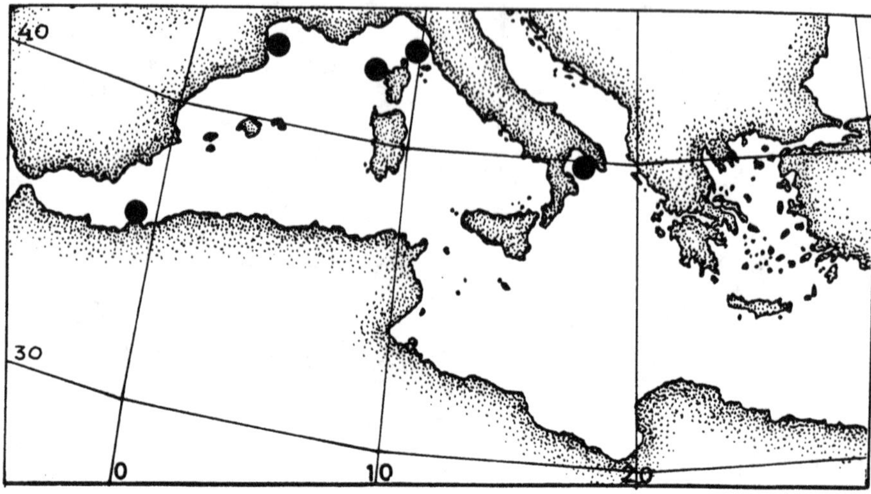

MAP 26. *Leptochiton (Parachiton) africanus* (Nierstrasz).

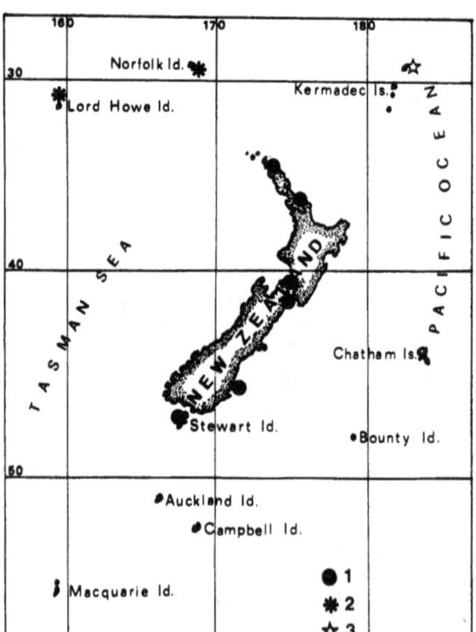

MAP 27 1, *Leptochiton (L.) inquinatus* (Reeve); 2, *Leptochiton (L.) norfolcensis norfolcensis* (Hedley & Hull); 3, *Leptochiton (L.) norfolcensis subtropicalis* (Iredale).

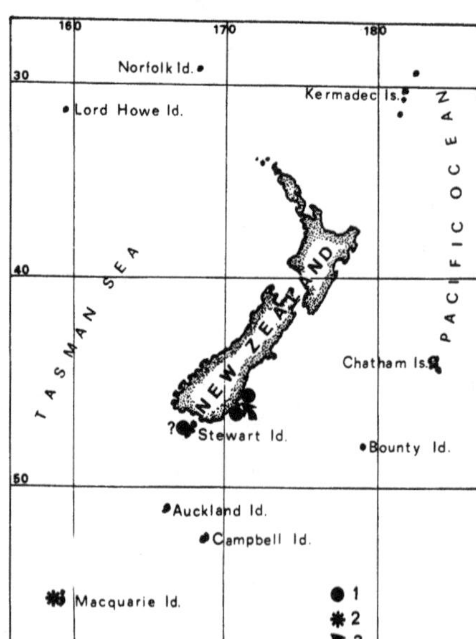

MAP 28. 1, *Leptochiton (L.) otagoensis* (Iredale & Hull); 2, *Leptochiton (L.) fairchildi* (Iredale & Hull); 3, *Leptochiton (L.) finlayi* (Ashby).

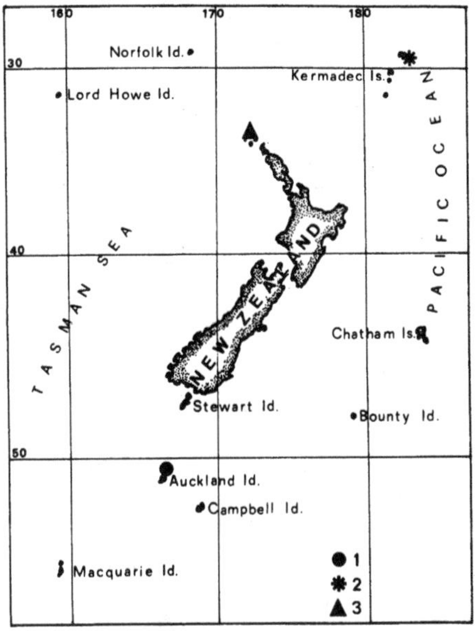

MAP 29. 1, *Leptochiton (L.) subantarcticus* (Iredale & Hull); 2, *Leptochiton (Parachiton) mestayerae* (Iredale); 3, *Leptochiton (Parachiton) textilis* (Powell).

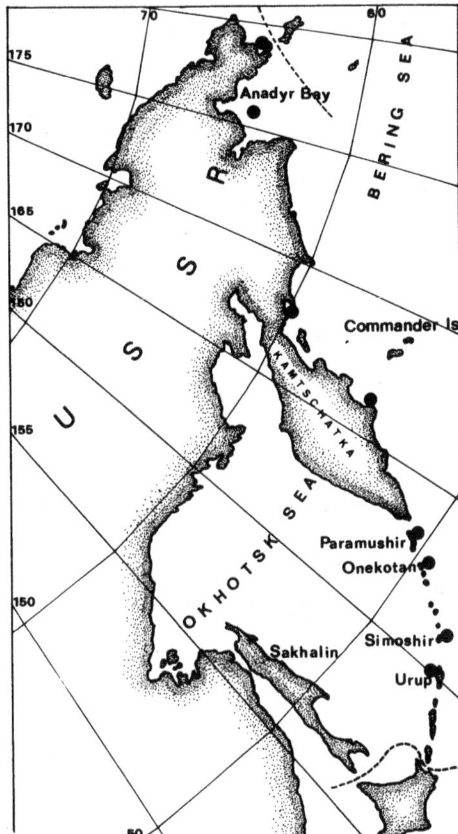

MAP 30. *Hanleyella asiatica* Sirenko.

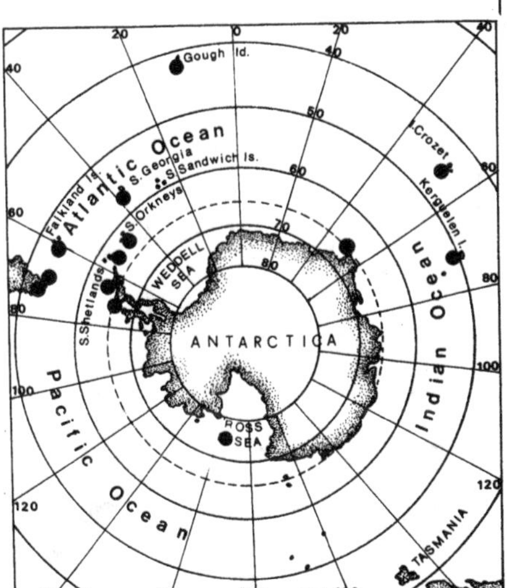

MAP 31. *Leptochiton (L.) kerguelensis* (Haddon).

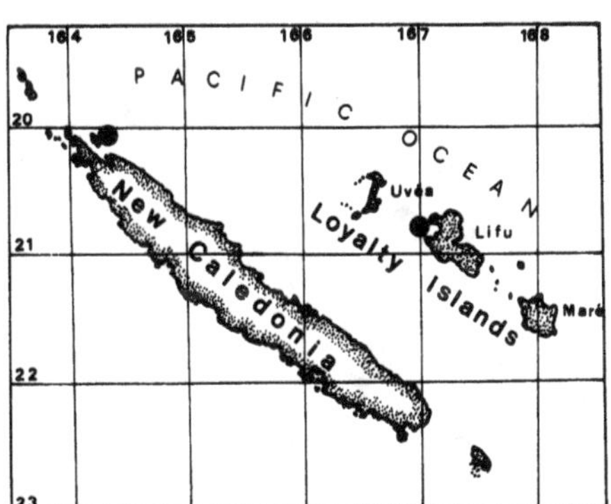

MAP 32. *Leptochiton (Parachiton) lifuensis* (Hull & Risbec).

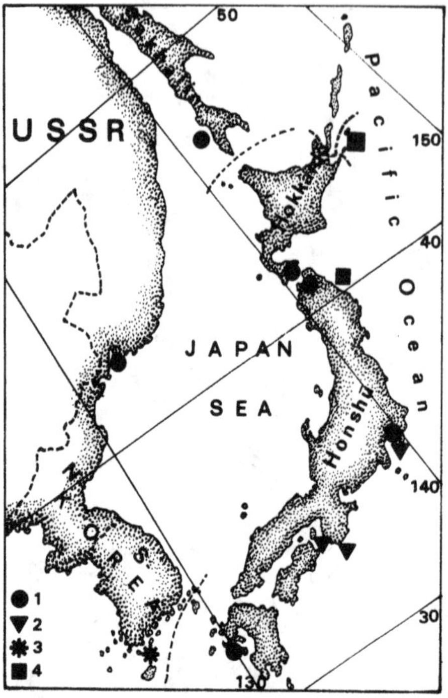

MAP 33. 1, *Leptochiton (L.) hakodatensis* (Thiele); 2, *Leptochiton (L.) hirasei* (Taki & Taki); 3, *Leptochiton (L.) fuliginatus* (Reeve); 4, *Leptochi-ton (L.) batialis* (Sirenko).

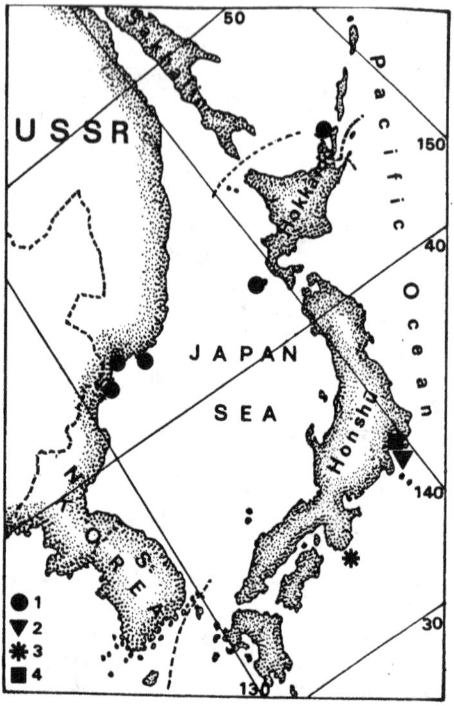

MAP 34. 1, *Leptochiton (L.) curvatus* (Carpenter MS, Pilsbry); 2, *Leptochiton (L.) latidens* (Bergenhayn); 3, *Leptochiton (L.) diomedeae* (Berry); 4, *Oldroydia bidentata* Taki.

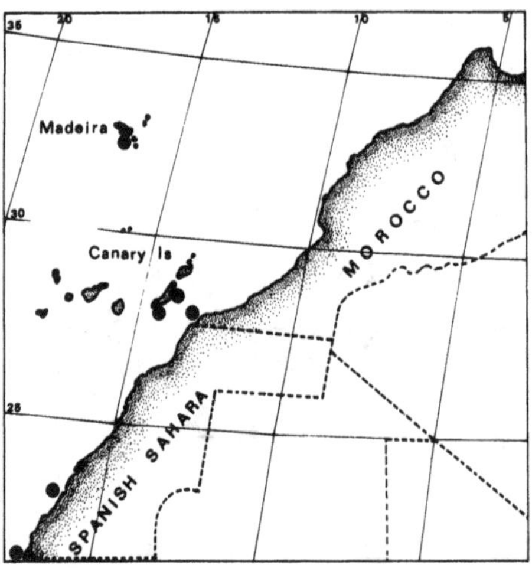

MAP 35. *Leptochiton (L.) odhneri* (Bergenhayn).

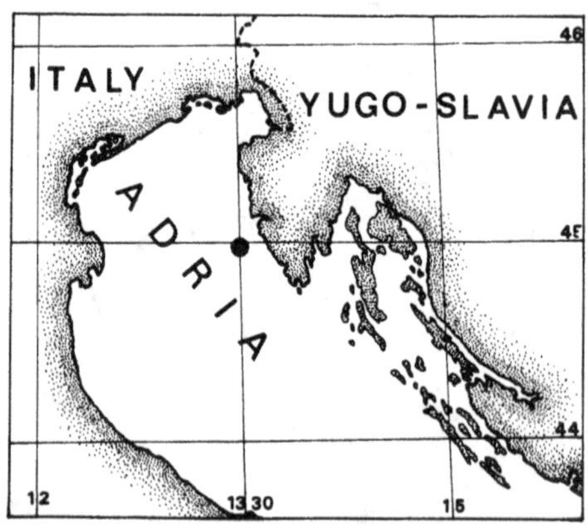

MAP 36. *Leptochiton (L.) intermedius* (von Salvini-Plawen).

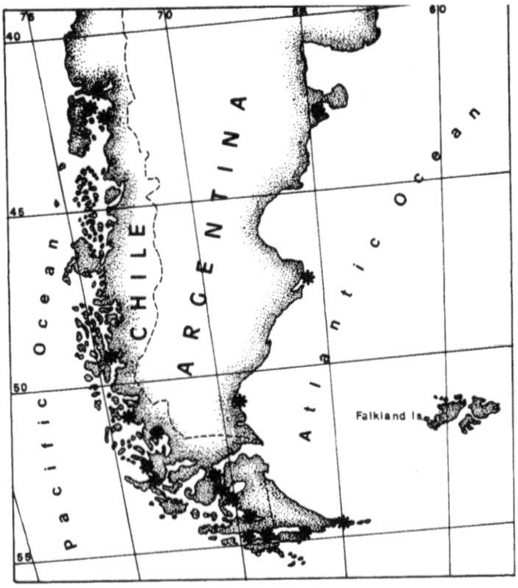

MAP 37. *Leptochiton (L.) medinae* (Plate).

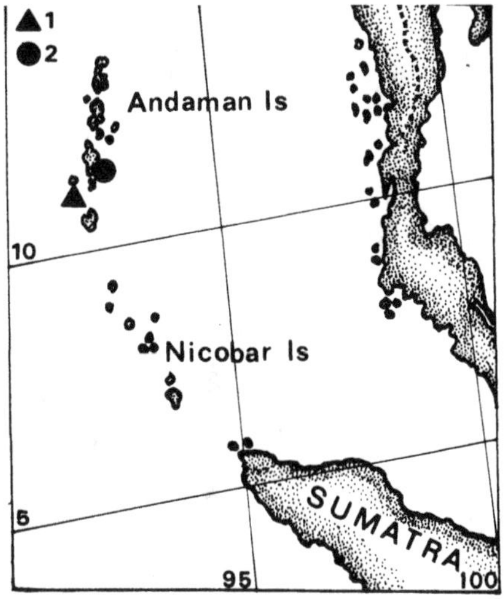

MAP 38. 1, *Leptochiton (L.) andamanicus* (Smith); 2, *Leptochiton (Parachiton) ronaldi* sp. nov.

MAP 39. *Oldroydia percrassa* (Dall).

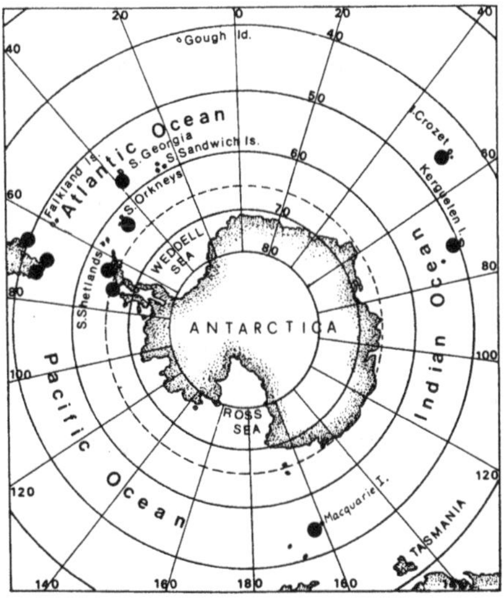

MAP 40. *Hemiarthrum setulosum* (Carpenter in Dall).

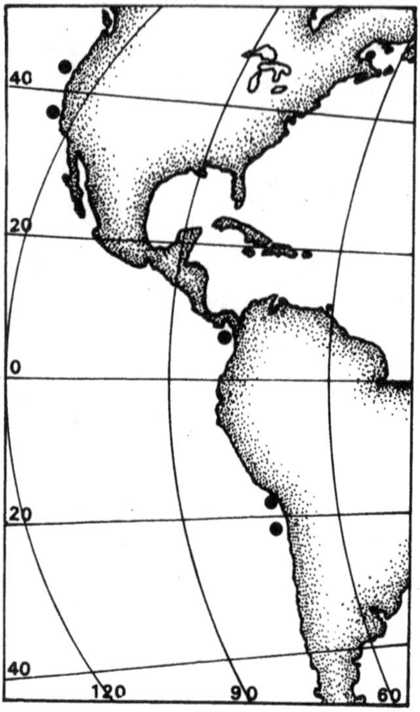

MAP 41. *Leptochiton (L.) americanus* sp. nov.

MAP 43. 1, *Leptochiton (L.) vitjazi* (Sirenko); 2, *Leptochiton (Parachiton) acuminatus* (Thiele); 3, *L. (P.) fornix* sp. nov.

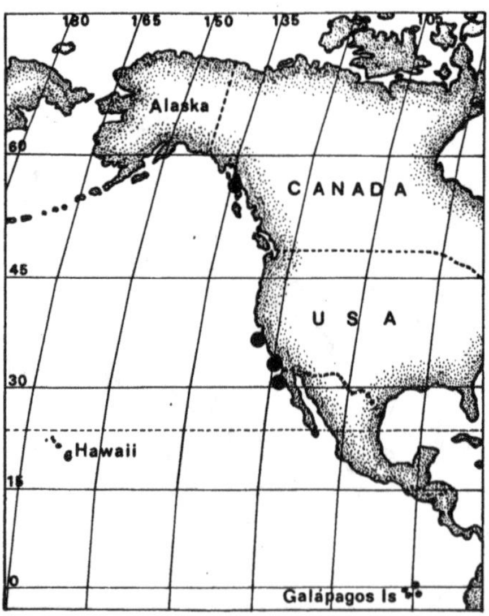

MAP 42. *Hanleyella oldroydi* (Bartsch MS, Dall).

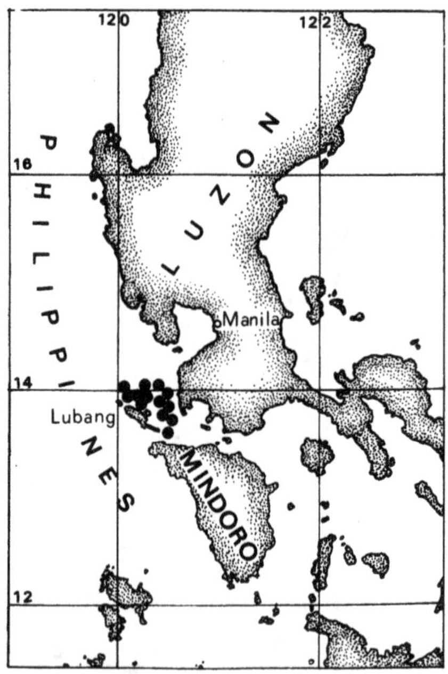

MAP 44. *Leptochiton (L.) lineatus* (Nierstrasz); *L. (L.) foresti* (Leloup); *L. (L.) juvenis* (Leloup); *L. (L.) philippinus* (Leloup); *L. (L.) cancelloides* Kaas.

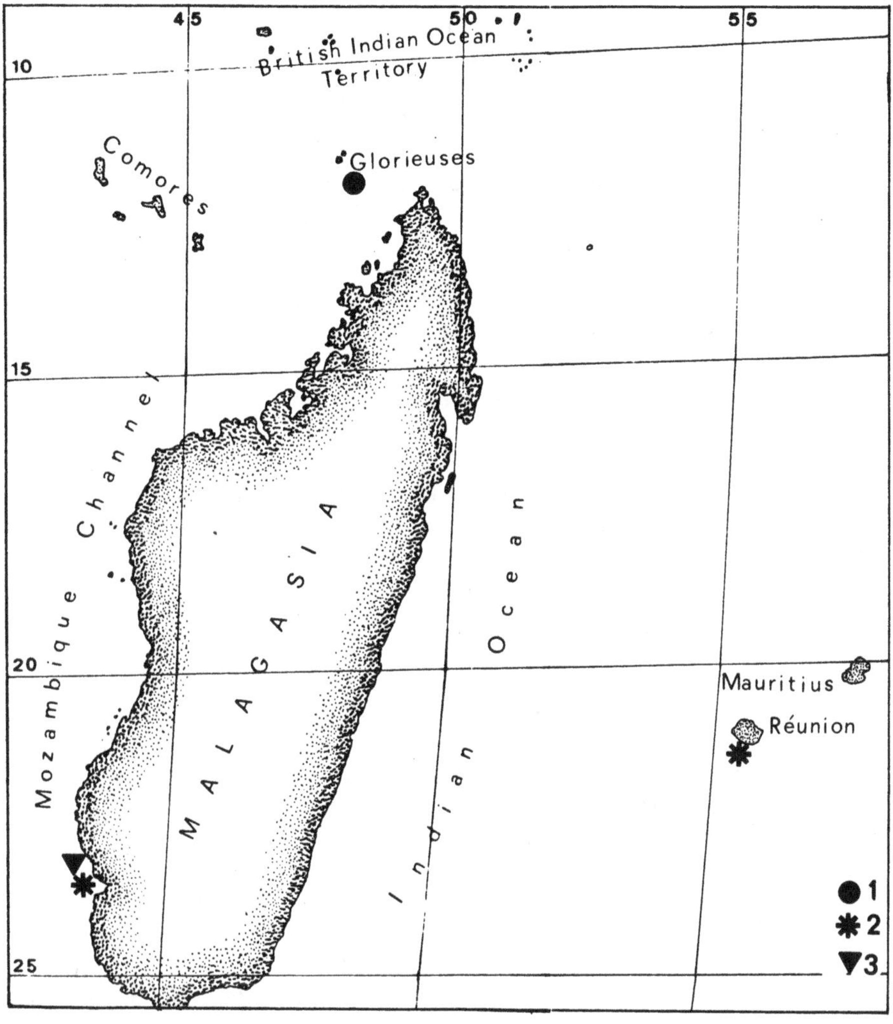

MAP 45. 1, *Leptochiton (L.) benthedi* (Leloup); 2, *L. (L.) nierstraszi* (Leloup); 3 *L. (Parachiton) eugenei* sp. nov.

Species insertae sedis

Lepidopleurus bottae de Rochebrune, 1882:192.

The original description of this Red Sea species leaves no doubt that it belongs to the genus *Chiton,* subgenus *Rhyssoplax*. Pilsbry (1893:98) suggested it might be the species he named *Callistochiton heterodon*. Thiele (1909:8; 1910:91), Pallary (1926:30) and Leloup (1952:27) considered it a synonym of *Chiton (Rhyssoplax) affinis* Issel, 1869. The exact status of *L. bottae* will be reconsidered later in this work.

Lepidopleurus cullieretі de Rochebrune, 1889:140, pl. 9 figs:9a-b.

This species, collected in Orange Bay, Patagonia, is certainly no *Lepidopleurus*. As already suggested by Pilsbry (1893:111) and Thiele (1908:13), it probably belongs to the genus *Chaetopleura*. Its correct identification will be studied when the genus is being revised.

Leptochiton belgicae Pelseneer, 1903:16, pl. 1:fig. 1.

Though Thiele (1908:10) suggested that this species from Tierra del Fuego might be identical with *Leptochiton kerguelensis* Haddon, 1886, Leloup (1956b:13) states that the insertion plates are apparent, and that it should be classed in the subfamily Ischnochitoninae. As he probably had the type specimen before him when he wrote this statement, it will be studied when that subfamily is under revision.

Leptochiton crassus Carpenter MS, Van Winkle Palmer, 1958:46.

The name is first mentioned on p. 46 in a list of nomina nuda, all manuscript designations of Carpenter. On p. 262 the author gives the following details: "There is a specimen in the Redpath Museum (no. 11) labelled by Carpenter '*Leptochiton crassus* Cpr. unique type S. Diego under rocks in sand extreme low tide Hemphill'. The writer has found no reference to such a species in literature. It apparently applied to a species which Carpenter intended to describe." It does not appear in the MSS of Carpenter's unpublished monograph either.
As no further study of that specimen has been published, the name has never been validated and remains a nomen nudum.

End of volume 1.
Volume 2 will comprise the suborder Ischnochitonina; family Ischnochitonidae: Schizoplacinae, Callochitoninae, Lepidochitoninae.

References

Adams, H. & G.F. Angas, 1864. Descriptions of new genera and species of Chitonidae from the Australian Seas, in the collection of George French Angas. – *Proc. zool. Soc. Lond.*:192-194.

Altimira, C., 1978. Moluscos marinos de las costas del NW de Africa (Expedicion "Atlor VII"). – *Result. Expedic. cientif.* 7:173-193.

Anadon, N., 1979. Poliplacoforos de las costas Asturianas. I, Estudios taxanomicos. -*Sup.Cien.Bol.Idea* 24:119-130.

Angas, G. F., 1865. Descriptions of ten new species of shells, chiefly from the Australian seas. -*Proc. zool. Soc. Lond.*:55-58, pl. 2.

– 1867b. A list of species of marine Mollusca found in Port Jackson Harbour, NSW, and on the adjacent coasts, with notes on their habits, etc, Part I. -*Proc. zool. Soc. Lond.*:185-233.

Annandale, N. & F.H. Stewart, 1909. *Illustrations of the zoology of the Royal Indian Survey Ship "Investigator". Mollusca,* pt. 6, pls 21-23. Calcutta.

Arey, L.B. & W.J. Crozier, 1918. The sensority responses of *Chiton*. -*J. exp. Zool.* 29(2):157-260.

Ashby, E., 1918c. Notes on South Australian Polyplacophora, with additions to the fauna; together with a list of Australian Polyplacophora, showing their distribution in the Australian States. -*Trans. R. Soc. S. Aust.* 42:79-87.

– 1921a. Description of Bracebridge Wilson's collection of Victorian chitons with description of two new species. -*Proc. R. Soc. Vict.* 33(n.s.):149-159, pl. 8.

– 1921c. The rediscovery of *Choriplax* (=*Microplax*) *grayi,* Adams and Angas (Order Polyplacophora), with notes on its true place in the natural system, and the description of a new subspecies. -*Trans. R. Soc. S. Aust.* 45:136-142, pl. 9.

– 1923a. Monograph on the Australian Lepidopleuridae (Order Polyplacophora), with a description of a new species. -*Trans. R. Soc. S. Aust.* 47:216-223, pls 16-18.

– 1924a. The chiton fauna (Polyplacophora) of Port Stephens, New South Wales. -*Trans. R. Soc. S. Aust.* 48:313-322, pl. 31.

– 1924c. Notes on the types of Australasian Polyplacophora in the British Museum. -*Trans. R. Soc. S. Aust.* 48:328-333.

– 1925. Monograph on Australian fossil Polyplacophora (Chitons). -*Proc. R. Soc. Vict.* 37 (n.s.)(2):170-205, pls 18-22.

– 1927a. Notes on, and additions to, the chiton fauna of North-West Tasmania, together with a brief review of the genus *Stenochiton*. -*Pap. Proc. R. Soc. Tasm.* (1926):92-117.

– 1928b. Further notes on Western Australian chitons (Polyplacophora). -*Trans. R. Soc. S. Aust.* 52:174-181, pl. 12.

– 1929a. New Zealand fossil Polyplacophora (Chitons). -*Trans. N.Z. Inst.* 60:366-369, pl. 32.

– 1929b. Further notes on New Zealand chitons (Polyplacophora). -*Trans. N.Z. Inst.* 60:370-378, pl. 32. (pars).

– 1929c. Contributions to the fauna of Rottnest Island. No. II. Polyplacophora. -*J. R. Soc. W. Aust.* 15:47-54, figs 10-16, map.

– 1931a. Monograph of the South African Polyplacophora (Chitons). -*Ann. S. Afr. Mus.* 30:1-59, figs 1-2, pls 1-7.

Ashby, E. & B.C. Cotton, 1930. Notes on Australian Polyplacophora. -*Trans. R. Soc. S. Aust.* 54:57-58.

Baluk, W., 1971. Lower Tortonian chitons from the Korytnica clays, southern slopes of the Holy Cross Mountains. -*Acta geol. pol.* 21(3):449-472, figs 1-3, pls 1-6.

Barash, A. & Z. Danin, 1977. Polyplacophora (Mollusca) from the eastern Mediterranean. -*J. Conch. Paris* 114(l-2):3-28, maps 1-3, figs 1-12.

Barnard, K.H., 1963. Contributions to the knowledge of South African marine Mollusca. IV. Gastropoda... Polyplacophora... -*Ann. S. Afr. Mus.* 47:201-360, figs 1-30.

Bean, W., 1844. *In:* C. Thorpe, *British marine conchology; being a descriptive catalogue, arranged according to the Lamarckian system, of the salt water shells in Great Britain. A supplement of new species:* 263-267. Lumley, London.

Bednall, W. T., 1897. The Polyplacophora of South Australia. -*Proc. malac. Soc. Lond.* 2:139-159, pl 12.

Bednall, W.T. & E.H.V. Matthews, 1906. On new species of Polyplacophora from South Australia. -*Proc. malac. Soc. Lond.* 7(2):91-92, pl. 9.

Belle, R.A. Van, 1972. Over de Polyplacophora van Bretagne. -*Gloria Maris,* Antwerpen 11:201-204, figs 1-2.

– 1975-1978. Sur la classification des Polyplacophora. I-VII. -*Inf. Soc. belg. Malac.* 4(5):121-131, pl. 1; 4(6):135-145, pls 2-3 (1975); 5(2):15-42, pls 4-5 (1977); 6(l):3-28, pls 6-8; 6(2):35-44, pl. 9; 6(3):65-82 (1978).

– 1975a. Sur la présence en Méditerranée de *Lepidopleurus alveolus* (Lovén, 1846) (Mollusca: Polyplacophora). -*Inf. Soc. belg. Malac.* 4(3):57-58, figs 1-4.

Bergenhayn, J.R.M., 1931a. Beiträge zur Malakozoologie der Kanarischen Inseln. Die Loricaten. -*Ark. Zool.* 23A(13):1-38, pls 1:3.

– 1933. Die Loricaten von Prof. Dr. Sixten Bocks Expedition nach Japan und den Bonin-Inseln 1914. -*K. svenska Vetensk. Akad. Handl.* 12(4):l-58, figs 1-17, pls 1-3.

– 1955. Die fossilen Schwedischen Loricaten nebst einer vorläufigen Revision der ganzen Klasse Loricata. -*K. Fysiogr. Sällsk. Handl., Lund* N.F.66(8):l-44, figs 1-10, pls 1-2.

Berry, S.S., 1907. Molluscan fauna of Monterey Bay, California. -*Nautilus* 21(2):17-22, (3):34-35, (4):39-47, (5):51-52.

– 1917b. Chitons taken by the United States Fisheries Steamer "Albatross" in the Northwest Pacific in 1906. -*Proc. U.S. natn. Mus.* 54(2223):1-18, pls 1-10.

– 1919a. Preliminary notices of some new West American chitons. -*Lorquinia* 2(6):44-47.

– 1919b. Notes on West American chitons II. -*Proc. Calif. Acad. Sci.* (4)9(1): 1-36, figs 1-6, pls 1-8.

Beu, A.G., 1977. New Zealand chitons. -*Cookia* 2(2):36-53, figs A-S.

Blainville, H.M.D. de, 1816. Prodrome d'une nouvelle distribution systématique du règne animal. -*Bull. Sci. Soc. philom. Paris* :113-124.

– 1825. *Oscabrelle. Oscabrion.* In: *Dictionnaire des sciences naturelles* (Ed, Levrault) 36:519:556.

Boone, L., 1928. Scientific results of the first Oceanographic Expedition of the "Pawnee", 1925. Mollusca from tropical East American seas. -*Bull. Bingham oceanogr. Coll.* 1(3):1-20, figs 1-4.

Boudet-Rommel, I., 1945. Los quitones Chilenos. -*Rvta chil. Hist. nat.* :122-140(1944).

Bullock, R.C. & K.J. Boss, 1971. *Lithophaga aristata* in the shell plates of chitons (Mollusca). -*Breviora* (369):1-10, figs 1-8.

Burne, R.H., 1896. Notes on the anatomy of *Hanleya abyssorum,* M. Sars. -*Proc. malac. Soc. Lond.* 2:4-13, figs 1-3, pl. 2.

Capellini, J., 1859. Catalogue des oscabrions de la Méditerranée, suivi de la description de quelques espéces nouvelles. -*J. Conch. Paris* 7:321-328, pl. 12.

Carcelles, A.R., 1950. Catálogo de los Moluscos marinos de la Patagonia. -*An. Mus. Nahuel Huapí* 2:41-99, 6 pls, map.

– 1953. Catálogo de la malacofauna antartica Argentina. -*An. mus. Nahuel Huapí* 3:155-250, pls 1-5, map.

Carcelles, A.R. & S.I. Williamson, 1951. Catálogo de los moluscos marinos de la Provincia Magellanica. Amphineura. -*Rvta Inst. nac. Invest. Cienc. nat. Mus. argent. Cienc. nat. Bernardino Rivadavia, Zool.* 2(5):225-383.

Carpenter, P.P., 1864a. Supplementary report on the present state of our knowledge with regard to the Mollusca of the West coast of North America. -*Rep. Br. Ass. Advmt Sci.* (1863):517-686.

Castellanos, Z.J.A. de, 1951. Poliplacoforos del museo Argentino de ciencias naturales. -*Com. Inst. nac. Invest. Cienc. nat. Mus. argent. Cienc. nat. Bernardino Rivadavia, Zool.* 1(15):1-31, figs 1-10.

– 1956. Catálogo de los Poliplacoforos Argentinos y de aguas vecinas al Estrecho de Magallanes. -*Revta Mus. La Plata, Zool.* (n.s.)6:465-486, pls 1-9.

Chemnitz, J.H., 1785. *Neues systematisches Conchylienkabinet. Muscheln mit sichtbaren Gelenken,* 8:252-293, pls 94-96 (Nürnberg).

Christiansen, M.E., 1954. The life history of *Lepidopleurus asellus* (Spengler) (Placophora). -*Nytt Mag. Zool.* 2:52-72, figs 1-26, 4 tabs.

Clench, W.J. & R.D. Turner, 1950. The Western Atlantic marine mollusks described by C.B. Adams. -*Occ. Pap. Mollusks, Harv.* 1(15):233-403, pls 29-49.

Cotton, B.C., 1937. Loricata. -*Rep. B.A.N.Z. Antarctic Res. Exp.* 1929-1931 (B)4(l):9-19, figs 1-29, 1 pl.

– 1964. *South Australian Mollusca. Chitons*:l-151, figs 1-139 (Adelaide).

Cotton, B.C. & F.K. Godfrey, 1940. *The molluscs of South Australia.* 2, *Scaphopoda, Cephalopoda, Aplacophora and Crepipoda. Crepipoda*:469-590, figs 451-589 (Adelaide).

– 1940a. A systematic list of the Pelecypoda, Scaphopoda, Cephalopoda & Crepipoda. The bivalve Mollusca, tusk-shells, Nautilus, squids, Octopi, and chitons of South Australia. Class Crepipoda (Chitons). *-Publ. malac. Soc. S. Aust.* 2:22-27.

Cotton, B.C. & B.J. Weeding, 1939. Flindersian loricates. *-Trans. R. Soc. S. Aust.* 63(2): 180-199, pl 7.

Crozier, W.J. & L.B. Arey, 1919. On the ethology of *Chiton tuberculatus. -Proc. Acad. Sci. Wash.* 5:496-498.

Dall, W.H., 1876. Mollusks. *In:* J.H. Kidder, Contributions to the natural history of Kerguelen Island, made in connection with the United States "Transit-of-Venus" Expedition, 1874-1875, II. *-Bull. U.S. natn. Mus.* 3:42-48.

– 1878. Descriptions of new forms of mollusks from Alaska contained in the collections of the National Museum. *-Proc. U.S. natn. Mus.* 1:1-3.

– 1879. Report on the limpets and chitons of the Alaskan and Arctic regions, with descriptions of genera and subspecies believed to be new. *-Proc. U.S. natn. Mus.* 1:281-344, figs A-E, pls 1-5.

– 1881. Reports on the results of dredging, under the supervision of Alexander Agassiz, in the Gulf of Mexico, and in the Caribbean Sea, 1877-79, by the United States coast survey steamer "Blake", Lieutenant-Commander C.D. Sigsbee, U.S.N., and Commander J.R. Bartlett, U.S.N., commanding. XV, Preliminary report on the Mollusca. *-Bull. Mus. comp. Zool. Harvard Coll.* 9(2):33-144.

– 1882b. On certain limpets and chitons from the deep waters off the eastern coasts of the United States. *-Proc. U.S. natn. Mus.* 4:400-414.

– 1889a. A preliminary catalogue of the shell-bearing marine mollusks and brachiopods of the southeastern coast of the United States, with illustrations of many of the species. *-Bull. U.S. natn. Mus.* 37:1-221, pls 1-74.

– 1889b. Report on the results of dredging, under the supervision of Alexander Agassiz, in the Gulf of Mexico (1877-78) and in the Caribbean Sea (1879-80), by the U.S. coast survey steamer "Blake", Lieutenant-Commander C.D. Sigsbee, U.S.N., and Commander J.R. Bartlett, U.S.N., commanding, 29, Report on the Mollusca. 2, Gastropoda and Scaphopoda. *-Bull. Mus. comp. Zool. Harvard Univ.* 18:1-492, pls 10-40.

– 1894. A new chiton from California. *-Nautilus* 8:90-91.

– 1902. Illustrations and descriptions of new, unfigured, or imperfectly known shells, chiefly American, in the U.S. National Museum. *-Proc. U.S. natn. Mus.* 24:499-566, pls 27-40.

– 1908. Report on the dredging operations off the West coast of Central America, to the Galapagos, etc ... by the U.S. Fish. Comm. Steamer "Albatross", 1891. 37, Reports on the scientific results. etc. 14, The Mollusca and Brachiopoda. *-Bull. Mus. comp. Zool. Harvard Univ.* 43(6):205-487, pls 1-22.

– 1919. Descriptions of new species of chitons from the Pacific coast of America. *-Proc. U.S. natn. Mus.* 55:499-516.

– 1927. Diagnoses of undescribed new species of mollusks in the collection of the United States National Museum. *-Proc. U.S. natn. Mus.* 70:1-11.

Dall, W.H. & C.T. Simpson, 1901. The Mollusca of Porto Rico. -*Bull. U.S. Fish. Comm.* (1900) 20(1):351-524, pls 53-58.

Daniel, F., 1883. Faune malacologique terrestre, fluviatile et marine des environs de Brest (Finistére) (Deuxiéme partie). -*J. Conch. Paris* 31:330-391.

Dautzenberg, P. & H. Fischer, 1912. *Résultats des campagnes scientifiques accomplies sur son yacht par Albert Ier.* 27, *Mollusques provenant des campagnes de l'Hirondelle et de la Princesse-Alice dans les Mers du Nord*: 1-630, pls 1-11 (Monaco).

Dell, R.K., 1964a. Antarctic and Subantarctic Mollusca: Amphineura, Scaphopoda, and Bivalvia. -*Discovery Rep.* 33:93-250, figs 1-4, pls 2-7.

– 1964b. Marine Mollusca from Macquarie and Heard Islands. -*Rec. Domin. Mus.* 4(20):267-301, figs.

– 1971. The marine Mollusca of the Royal Society Expedition to southern Chile 1958-1959. -*Rec. Domin. Mus.* 7(17):155-233, pls 1-5.

Dons, C., 1933. Zoologiske Notiser XVIII. Om utbredelsen av *Hanleya nagelfar.* -*Forh. K. norsk. Vidensk. Selsk. Trondhjem* 5(39):151-153, figs 1-3.

Ferreira, A.J., 1979b. The family Lepidopleuridae (Mollusca: Polyplacophora) in the eastern Pacific. -*Veliger* 22(2):145-165, figs 1-40.

– 1980. A new species of *Lepidopleurus* Risso, 1826 (Mollusca: Polyplacophora) in the deep waters of the eastern Pacific. -*Veliger* 23(1):55-61, figs 1-11.

– 1981b. *Laminoplax,* a new genus of chitons and the taxonomic position of *Hanleya dalli* Kaas, 1957 (Polyplacophora: Afossochitonidae). -*Nautilus* 95(4): 189-193, figs 1-6.

– 1981c. Observations on the deep-water chiton, *Leptochiton rissoi* (Nierstrasz 1905) in the eastern Pacific. -*Bull. S. Calif. Acad. Sci.* 80(1):36-41, figs 1-2.

Fischer, P., 1885. *Manuel de conchyliologie et de paléontologie conchyliologique.* Paris.

Fretter, V., 1937. The structure and function of the alimentary canal of some species of Polyplacophora. -*Trans. R. Soc. Edinb.* 59(1):119-164, figs.

Gatliff, J.H. & C. J. Gabriel, 1917. Additions to and alterations in the catalogue of the marine shells of Victoria. -*Proc. R. Soc. Vict.* 30(n.s.) (1):21-31, pl. 3.

Gmelin, J.F., 1791. *C. von Linnaeus' Systema Naturae,* ed. 13. tome 1, pars 6, *Vermes Testacea, Genus 300, Chiton*:3202-3207. Holmiae.

Götting, K.-J., 1973. Die Polyplacophora der karibischen Küste Kolumbiens. -*Arch. Molluskenk.* 103(4/6):243-261, figs 1-4, pls 8-11.

– 1974. *Malakozoologie*:1-321, figs 1-160 (Stuttgart).

Gould, A.A., 1852. *United States Exploring Expedition during the years 1838-42, under the command of Charles Wilkes, U.S.N.* 13, *Mollusca and shells*: i-xv, 1-510, atlas.

– 1859. Description of shells collected by the North Pacific Exploring Expedition. -*Proc. Boston Soc. nat. Hist.* 7(11):161-165.

– 1862. *Otia Conchologica. Descriptions of shells and mollusks, from 1839 to 1862*:i-iv, 1-256 (Boston).

Gray, J.E. 1821. A natural arrangement of Mollusca according to their internal structure. -*Lond. med. Repos.* 15:221-239.

– 1847a. On the genera of the family Chitonidae. -*Proc. zool. Soc. Lond.* 15:63-70.

– 1847b. Additional observations on Chitones. -*Proc. zool. Soc. Lond.* 15:126-127.

– 1847c. A list of the genera of recent Mollusca, their synonymy and types. -*Proc. zool. Soc. Lond.* 15:129-206.

– 1857. *Guide to the systematic distribution of Mollusca in the British Museum.* I, *Polyplacophora*:176-188. British Museum (Nat. Hist.), London.

Guilding, L., 1829. Observations on the Chitonidae. -*Zool. J. Lond.* 5:25-35.

Haddon, A.C., 1886. Report on the Polyplacophora collected by H.M.S. "Challenger" during the years 1873-1876. -*Challenger Rep.* 15(43):1-50, pls 1-3.

Hedley, C., 1916. Mollusca. *In:* Australian Antarctic Expedition, 1911-14, under the leadership of Sir Douglas Mawton, D. Sc., B. E. -*Sci. Rep.,* C, *Zoology and Botany* 4(1): 1-80, pls 1-9 (Adelaide).

Hedley, C. & A.F.B. Hull, 1909. Descriptions of new and notes on other Australian Polyplacophora. -*Rec. Aust. Mus.* 7(4):260-266, pls 73-74.

– 1912. The Polyplacophora of Lord Howe and Norfolk Islands. -*Proc. Linn. Soc. N.S.W.* 37(2):271-281, pls 11-13.

Hedley, C. & W.L. May, 1908. Mollusca from one hundred fathoms seven miles East of Cape Pillar, Tasmania. -*Rec. Aust. Mus.* 7:108-145, pls 22-25.

Hoffmann, H., 1929-1930. *In:* H.G., Bronn. *Klassen und Ordnungen des Thier-Reichs.* 3(1), Nachtrag 2, *Polyplacophora, Amphineura, Scaphopoda* (1930): 369-511, 57 figs (Leipzig).

Hull, A.F.B., 1923a. New Australian Polyplacophora, and notes on the distribution of certain species. -*Aust. Zool.* 3:157-166, pls 24-26.

– 1923b. New Australian Loricata and notes on the distribution of certain species. 1. -*Aust. Zool.* 3:195-201, pls 27-28.

Hull, A.F.B. & J. Risbec, 1930-1931. The Loricates of the New Caledonian region (Class Mollusca – Order Loricata). -*Aust. Zool.* 6:277-286, figs 1-25, pl. 20 (1930);372-386, figs 1-25, pl. 31 (1931).

Hyman, L.H., 1967. *Class Polyplacophora.* In: *The Invertebrates,* 6, *Mollusca* 1:70-142, figs 29-57 (New York).

Iredale, T., 1914a. The chiton fauna of the Kermadec Islands. -*Proc. malac. Soc. Lond.* 11:25-51, pls 1-2.

Iredale, T. & A.F.B. Hull, 1923-1926. A monograph of the Australian loricates, I-VIII. -*Aust. Zool.* 3:186-194, pls 26-27 (1923); 227-238, pls 33-34; 277-297, pls 35-37 (1924); 339-362, pls 39-40; 4:75-111, pls 9-12 (1925); 164-185, pls 18-20; 256-276, pl. 37; 324-328, pl. 49 (1926).

– 1929-1932. The loricates of the Neo-Zelanic region. -*Aust. Zool.* 5:305-323, pls 30-34; 6:75-95, pls 9-10 (1929); 157-166, pl. 16 (1930); 7:59-76, pl. 3 (1931); 119-164, pls 7-10 (1932).

Jakovleva, A.M., 1952. *Shell-bearing Mollusks (Loricata) of the seas of the U.S.S.R.* -*Fauna USSR* 45:1-107, figs 1-53, pls 1-11 (Zool. Inst. Acad. Sci.

U.S.S.R., Moskwa and Leningrad). Translated into English by the Israel Program for Scientific Translations, Jerusalem, 1965.

Jeffreys, J.G., 1839. A list of marine Mollusca taken during a few days stay at Oban, in Argyleshire, in the autumn of 1838. -*Malac. conch. Mag.* 2:33-54.

– 1865-69. *British Conchology, or an account of the Mollusca which now inhabit the British Isles and the surrounding seas,* 3, Family Chitonidae:203-229 (1865); 5:197-199, pls 55-56 (1869). Van Voorst, London.

– 1880. On a new species of chiton lately found on the British coasts. -*Ann. Mag. nat. Hist.* 6:33-35.

– 1882. On the Mollusca procured during the "Lightning" and "Porcupine" expeditions, 1868-70 (part V). -*Proc. zool. Soc. Lond.*:656-687, pls 49-50.

Johnson, C.W., 1934. List of marine Mollusca of the Atlantic coast from Labrador to Texas. -*Proc. Boston Soc. nat. Hist.* 40(1): 1-204.

Kaas, P., 1957. Notes on Loricata. 5, On some preoccupied names. -*Basteria* 21(4-5):83-87.

– 1972. Polyplacophora of the Caribbean region. -*Stud. Fauna Curaçao* 41(137):1-162, figs 1-247, pls 1-9.

– 1977. Notes on Loricata. 9, On the rediscovery of *Lepidopleurus africanus* Nierstrasz, 1906 and the systematic position of the taxon *Parachiton* Thiele, 1909. -*Basteria* 41:81-86, figs 1-6.

– 1979a. On a collection of Polyplacophora (Mollusca, Amphineura) from the Bay of Biscay. -*Bull. Mus. natn. Hist. nat. Paris* 4(A)A(1):13-31, pls 1-5.

– 1981. Scandinavian species of *Leptochiton* Gray, 1847 (Mollusca, Polyplacophora). -*Sarsia* 66:217-229, figs 1-10.

– 1982b: *Leptochiton* species (Polyplacophora: Leptochitonidae) of the Musorstom 1 (1976) and 2 (1980) – Philippines-Expeditions. -*Basteria* 46:87 -92, figs 1-13.

Knudsen, J., 1949. *Amphineura.* -*The zoology of Iceland,* vol. IV, part 59:1-11 (Copenhagen and Reykjavik).

– 1970. Amphineura. -*Zool. Faroes* 51:1-8.

Laghi, G.F., 1977. Polyplacophora (Mollusca) neogenici dell'Appennino Settentrionale. -*Boll. Soc. paleont. ital.* 16(1):87-115, figs 1-3, pls 1-4.

Laghi, G.F., F. Russo, & B. Dell'Angelo, 1981. Recenti ritrovamenti di *Lepidopleurus (Parachiton) africanus* Nierstrasz, 1906, (Polyplacophora, Mollusca). -*Atti Soc. Nat. Mat. Modena* 111 (1980):l-7, figs 1-9.

Leach, W.E., 1852. *A synopsis of the Mollusca of Great Britain, arranged according to their natural affinities and anatomical structure*:i-viii, 1-376, pls 2-13 (London).

Leloup, E., 1934a. Contributions á l'étude de la faune belge. IV, Les polyplacophores de la côte belge. -*Bull. Mus. r. Hist. nat. Belg.* 10(17):l-23, 1 map.

– 1934b. Contribution à la connaissance de la faune des chitons de Villefranche-sur-Mer et des environs (France, Méditerranée). -*Bull. Mus. r. Hist. nat. Belg.* 10(47):1-20, figs 1-16.

– 1956b. Polyplacophora. Reports of the Lund University Chile expedition 1948-

1949, 27. -*Acta Univ. Lund.* (n.s.)52:1-94, figs 1-53.

– 1968a. Chitons de la côte Africaine Occidentals. –*Atlantide Rep.* 10:7-31, figs 1-9, pls 1-2.

– 1981a. Résultats des Campagnes Musorstom I – Philippines (18-28 Mars 1976), 15. Mollusques: Polyplacophores. -*Mém. Orstom* 91:317-323, figs 1-6, pls 1-2.

– 1981b. Chitons du Sud-Ouest de l'Océan Indien. -*Bull. Inst. r. Sci. nat. Belg.* 53 (10):1-4, figs 1-3, pl. 1.

– 1981d. Chitons de Tuléar, Réunion, Maurice et Tahiti. -*Bull. Inst. r. Sci. nat. Belg.* 53(3):1-46, 22 figs, 4 pls.

– & P. Volz, 1938. Die Chitonen (Polyplacophoren) der Adria. -*Thalassia* 2(10):1-63, figs 1-57, map.

Linnaeus, C., 1758. *Systema Naturae. Regnum Animale,* ed. 10, vol. 1, *Vermes. Testacea. Chiton,* p. 667. -Photographic facsimile, British Mus. (Nat. Hist.) (1956).

– 1764. *Museum S:ae R:ae M:tis Ludovicae Ulricae Reginae, etc. Chiton,* p. 465. Holmiae.

– 1767. *Systema Naturae,* Tom. 1, Pars 2, ed. duodecima reformata. *Vermes* iii *Testacea,* genus 300. *Chiton*: 1106-1107. Holmiae.

Lovén, S.L., 1846. Index molluscorum litora Scandinaviae Occidentalia habitantium. -*Öfvers. K. Vetensk. Akad. Förh. Stockh.* 3:134-160.

– 1847. Om tungans beväpning hos mollusker. -*Öfvers. K. Vetensk. Akad. Förh. Stockh.* 4:175-199, pls 2-6.

Lyngnes, R., 1924. Zur Kenntnis der Eihüllen der Chitoniden. -*Vid. Selsk. Skrift. I Matem. Naturvid. Kl.* 20: 1-12, figs (1923).

Mackay, J.S., 1930. Notes on loricates (Chitons) collected on the coast of Queensland in 1928, 1929 and 1930. -*Aust. Zool.* 6(3):287-300, figs 1-11.

Macpherson, J. Hope & C.J. Gabriel, 1962. *Marine molluscs of Victoria*:1-475, figs 1-486 (Melbourne).

Malatesta, A., 1962. Mediterranean Polyplacophora cenozoic and recent. -*Geologica romana* 1:145-171, figs 1-25.

Marshall, J.T., 1893. Additions to "British Conchology". -*J. Conch., Lond.* 7:241-265.

Matthews, G., 1967. The identification of British chitons. -*Pap. Stud. Conch. Soc. Gr. Brit. Irel.* 9:1-4.

May, W.L., 1923. *An illustrated index of Tasmanian shells*:1-54, pls 1-50. Revised edition, with appendix, by J. Hope Macpherson (Government Printer, 1958).

McKay, D.W. & S.M. Smith, 1979. *Marine Mollusca of East Scotland:* 1-185, maps (The Royal Scottish Museum, Edinburgh).

Middendorff, A.T. von, 1847b. Beiträge zu einer Malakozoologia Rossica. -*Mém. Sci. nat. Acad. imp. St Pétersb.* 6(1):1-151, pls 1-14.

Mighels, J.W. & C.B. Adams, 1842. Descriptions of twentyfour species of the shells of New England. -*Boston J. nat. Hist.* 4:37-54.

Montagu, G., 1803. *Testacea Britannica or Natural History of British shells, marine, land and fresh water, including the most minute, systematically arranged and embellished with figures,* 1:1-291; 2:292-606, pls 1-16 (London).

Muus, B., 1959. *Skallus, søtænder, blæksprutter. Danmarks fauna* 65: 1-239, figs 1-117. (København).

Nierstrasz, H.F., 1905. Die Chitonen der Siboga-expedition. -*Siboga Exp.* 48:1-112, pls 1-8.

– 1906. Remarks on the Chitonidae. -*Tijdschr. ned. dierk. Vereen.* (2)10:141-172.

Nierstrasz, H.F. & H. Hoffmann, 1929. *Die Tierwelt der Nord- und Ostsee.* 9a, *Aculifera*: 1-64, figs 1-58 (Leipzig).

Nobre, A., 1931-1936. *Moluscos marinhos de Portugal,* 1:1-463, pls. 1-80 (Inst. Zool. Univ. Pôrto) (1931): 2:1-378, pls 81-86 (Comp. Edit. do Minho, Barcelos) (1936).

– 1937. Moluscos testáceos marinhos do arquipélago da Madeira. -*Mem. Mus. zool. Univ. Coimbra* (1)98:1-101.

– 1940. *Moluscos marinhos e das águas salobras*:i-xxxii, 1-806 (Porto).

Nordgard, O., 1907. Mofjordens naturforhold. -*Skr. Trondh. Selsk.* (1906) 9:1-42, pl. 1.

Norman, A.M., 1902. Notes on the natural history of East Finmark. -*Ann. Mag. nat. Hist.* (7)10:341-361.

Nowikoff, M., 1907. Über die Rückensinnesorgane der Placophoren, nebst einige Bemerkungen über die Schale derselben. -*Z.wiss. Zool.* 88:153-186, pls 10,11.

Oliver, W.R.B., 1915. The Mollusca of the Kermadec Islands. -*Trans. Proc. N.Z. Inst.* 47(1914):509-559, pls 9-12.

Óskarsson, I., 1944. Saeskeld?rarannsóknir i Eyjafir?i. -*Nattúrufraedingurinn* 14:1-21, 1 fig. (in Icelandic).

– 1960. Note on some rare and new species of Mollusca off the coast of Iceland. -*Nattúrufraedingurinn* 30:176-187, figs 1-9 (In Icelandic).

Pallary, P., 1926. Explications des planches de J.G. Savigny avec la collaboration de Ph. Dautzenberg, Major Dupuis, L. Joleaud, A. Koch, E. Lamy, A. Vayssière, et H. Vignel. -*Mém. Inst. Egypte* 11:1-138, pls 1-18.

Pelseneer, P., 1899. Sur la morphologie des branchies et des orifices rénaux et génitaux des chitons. -*Bull. scient. Fr. -Belg.* 31:23.

– 1903. *Résultats du voyage du S. Y. Belgica ... Zoologie, Mollusques (Amphineures, gastropodes et lamellibranches)*:1-85, pls 1-9 (Antwerp).

Pfeffer, G., 1886. Polyplacophora. *In:* E. von Martens & G. Pfeffer, Die Mollusken von Süd-Georgien. -*Jb. hamb. wiss. Anst.* 3:103-109, pl 3.

Pilsbry, H.A., 1892-94. *Monograph of the Polyplacophora. In:* G.W. Tryon, *Manual of Conchology,* 14:1-128, pls 1-30 (1892); i-xxxiv, 129-350, pls 31-68 (1893); 15:1-64, pls 1-10 (1893); 65-133, pls 11-17 (1894) (Academy of Natural Sciences, Philadelphia).

– 1894b. Notices of new chitons, III. -*Nautilus* 7:138-139.

Plate, L.H., 1898-1901. Die Anatomie und Phylogenie der Chitonen. Fauna Chilensis. -*Zool. Jb. Syst.* 1 (Suppl. 4):1-243, pls 1-12 (1898); 2 (Suppl. 5):15-216, pls 2-11 (1899); 3 (Suppl. 5):281-600, pls 12-16 (1901).

Poli, J., 1791. *Testacea utriusque Siciliae corumque historia et anatome.* 1, *Multivalvia:* 1-50, pls 3-6 (Parma).

Posselt, H.J., 1898. Placophora. *In:* Conspectus faunae Groenlandicae. Brachiopoda et Mollusca. -*Meddr. Grönland* 23:107-113.

Powell, A.W.B., 1937. New species of marine Mollusca from New Zealand. – *Discovery Rep.* 15:153-222, pls 45-56.

– 1955. Mollusca of the southern islands of New Zealand. – *Cape Exped. series, Bull.* 15:1-138, pls 1-5.

– 1979. *New Zealand Mollusca. Marine, Land and Freshwater shells.* -Collins, Auckland, Sydney, London: i-xiii, 1-500, 121 figs, 82 pls, maps.

Rainer, S.F., 1981. Soft-bottom benthic communities in Otago Harbour and Blueskin Bay, New Zealand. -*N.Z. Oceanogr. Inst. Mem.* 80:16-18.

Reeve, L., 1847a. Descriptions of new species of shells collected in the eastern Archipelago by Capt. Sir Edward Belcher and Mr. Adams during the voyage of H.M.S. Samarang. -*Proc. zool. Soc. Lond.* 15:24-26.

– 1847b-1848. *Conchologia iconica, or illustrations of the shells of molluscous animals.* 4, *Monograph of the genus Chitonellus:* pl. 1; *Monograph of the genus Chiton:* pls 1-27 (1847), pl. 28 (1848).

Ricketts, E.F. & J. Calvin, 1939. *Between Pacific tides:* i-xxii, 1-320, figs 1-112, pls 1-96.

Righi, G., 1973a. Moluscos da Baía da Ilha Grande, Rio de Janeiro, Brasil. II, Polyplacophora. -*Papéis avuls. Zool. S. Paulo* 26(19):237-245, figs 1-2, map.

– 1973b. Adicões aos Polyplacophora brasileiros (Mollusca). -*Papéis avuls. Zool. S. Paulo* 26(22):259-273, figs 1-46.

Risso, A., 1826. *Histoire naturelle des principales productions de l'Europe Méridionale et particulièrement de celles des environs de Nice et des Alpes Maritimes* 4:i-vii, 1-439, pls 1-12 (Paris).

Rochebrune, A.T. de, 1882. Diagnoses d'espéces nouvelles de la famille des Chitonidae. -*Bull. Soc. philom. Paris* (7)6:190-197.

– 1889. *Polyplacophores. In:* A.T. de Rochebrune, & J. Mabille, *Mission scientifique du Cap Horn.* 4, *Zoologie*: 131-143, pl. 9 (Paris).

Sabelli, B.A., 1972. Sulla presenza e distribuzione di *Hanleya hanleyi* in Mediterraneo. -*Conchiglie* 8(7-8):97-100, figs 1-6.

– 1974a. Origine e distribuzione dei Polyplacophora viventi in Mediterraneo. -*Quad. civ. Soc. idrobiol. Milano* 5:71-78.

– 1974b. *Hanleya hanleyi* (Bean, 1844). -*Schede malacologische del Mediterraneo* 44, figs 1-13.

Sabelli, B.A. & G. Spada, 1971a. *Lepidopleurus cajetanus* (Poli, 1791). -*Schede malacologische del Mediterraneo* 35, figs 1-17.

Salvini-Plawen, L. von, 1968. Neue Formen im marinen Mesopsammon. -*Ann.*

naturh. Mus. Wien 72:231-272, figs 1-17, pls 1-9.

Sars, G.O., 1878. *Mollusca regionis Arcticae Norvegiae*:i-xiii, 1-466 (Christiania).

Schröter, J.S., 1786. *Einleitung in die Conchylienkenntniss nach Linné*. 3, *Das fünfzehende Geschlecht, Chiton*:490-520, pl. 9 (Halle).

Sirenko, B.I., 1973b. Amphipacific distribution of chitons (Loricata) and their new species in the North-West Pacific. -*Zool. Zh.* 52(5):659-666, figs 1-3 (In Russian). English translation by I. McT. Cowan & A.G. Smith in Of Sea and Shore, summer 1974:59-63.

– 1973c. A new genus of the family Lepidopleuridae (Neoloricata). -*Zool. Zh.* 52(10):1569-1571, figs 1-16 (In Russian).

– 1976a. Chitones from the Vostok Bay (Sea of Japan). -*Trans.* 5 *Acad. Sci. U.S.S.R., Far East Centre, Inst. mar. Biol.*:87-92 (In Russian).

– 1977. Vertical distribution of chitons of the genus *Lepidopleurus* (Lepidopleuridae) and its new ultraabyssal species. -*Zool. Zh.* 56(7):1107-1110, figs 1-13 (In Russian).

– 1979. On the composition of the family Leptochitonidae Dall, 1889 (=Lepidopleuridae Pilsbry, 1892) (Polyplacophora) with description of a new bathyal species. -*Proc. zool. Inst. Acad. Sci. U.S.S.R.* 80:116-121, figs 1-12 (In Russian).

Smith, A.G., 1960a. *Amphineura. In:* R.C. Moore, *Treatise on invertebrate paleontology.* I, *Mollusca* 1:141-176, figs 31-45 (University of Kansas Press).

– 1960b. The type species of *Lepidopleurus* Leach in Risso, 1826. -*Veliger* 2(4):75-76, pl. 17.

Smith, A.G. & A.J. Ferreira, 1977. Chiton fauna of the Galápagos Islands. -*Veliger* 20(2):82-97, 4 pls.

Smith, E.A., 1894. Report on some Mollusca, dredged in the Bay of Bengal and the Arabian Sea. -*Ann. Mag. nat. Hist.*: 157-174, 366-368, pls 3-5.

– 1906. Natural history notes from H.M. marine survey steamer "Investigator", commander T.H. Heming, R.N. Series 3, no 10, On Mollusca from the Bay of Bengal and the Arabian Sea. -*Ann. Mag. nat. Hist.* (7)18:245-264.

Sowerby, G.B. II, 1832-1840. *The conchological illustrations, or coloured figures of all the hitherto unfigured recent shells,* parts 38-45 (1833), 159-177 (1839-1840) (London).

– 1840a. *Descriptive catalogue of British chitones* (Appendix to the Conchological illustrations):1-7 (London).

Sowerby, G.B. III, 1903. Mollusca of South Africa. -*Mar. Invest. S. Afr.* 2(3):213-232, pls 3-5.

Sparre Schneider, J., 1886. Undersøgelser af dyrelivet i de arktiske fjorde. 3. Tromsøsundets Molluskfauna. – *Tromsø Mus. Aarsh.* 8:44-112 (1885).

Spengler, L., 1797. Udförlig beskrivelse over det mangeskallede konkylie-slaegt, af Linneus kaldet *Chiton;* med endeel nye arter og varieteter. -*Skrivt. naturh. Selsk.* 4:62-103, pl. 6.

Starobogatov, Ys.I. & B.I. Sirenko, 1975. On the systematics of the Polyplacophora. (Fifth meeting on mollusks, Leningrad, 1975). -*Malac. Review* 11:73-74.

Sulc, J., 1936. Studie über die fossilen Chitonen. 1, Die fossilen Chitonen im Neogen des Wiener Beckens und der angrenzenden Gebiete. -*Annln naturh. Mus. Wien* 47:1-31, figs 1-4, pls 1-2.

Sykes, E.R., 1894. Notes on the British chitons. -*Proc. malac. Soc. Lond.* 1(2):35-37, pl. 3.

– 1896. Report on a collection of Polyplacophora from Port Phillip, Victoria. -*Proc. malac. Soc. Lond.* 2:84-93, pl. 6.

Taki, Is., 1938a. Report of the biological survey of Mutsu Bay. 31, Studies on chitons of Mutsu Bay with general discussion on chitons of Japan. -*Sci. Rep. Tôhoku imp. Univ.* (4) *Biol.* 12(3):323-423, figs 1-7, pls 14-34.

– 1938b. Taxonomic study of Japanese chitons (2). -*Venus* 8(2):81-84, pl. 1.

– 1962. A list of the Polyplacophora from Japanese islands and vicinity. -*Venus* 22(1):29-53.

– & Iw. Taki, 1929c. Studies on Japanese chitons (3). -*Venus* 1(5):157-164, figs 44-58.

Taki, Iw., 1964b. Classification of the class Polyplacophora, with a list of Japanese chitons. -*Venus* 22(4):401-414 (In Japanese).

Taslé, M., 1867. *Histoire naturelle du Morbihan. Zoologie. Catalogue des mollusques marins, terrestres et fluviatiles observés dans le département*:1-72 (Société polymathique, Vannes).

Tate, R. & W.L. May, 1901. A revised census of the marine Mollusca of Tasmania. -*Proc. Linn. Soc. N.S.W.* 26(3):344-472, pls 23-27.

Thiele, J., 1893. *Lepidoglossa. In:* F.H. Troschel, *Das Gebiss der Schnecken* 2:353-401, pls 30-32.

– 1902. Die systematische Stellung der Solenogastren und die Phylogenie der Mollusken. -*Z. wiss. Zool.* 72:249-466, figs 1-21, pls 18-27.

– 1906. Ueber die Chitonen der deutschen Tiefsee-Expedition. *Wiss. Ergebn. dt. Tiefsee Exp.* 9(2):327-334, pl. 29.

– 1908. Die antarktischen und subantarktischen Chitonen. *Deutsche Südpolar Expedition* 1901-1903. 10, Zoologie 2:9-23, pl. 1.

– 1909-1910. Revision des Systems der Chitonen. -*Zoologica Stuttg.* 22:1-70, pls 1-6 (1909); 71-132, pls 7-10 (1910).

– 1929. *Handbuch der systematischen Weichtierkunde.* 1, *Classis Loricata*: 1-22, figs 1-22 (Jena).

Thorpe, S., 1971. *Class Polyplacophora. In:* A.M. Keen, *Sea shells of tropical West America. Marine mollusks from Baja California to Peru,* ed. 2:861-882, figs 1-50.

Torr, W.G., 1911. Western Australian Polyplacophora. -*Trans. R. Soc. S. Aust.* 35:94-107, pls 24-25.

– 1912b. South Australian Polyplacophora. -*Trans. R. Soc. S. Aust.* 36:140-170, pls 5-7.

Warén, A., 1980. Marine Mollusca described by John Gwyn Jeffreys, with the location of the type material. -*Special Publ. Conch. Soc. Gr. Brit. Irel.* 1:1-60, pls 1-8.

Watters, G.T., 1981a. A note on the occurrence of *Lithophaga (Leiosolenus) spatiosa* (Carpenter, 1857) in the shell-plates of *Acanthochitona hirudiniformis* (Sowerby, 1832). -*Veliger* 24(1):77.

Wells, F.F., 1977. Type specimens in the department of molluscs, Western Australian Museum. -*Rec. W. Aust. Mus.* 6(1):33-61.

Winckworth, R., 1926. Notes on British Mollusca. I. -*J. Conch. Lond.* 18(1):13-15, pl. 1.

– 1940. A systematic list of the "Investigator" Mollusca. -*Proc. malac. Soc. Lond.* 24:19-29.

Winkle Palmer, K. van, 1958. Type specimens of marine Mollusca described by P.P. Carpenter from the West coast (San Diego to British Columbia). -*Mem. geol. Soc. Am.* 76:i-vi, 1-376, pls 1-35.

Wood, W., 1815. *General conchology; or, a description of shells arranged according to the Linnean system*:i-lxi, 1-7, i-iv, 1-246, pls 1-60 (London).

Index

Valid names of taxa belonging to the suborder Lepidopleurina are in italics. Numbers in bold refer to the main treatment of the taxon in question.

www.ingramcontent.com/pod-product-compliance
Lightning Source LLC
LaVergne TN
LVHW010626300926
849162LV00001B/78